AF389497

Phenolic Compounds in Food: Characterization and Health Benefits

Phenolic Compounds in Food: Characterization and Health Benefits

Editor

Mirella Nardini

MDPI • Basel • Beijing • Wuhan • Barcelona • Belgrade • Manchester • Tokyo • Cluj • Tianjin

Editor
Mirella Nardini
CREA Research Centre for
Food and Nutrition
Council for Agricultural
Research and Economics
(CREA)
Rome
Italy

Editorial Office
MDPI
St. Alban-Anlage 66
4052 Basel, Switzerland

This is a reprint of articles from the Special Issue published online in the open access journal *Molecules* (ISSN 1420-3049) (available at: www.mdpi.com/journal/molecules/special_issues/phenolic_food).

For citation purposes, cite each article independently as indicated on the article page online and as indicated below:

LastName, A.A.; LastName, B.B.; LastName, C.C. Article Title. *Journal Name* **Year**, *Volume Number*, Page Range.

ISBN 978-3-0365-6116-5 (Hbk)
ISBN 978-3-0365-6115-8 (PDF)

Contents

Editorial

Phenolic Compounds in Food: Characterization and Health Benefits

Mirella Nardini

CREA, Research Centre for Food and Nutrition, Via Ardeatina 546, 00178 Rome, Italy; mirella.nardini@crea.gov.it

Citation: Nardini, M. Phenolic Compounds in Food: Characterization and Health Benefits. *Molecules* 2022, 27, 783. https://doi.org/10.3390/molecules27030783

Received: 19 January 2022
Accepted: 24 January 2022
Published: 25 January 2022

Publisher's Note: MDPI stays neutral with regard to jurisdictional claims in published maps and institutional affiliations.

Oxidative stress is involved in the onset and development of several human diseases, such as cardiovascular diseases, diabetes, ageing, cancer, and neurodegenerative diseases [1]. Antioxidants from one's diet, which counteract oxidative stress, may offer protection toward these pathological conditions. Polyphenolic compounds are by far the most abundant antioxidants in the human diet, being largely present in plant-based food and beverages [2,3]. Polyphenols play many roles in plants. They protect against stresses such as UV light, attacks from pests, and provide color to attract insects. Evidence from epidemiological studies suggest that the long-term consumption of polyphenolic-rich foods affords protection against the development of cardiovascular and degenerative diseases, cancer, and diabetes [4–6]. The absorption and metabolism of polyphenols have been extensively described. Polyphenols from foods and beverages are quickly absorbed and metabolized in humans. Gut microbiota play a critical role in the absorption process. For individuals who regularly consume wine, beer, coffee, and tea, these beverages represent the main sources of dietary polyphenols [7–10]. Polyphenols content in foods and beverages strongly depends on cultivation, technology processes, and transformation. This Special Issue has collected manuscripts regarding the composition of polyphenols in food, with special emphasis to extractive and analytical aspects. The role of technological processes in the nutritional quality of foods was also considered. Research studies dealing with biological activity and healthy effects of polyphenols were also presented.

The first research article of this Special Issue by Johnson et al. [11] deals with the characterization of phenolic profiles of Australian faba bean (*Vicia faba* L.) varieties. The phenolic acids and flavonoids composition of ten commercial Australian faba bean varieties grown at two different locations was presented. Phenolic profiling by HPLC revealed catechin and rutin as the most abundant flavonoids. Among phenolic acids, syringic acid was found in high concentrations. The levels of phenolics varied significantly with the varieties. Some effects of the growing location on phenolics content were also observed.

The study of El Cadi et al. [12] describes the phytochemical content and the antioxidant activity of the Moroccan species *Chamaerops humilis* L. fruits. The n-hexane fraction analyzed using GC/MS exhibited 69 compounds belonging to different chemical classes. The polyphenolic profile obtained using HPLC-PDA/MS led to the identification of 16 and 13 compounds in two different extracts obtained by ethyl acetate or methanol extraction, with ferulic acid and chlorogenic acid as major compounds, respectively. The ethyl acetate extract showed the highest antioxidant activity measured using the DPPH method. The results obtained highlighted *Chamaerops humilis* L. fruits as important sources of bioactive compounds.

Another study by Pinto et al. [13] reports the polyphenol profiling, obtained using MALDI-TOF-MS and ESI-qTOF-MS, of chestnut pericarp, integument, and curing water extracts to qualify these food by-products as a source of antioxidants. The study provides useful indications of the molecular processes associated with the traditional practice of the water curing of chestnut, which aims to prevent insect and mold development during fruit storage. This research provides a rationale to traditional processing practices on fruit appearance and qualifies the corresponding wastes as a source of bioactive compounds for other nutraceutical applications.

The review of Serna-Vazquez et al. [14] describes the latest insights into novel deep eutectic solvents (DES) for the sustainable extraction of phenolic compounds from natural sources. DES are green alternatives for the extraction processes, given their low or non-toxicity, biodegradability, and reusability. The latest studies employing DES for phenolic extraction, solvent components, extraction yield, and extraction techniques were reviewed. Moreover, the most relevant DES-based studies for phenolic extraction from natural sources and potential applications were reported.

In the research article by La Torre et al. [15], the effect of the long-term storage of Kombucha from black tea on phenolics content and radical scavenging properties is reported. Kombucha is a beverage obtained by fermenting tea with the addition of sugars. It is a highly commercialized drink produced industrially. The novel finding of this pilot study revealed that kombucha from sugared black tea can be stored at refrigerator temperature for four months. After this period, the antioxidant properties of kombucha are no longer retained.

In the article of Adebo et al. [16], the kinetic of phenolic compounds' modification during maize flour fermentation over different fermentation times is studied. The flavonoids apigenin, kaempferol, luteolin, quercetin, and taxifolin and the phenolic acids caffeic, ferulic, gallic, *p*-coumaric, sinapic, and vanillic acids were investigated. The results obtained showed that flavonoids were significantly reduced after fermentation, while phenolic acids gradually increased in prevalence. The modification of phenolics during fermentation is compound-specific, and the modification rate depends on their forms of existence in the fermented products.

The study of Salamatullah et al. [17] describes the effect of boiling techniques on the bioactive and antimicrobial properties of basil (*Ocimum sanctum*) and rosemary (*Rosmarinus officinalis*). The duration of the boiling time has a significant influence on total polyphenols and flavonoids content and antioxidant activity. Basil showed the highest antioxidant activity, total polyphenols, and total flavonoids content when it was boiled for 5 min, while rosemary exhibited the highest antioxidant activity, total polyphenols, and total flavonoids content when it was boiled for 15 min. Rosemary extracts showed high growth inhibition against Gram-positive bacteria. Salicylic acid was the most abundant phenolic compound in the rosemary sample boiled for 5 min, while acetyl salicylic acid was the most abundant phenolic compound in the basil sample boiled for 15 min.

The review of Ambra et al. [18] analyzes the main experimental reports on the fate, accessibility and bioavailability of phenolic compounds present in cooking oils and cooked vegetables. The authors considered different cooking techniques (deep-fat frying, sautéing, roasting, air-frying, microwaving, and boiling with oil), the types of oil, and the type of food, using oil alone or in combination with vegetables and how the protective effect of phenolic compounds may be improved to counteract harmful effects of oil cooking. The study of Ambra et al. [18] indicates that incomplete and contradictory observations have been published in the last few years and suggests further research is necessary to clarify the impact of cooking techniques on phenolic compounds in oil.

The article of Demir and Agaoglu deals with the identification of bioactive compounds of artichoke (*Cynara scolymus*) powder and the characterization of its antioxidant-, antimicrobial-, and metmyoglobin-reducing activity when added to minced meat during frozen storage [19]. The phenolics and flavonoids content was determined using LC-QTOF-MS. Nineteen phenolic compounds were identified via LC-QTOF-MS, with quercetin, chlorogenic acid, luteolin, catechin, and caffeic acid as the most abundant phenolics. The antioxidant activity was measured using DPPH, FRAP, and ABTS assays, while the antimicrobial activity of artichoke extract was evaluated on five different food pathogens (*S. typhimurium*, *L. monocytogenes*, *E. faecalis*, *S. aureus*, and *E. coli*) by using the disc diffusion method. In conclusion, *C. scolymus* extract exhibited good antimicrobial and antioxidant effects, stabilized the color of minced meat, and had a significant impact on sensory characteristics during the storage period.

Another interesting review of this Special Issue by Tabari et al. [20] focuses on flavonoids as promising antiviral agents against SARS-CoV-2 infection. This work provides a comprehensive review of the benefits of flavonoids in relation to COVID-19 disease. The previously reported effects of flavonoids on five RNA viruses, including influenza, Human Immunodeficiency Virus (HIV), Severe Acute Respiratory Syndrome (SARS), Middle East Respiratory syndrome (MERS), and Ebola, were considered. Flavonoids act via direct antiviral activity (via the inhibition of viral proteases, RNA polymerase, and mRNA, virus replication, and infectivity) and indirectly through the modulation of host responses to viral infection and subsequent complications (the regulation of interferons, pro-inflammatory cytokines, and inflammatory pathways).

The research article of Shafreen et al. [21] investigates the in vitro and in silico interactions of red wine polyphenols with different serum proteins: human serum albumin, fibrinogen, glutathione peroxidase 3, and C-reactive protein. The study indicates that polyphenols from red wine can interact with the key regions of serum proteins to enhance their biological activity. Among them, rutin, resveratrol, and tannic acid showed good binding affinity. Particularly, the flavonoid rutin showed the highest binding affinity with all target proteins under study. In conclusion, red wine polyphenols possess beneficial properties that can exalt their role in clinical applications.

Funding: This research received no external funding.

Institutional Review Board Statement: Not applicable.

Informed Consent Statement: Not applicable.

Data Availability Statement: Not applicable.

Acknowledgments: All the authors who kindly contributed to this Special Issue are gratefully acknowledged.

Conflicts of Interest: The authors declare no conflict of interest.

References

1. Aruoma, O. Free radicals, oxidative stress and antioxidants in human health and diseases. *J. Am. Oil Chem. Soc.* **1998**, *75*, 199–212. [CrossRef] [PubMed]
2. Scalbert, A.; Williamson, G. Dietary intake and bioavailability of polyphenols. *J. Nutr.* **2000**, *130*, 2073S–2085S. [CrossRef] [PubMed]
3. Pulido, R.; Hernandez-Garcia, M.; Saura-Calixto, F. Contribution of beverages to the intake of lipophilic and hydrophilic antioxidants in the Spanish diet. *Eur. J. Clin. Nutr.* **2003**, *57*, 1275–1282. [CrossRef] [PubMed]
4. Rienks, J.; Barbaresko, J.; Nothlings, U. Association of polyphenol biomarkers with cardiovascular disease and mortality risk: A systematic review and meta-analysis of observational studies. *Nutrients* **2017**, *9*, e415. [CrossRef] [PubMed]
5. Grosso, G.; Micek, A.; Godos, J.; Pajak, A.; Sciacca, S.; Galvano, F.; Giovannucci, E.L. Dietary flavonoid and lignan intake and mortality in prospective cohort studies: Systematic review and dose-response meta-analysis. *Am. J. Epidemiol.* **2017**, *185*, 1304–1316. [CrossRef]
6. Del Rio, D.; Rodriguez-Mateos, A.; Spencer, J.P.; Tognolini, M.; Borges, G.; Crozier, A. Dietary (poly)phenolics in human health: Structures, bioavailability, and evidence of protective effects against chronic diseases. *Antiox. Redox Signal.* **2013**, *18*, 1818–1892. [CrossRef] [PubMed]
7. Gronbaek, M.; Deis, A.; Sorensen, T.I.; Becker, U.; Schnohr, P.; Jensen, G. Mortality associated with moderate intakes of wine, beer and spirits. *Br. Med. J.* **1995**, *310*, 1165–1169. [CrossRef]
8. Gorinstein, S.; Caspi, A.; Libman, E.; Leontowicz, M.; Tahsma, Z.; Katrich, E.; Jastrzebski, Z.; Trakhtenberg, S. Bioactivity of beer and its influence on human metabolism. *Int. J. Food Sci. Nutr.* **2007**, *58*, 94–107. [CrossRef]
9. Kaplan, N.M.; Palmer, B.F. Nutritional and health benefits of beer. *Am. J. Med. Sci.* **2000**, *320*, 320–326. [CrossRef]
10. Costanzo, S.; Di Castelnuovo, A.; Donati, M.B.; Iacoviello, L.; de Gaetano, G. Wine, beer or spirit drinking in relation to fatal and non-fatal cardiovascular events: A meta-analysis. *Eur. J. Epidemiol.* **2011**, *26*, 833–850. [CrossRef]
11. Johnson, J.B.; Skylas, D.J.; Mani, J.S.; Xiang, J.; Walsh, K.B.; Naiker, M. Phenolic profile of ten Australian Faba bean varieties. *Molecules* **2021**, *26*, 4642. [CrossRef] [PubMed]
12. El Cadi, H.; El Bouzidi, H.; Selama, G.; Ramdan, B.; El Majdoub, Y.O.; Alibrando, F.; Arena, K.; Palma Lovillo, M.; Brigui, J.; Mondello, L.; et al. Elucidation of antioxidant compounds in Moroccan *Chamaerops humilis* L. fruits by GC/MS and HPLC/MS techniques. *Molecules* **2021**, *26*, 2710. [CrossRef] [PubMed]

13. Pinto, G.; De Pascale, S.; Aponte, M.; Scaloni, A.; Addeo, F.; Caira, S. Polyphenol profiling of chestnut pericarp, integument and curing water extracts to qualify these food by-products as a source of antioxidants. *Molecules* **2021**, *26*, 2335. [CrossRef] [PubMed]
14. Serna Vazquez, J.; Ahmad, M.Z.; Boczkaj, G.; Castro-Munoz, R. Latest insights on novel deep eutectic solvents (DES) for sustainable extraction of phenolic compounds from natural sources. *Molecules* **2021**, *26*, 5037. [CrossRef]
15. La Torre, C.; Fazio, A.; Caputo, P.; Plastina, P.; Caroleo, M.C.; Cannataro, R.; Cione, E. Effects of long-term storage on radical scavenging properties and phenolic content of Kombucha from black tea. *Molecules* **2021**, *26*, 5474. [CrossRef] [PubMed]
16. Adebo, O.A.; Oyedeji, A.B.; Adebiyi, J.A.; Chinma, C.E.; Oyeyinka, S.B.; Olatunde, O.O.; Green, E.; Njobeh, P.B.; Kondiah, K. Kinetics of phenolic compounds modification during maize flour fermentation. *Molecules* **2021**, *26*, 6702. [CrossRef] [PubMed]
17. Salamatullah, A.M.; Hayat, K.; Arzoo, S.; Alzahrani, A.; Ahmed, M.A.; Yehia, H.M.; Alsulami, T.; Al-Badr, N.; Al-Zaied, B.A.M.; Althbiti, M.M. Boiling technique-based food processing effects on the bioactive and antimicrobial properties of basil and rosemary. *Molecules* **2021**, *26*, 7373. [CrossRef] [PubMed]
18. Ambra, R.; Lucchetti, S.; Pastore, G. A review of the effects of olive oil-cooking on phenolic compounds. *Molecules* **2022**, *27*, 661. [CrossRef]
19. Demir, T.; Agaoglu, S. Antioxidant, antimicrobial and metmyoglobin reducing activity of artichoke (*Cynara scolymus*) powder extract-added minced meat during frozen storage. *Molecules* **2021**, *26*, 5494. [CrossRef]
20. Tabari, M.A.K.; Iranpanah, A.; Bahramsoltani, R.; Rahimi, R. Flavonoids as promising antiviral agents against SARS-CoV-2 infection: A mechanistic review. *Molecules* **2021**, *26*, 3900. [CrossRef]
21. Shafreen, R.M.B.; Lakshmi, S.A.; Pandian, K.S.; Kim, Y.M.; Deutsch, J.; Katrich, E.; Gorinstein, S. In vitro and in silico interaction studies with red wine polyphenols against different proteins from human serum. *Molecules* **2021**, *26*, 6686. [CrossRef] [PubMed]

molecules

Review

A Review of the Effects of Olive Oil-Cooking on Phenolic Compounds

Roberto Ambra *, **Sabrina Lucchetti and Gianni Pastore**

CREA—Research Center for Food and Nutrition, Via Ardeatina 546, 00178 Rome, Italy;
sabrina.lucchetti@crea.gov.it (S.L.); giovanni.pastore@crea.gov.it (G.P.)
* Correspondence: roberto.ambra@crea.gov.it

Abstract: The fate of phenolic compounds in oil and food during cooking vary according to the type of cooking. From a nutritional point of view, reviews largely suggest a preference for using extra-virgin olive oil at a low temperature for a short time, except for frying and microwaving, for which there appears to be no significant advantages compared to olive oil. However, due to the poorly pertinent use of terminology, the different protocols adopted in studies aimed at the same objective, the different type and quality of oils used in experiments, and the different quality and quantity of PC present in the used oils and in the studied vegetables, the evidence available is mainly contradictory. This review tries to reanalyse the main experimental reports on the fate, accessibility and bioavailability of phenolic compounds in cooking oils and cooked vegetables, by considering different cooking techniques and types of oil and foods, and distinguishing experimental findings obtained using oil alone from those in combination with vegetables. The re-analysis indicates that incomplete and contradictory observations have been published in the last few years and suggests that further research is necessary to clarify the impact of cooking techniques on the phenolic compounds in oil and vegetables during cooking, especially when considering their nutritional properties.

Keywords: EVOO; vegetables; phenolic compounds; thermal treatment; processing techniques; bioaccessibility; bioavailability

Citation: Ambra, R.; Lucchetti, S.; Pastore, G. A Review of the Effects of Olive Oil-Cooking on Phenolic Compounds. *Molecules* **2022**, *27*, 661. https://doi.org/10.3390/molecules27030661

Academic Editor: Eun Kyoung Seo

Received: 19 November 2021
Accepted: 17 January 2022
Published: 20 January 2022

1. Introduction

Different culinary practices are involved in heating food with vegetable oils. Among the different techniques mentioned below (Section 4), the most widespread is deep-frying, where food is entirely submerged in hot oil, inducing a process that rapidly modifies the nutritional and organoleptic properties of oil and food, normally improving those of the latter and worsening those of the former. This cooking technique has been widely used since Ancient Egypt, and oral tradition suggests that the practice is as old as the invention of pots, more than six thousand years ago [1]. Its success, historically linked to the usage of oil obtained from Olea europaea var. sativa L. fruits in other practices (lighting, religion and aesthetics), lies in the fact that it produces tastier and more durable properties in the resulting food, compared to raw food or other cooking procedures. Nowadays, even if some limitations have been introduced, especially by Mediterranean countries due to a possible link between fat consumption and obesity, oil-cooking is one of the most widely used food preparation processes worldwide [2].

Cooking with oil triggers a series of physico-chemical reactions, favoured by high temperatures, oxygen, water, and metals, which have lipids as their main targets, but also other fat components i.e., tocopherols, pigments, sterols, vitamins and phenols. Cooking with oil can yield molecules with cytotoxic, mutagenic and carcinogenic activities, affecting the cardiovascular system or generating systematic inflammation [3,4]. Interestingly, such effects are influenced by the interaction between ingredient molecules, antioxidants, and other bioactive molecules. Among those, and besides monounsaturated fatty acids, tocopherols and tocotrienols are mainly present in seed oils, while phenolic compounds (PC) may be particularly abundant in unrefined olive oils (simple phenols, secoiridoids, lignans, flavonoids, phenolic acids and alcohols), able to confer specific organoleptic features to foods and induce different reactivity to cooking oils. The ability of PC to prevent oxidative deterioration occurring in olive oil (OO) during heating was hypothesized more than 20 years ago [5]. More recently, PC in OO were shown to reduce the formation of acrylamide during potato frying [6].

Contradictory evidence is available on the fate and role of PC in oil and food during oil cooking procedures. A recent meta-analysis indicated that frying was the only cooking technique, along with steaming, that was not associated with statistically significant losses of PC from vegetables [7]. Some studies indicate losses from both food and oil, whereas others suggest some migration between oil and food. A partial explanation is that experimental cooking conditions, mimicking home ones, are hardly repeatable in and comparable between laboratories. Moreover, most works have studied the effects of heating at very long durations (repeated for even days), which are poorly representative of domestic conditions (short and unrepeated heating).

Additionally, insufficient data are available on the physical aspects of food heating. Research on other vegetable molecules indicates that heating could induce a net increase in PC as a result of several mechanisms i.e., partial evaporation of moisture that increases concentration in the food matrix, migration toward media with different polarity, changes in the microstructure of vegetable tissues, breakage or softening of the rigid cell walls and other components of the vegetable cells, and the decomposition of molecules linked to fibre. A recent review [8] reported that total frying time is normally limited to the first phase of the water cycle, evaporation, with the inner temperature of food not exceeding 100 °C. However, to our knowledge, this assumption, made 20 years before [9], was never corroborated. Other experiments with different kinds of food products indicate that the main effects on PC content rely on food cooked in oil rather than on oil itself. The situation is complicated by the fact that different oil cooking techniques exist, each with a different influencing factor that, if not considered, can result in misleading conclusions regarding health effects, as recently recapitulated by Garcimartín and collaborators in a book chapter [10].

This review tries to reanalyse the reports on the fate and on the bioavailability of PC in cooking oils and cooked vegetables, considering the different culinary approaches and the types of oil and foods and focusing on experimental findings obtained by both using oil alone or in combination with vegetables.

2. Phenolic Compounds Classification

PC, very widespread in the plant world, are bioactive molecules essential in numerous functional processes [11]. Their common structural characteristics include at least a hydroxyl group linked to an aromatic ring, but PC can also be polymerized compounds with functional groups such as mono or polysaccharides, esters and methyl esters with one or more phenolic group [12].

Historically, PC have been classified taking into account various parameters. For example, in 1962 Swain and Bate-Smith classified PC considering their spread [13], in 2017 Basli considered uniquely their carbon skeleton [14], and in 2019 Tsimogiannis and Oreopoulou classified phenols in a much more specific way, subdividing them into six classes [15]: (1) molecules with a single benzene ring (C6 class); (2) molecules with a C6 ring and a linked chain composed of one to four or seven carbon atoms (C6-Cn class); (3) and molecules belonging to the previous class with the addition of one more benzene ring linked to the carbon chain (C6-Cn-C6 class). In respect to these groups, complex phenols are organized as follows: (4) dimers [(C6-C1), $(C6-C3)_2$, $(C6-C2-C6)_2$, $(C6-C3-C6)_2$], (5) oligomeric compounds due to the condensation of dimers, (6) polymers, normally with the following structure [$(C6)_n$, $(C6-C3)_n$, $(C6-C3-C6)_n$] (7) hybrid phenolics constituted of phenolics connected to other types of molecules, such as terpenes or lipids [15]. Below are presented only categories containing molecules that are discussed in the following paragraphs.

Gallic, *p*-hydroxybenzoic, 3,4-dihydroxybenzoic, vanillic, salicylic, ellagic and syringic, caffeic and ferulic acids are some examples of the C6-C1 group typically found in fruits, cereals, vegetables and teas, and interestingly, many of them are also characteristic of OO (Figure 1). The C6-C2 group (phenylethanoids) is scarcely present in plants but contains two compounds very characteristic of OO, namely tyrosol (Ty) and hydroxytyrosol (Hy) [16] (Figure 1). Largely distributed in the vegetal world are hydroxycinnamic acids, the main molecules of C6-C3 group (phenylpropanoids), and the more diffused are coumaric, caffeic and ferulic acids (Figure 1), often esterified with other molecules. Coumarins, also belonging to the C6-C3 group, can be found in a free status or coupled with sugars, such as heterosides and glycosides in many plant families. Among the different molecules included in the C6-Cn-C6 group worth mentioning are flavonoids (C6-C3-C6) (Figure 1), the most widespread PC in the vegetal world and largely present in the human diet. Flavonoids are mainly constituted by two aromatic rings linked through a 3−carbon bridge that often forms a heterocyclic ring. Of note are quercetin and kaemferols, widespread and very concentrated in capers and in red onion, the former is also present in OO, while the latter in cabbage, beans, tea, spinach and broccoli. Apigenin and luteolin, present in OO, are within flavones, while naringenin, abundant in the genus citrus, is among flavanones [17]. Within dimers of the C6-C3 group are lignans (Figure 1). Pinoresinol is a lignan found in sesame seeds and in OO [18]. Finally, worth mentioning are secoiridoids, olive oil-specific PC [19] originating from oleuropein and ligstroside, i.e., the oleuropein aglicone mono-aldehyde (3,4-DHPEA-EA), the ligstroside aglicone mono-aldehyde (*p*-HPEA-EA), the dialdehydic form of elenolic acid linked to Hy (3,4-DHPEA-EDA, or oleacin) and the the dialdehydic form of elenolic acid linked to Ty (*p*-HPEA-EDA, or oleocanthal) (Figure 1).

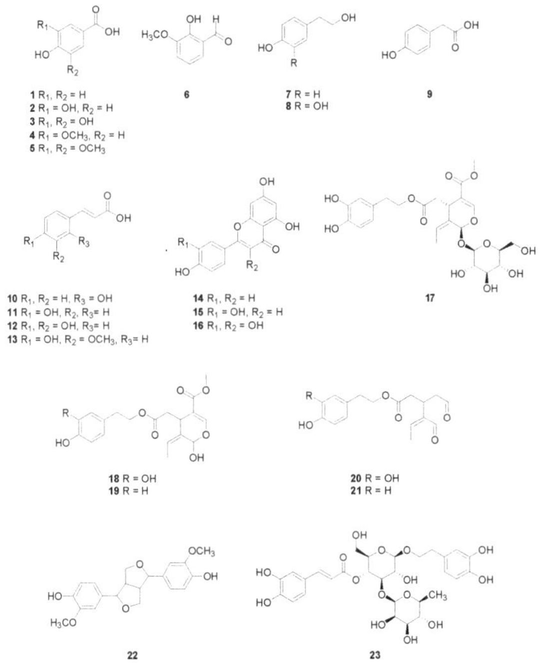

Figure 1. Phenolic compounds mentioned in figures. C6-C1 group: **1** *p*-hydroxybenzoic acid; **2** 3,4-dihydroxybenzoic acid; **3** gallic acid, **4** vanillic acids; **5** syringic acid; **6** *o*-vanillin. C6-C2 group: **7** tyrosol; **8** hydroxytyrosol; **9** *p*-hydroxyphenyl acetic acid. C6-C3 group: **10** *o*-coumaric acid; **11** *p*-coumaric acid; **12** caffeic acid; **13** ferulic acid. C6-C3-C6 group: **14** apigenin; **15** luteolin; **16** quercetin. Secoiridoids: **17** oleuropein, **18** 3,4-DHPEA-EA; **19** pHPEA-EA; **20** 3,4-DHPEA-EDA; **21** *p*-HPEA-EDA. Lignans: **22** pinoresinol; **23** verbascoside.

3. Origin and Functions of Phenolic Compounds

Despite the enormous differences found among the group, all PC originate from only two distinct pathways, namely the shikimimate/phenylpropanoids and the acetate/malonate (polyketide) pathways [20]. Different approaches, such as knockout mutations and small RNA silencing in plant models, have determined the identification of regulatory genes, biosynthetic enzymes and metabolites [21]. PC constitute an important evolutionary and diversification strategy for plant terrestrial colonization, ensuring safety from both abiotic (hydric, thermal, ultraviolet radiation, heavy metals) and biotic (predators and pathogens) seasonal stresses [22], and functional roles in growth and reproduction (for example acting as attractors for pollinating insects through colours and sensory characteristics) [23].

Many in vitro and in vivo studies have demonstrated a positive relation between PC and human health as anti-inflammatory, antibacterial and, above all, as antioxidant compounds [24]. In particular, the antioxidant capacity of PC was related to their structural characteristics, through which they can donate hydrogen or electrons, or also chelate iron and copper metals, therefore suppressing radical formation due to the presence of metals. Each group of molecules express a different antioxidant capacity related to the number and position of the hydroxyl groups linked to the carboxyl functional group. Due to their widespread presence in fruits, vegetables and beverages, PC have received significant attention in the last few decades as sources of healthy molecules in a correct and healthy diet [25]. Worth mentioning are some functions of PC present in oils produced from olives, the main subject of this review, in particular the four secoiridoids that yield Ty and Hy and that originate from oleuropein and ligstroside, the oleuropein aglicone mono-aldehyde (3,4-DHPEA-EA), the ligstroside aglicone mono-aldehyde (*p*-HPEA-EA), the dialdehydic form of elenolic acid linked to Hy (3,4-DHPEA-EDA, or oleacin) and the the dialdehydic form of elenolic acid linked to Ty (*p*-HPEA-EDA, or oleocanthal) (Figure 1) [26].

With respect to Ty, Hy, syringic acid and verbascoside, the effects of which on human health have been recently reviewed [27], animal and cell models indicate the involvement of caspase 3 in antioxidant, antigenotoxic and proapoptotic activities [28,29]. Secoiridoids anticancer effects involve the modulation of redox, inflammatory and cell-regulatory factors (reviewed in [30]). In particular, pHPEA-EDA was shown to be able to selectively block cyclooxygenases [31], demonstrating that, together with inflammatory transcription factors, kinases and cytokines are also present in antimutagenic and anticancer downstream pathways of gallic, caffeic and coumaric acids, verbascoside and pinoresinol [32–36]. Using cell models, the anticancer activities of apigenin and luteolin have been linked to the microRNA-dependent downregulation of E2F1/3 transcription factors [37], and the use of rats confirmed luteolin's properties in limiting the adverse effects of antitumorals [38,39]. Free radical scavenging or pro-apoptotic properties and ferulic, vanillic, syringic and 3,4-dihydroxybenzoic acids were demonstrated by in vitro studies [29,40], while animal models showed that restoration properties (e.g., on gut microbiota or asthma) by these molecules involved the modulation of several inflammation parameters and pro-inflammatory cytokines [41–43].

4. Are the Phenolic Compounds Influenced by Cooking?

During cooking, PC present in oils and foods are subjected to different chemical reactions that obviously depend on the cooking method technique used: deep-fat frying, sautéing (shallow/pan frying of vegetables in little amount of oil), roasting (air-baking of superficially oiled vegetables), air-frying with a small amount or no oil at all and microwaving and boiling with oil (cooking with water/oil mixes). Oil cooking reduces vegetable content in phytosterols, and reduction was linked to several variables, from length and type of technique used to oil unsaturation. Final products in deep-fat frying and roasting consist mainly in superficially browned food. In shallow frying, products range from flavoured tomato sauces for first dishes to eggplant Parmigiana, in which raw vegetables are greatly improved from a flavour point of view. As described below, on one hand, the presence of vegetables affects the content of PC in EVOO, on the other, cooking with EVOO may improve the uptake of PC, thanks to the transfer of molecules from oil to raw vegetables. Different cooking procedures are also related to the level of production of fried items: home, restaurant and industry. While domestic choices mainly favour tradition and health vs. efficiency and shelf life, industries are mainly oriented towards efficiency and price, while restaurateurs' choices are more variable due to the huge diversity in the quality of food offered and the type of consumers targeted. For example, detrimental repeated frying is avoided in shallow frying as oil is used just once, with consequences on oil uptake. On the other hand, while restaurants tend to reuse and/or partially replenish deep-fat frying after oil cooling due to economic reasons, with effects on oil quality, industries mainly perform deep-fat frying in a continuous, but controlled, manner, in order to implement frying for consecutive hours or even days [44]. Finally, home deep-fat frying tends both to discard oil every time but also to reuse it, based on arbitrary criterion [2]. Very little is known on the effects on PC in industrial-level cooking, as practically all studies have been performed in laboratory simulations.

4.1. Non-Olive Oils

Very few experiments have analysed the fate of PC in oils different from olive ones during heating. Nogueira-de-Almeida and Castro applied the same domestic conditions used for OO (200 °C for 6 min, in absence of food) to soybean and sunflower seed oils and evaluated changes in total PC (TPC) concentration [45]. TPC losses were higher for EVOO (20% loss) than for refined OO (8% loss) and complete (undetectable TPC) in both soybean and sunflower seed oils [45].

In the presence of food, the loss of TPC was complete following the deep frying (3 h at 170 °C) of fresh potatoes with a refined vegetable oil blend, consisting mostly of sunflower oil [46]. However, differently from the above publication, the loss of TPC was lower (by 42%) for EVOO than for a blend of refined and VOOs (53% loss) [46]. Some authors used the technique of spiking vegetable oils with defined amounts of PC (see also Section 4.3.1). This approach was used in a set of experiments published over 7 years by the Greek research group headed by Andrikopoulos, aimed at studying the impact of PC on the preparation of fried potatoes. Specifically, the authors enriched different oils (olive, sunflower, palm, soybean) with increasing the amounts of olive leaf extract [47–50]. Authors made an observation that still questions the validity of studies comparing the effects of cooking on oil PC. In fact, they observed that the efficiency of recovery of TPC from freshly spiked oils-before cooking-was very low and variable (from 38 to 72%), and also depended on the type of oil, i.e., higher for sunflower oil and lower, and not proportional with the amount spiked, for palm oil [47]. Additionally, the recovery of individual PC present in the leaf extract was affected. As a consequence, the recovery of TPC from oils, following the pan-frying of potatoes, barely correlated with the amount of spiked PC. Authors hypothesized a role of different chemical and physical properties of oils, but a clear corroboration is still lacking. Nonetheless, potatoes fried in phenol-spiked sunflower oil (but not palm oil) were more enriched in TPC (and individual PC, especially oleuropein) than those fried in phenol-spiked OO using the same frying conditions (201 gr of sliced potatoes fried at 175 °C for 6 min) [47], suggesting that sunflower oil could be more indicated than OO if one wants to enrich food with PC. However, the migration of PC was due to the food's absorption of the cooking oil, which calls for nutritional considerations. Later, the authors reported that supplementation of oils with TPC also increased tocopherols transfer to potatoes by frying, which was again higher for sunflower vs. OO [48]. Unfortunately, even if the authors expressly declared that oils were chosen based on their different saturation levels (reporting the chemical analysis), they did not correlate nor discuss this with respect to this characteristic. Moreover, looking at their following publication analysing longer and repeated frying sessions effects [49], it appears that the higher enrichment achievable with sunflower, compared to OO, was temporary. Unfortunately, in this last publication, the authors did not test the effects on TPC.

Santos and collaborators recently compared the enrichment in PC of potatoes fried by air-frying with different vegetable oils (canola, soybean, sunflower and olive) [51]. In contrast to the above set of experiments, sunflower oil was the only oil that did not increase the amount of PC in air-fried potatoes. In particular, even if potatoes cooked in OO had higher absolute content in PC (thanks to higher PC content of OO), the PC increase in air-fried potatoes was higher for canola, followed by soybean and olive oils, possibly because of different fatty acid content, suggesting again the opportunity to enrich with phenolic extracts non-OO oils rather than OO ones. However, even if the authors performed air-frying for longer cooking times (15–25 vs. 6 min for deep-frying), air-fried potatoes did not retain more fat and neither fats were more oxidized (as indicated by lower content of trans fatty acid in potatoes [51]), which is quite unbelievable, especially for those PC whose concentration was reported reduced by other authors (see Section 4.3.4).

4.2. Effects of Oil Heating without Food

Unfortunately, only a few studies have analysed the effects of heating in the absence of foods that can provide their own PC to EVOO or absorb PC of EVOO. Moreover, as mentioned above, cooking with oil at home involves heating using different processes that affect the fate of PC, possibly to different extents. The work by Goulas and co-workers is one of the few that recently tried to recapitulate the effects occurring in different heating processes: boiling (60 mL of oil in 400 mL of water for 40, 60 and 80 min), frying (actually heating, as 50 mL oil samples were poured into glass beakers and heated on heating plates at 180 °C for 1 or 5 h), hot-air oven baking (50 mL at 180 °C for 45 or 90 min) and microwaving (50 mL for 5 min at 500 W) (final temperature was not reported) [52]. The authors concluded that the only heating procedure that did not induce loss of TPC (estimated both by Folin–Ciocateau and by fluorometric assay) was microwaving, while the loss for baking was 11%, and boiling and frying induced relevant losses, up to near 75% for heating, time dependent for boiling but not for heating [52]. However, unfortunately, as can be noticed already from Figure 2 (reproduced from authors publication), the duration of the heat treatments was very different, even for similar procedures (boiling, frying and baking, please see the footnote), making it very difficult to compare the effects of different cooking methods.

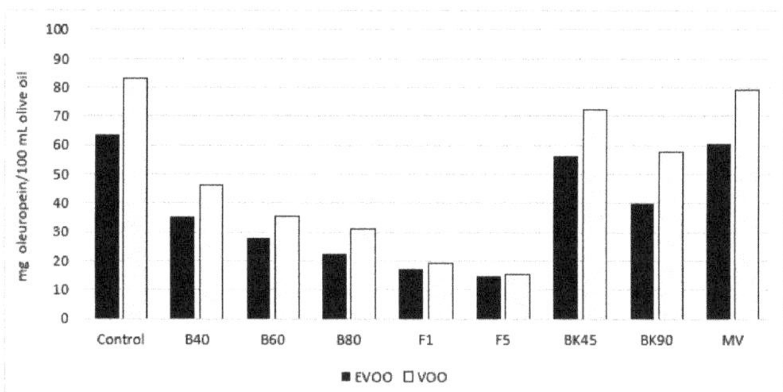

Figure 2. Impact of different heating processes on TPC of EVOO and OO in absence of foodstuff. Folin–Ciocalteu absolute quantification of TPC (mg oleuropein/100 mL of oil) in EVOO (black) and OO (white) subjected to boiling at 100 °C for 40, 60 and 80 min (B40, B60 and B80), heating at 180 °C for 1 or 5 h (F1 and F5), baking at 180 °C for 45 and 90 min (BK45 and BK90) and microwaving for 5 min at 500 W (MW). Picture from Goulas et al. [52].

Most studies that have analysed the effects of heating on PC in EVOO in absence of foods used HPLC or MS. One significant exception is the work by Campanella and co-workers that, using a biosensor to monitor changes in oxygen concentration during the tyrosinase enzyme-catalysed oxidation of PC to quinone, studied the kinetics of the decay of TPC in EVOO (25 mL), artificially oxidized by exposure to heat (98, 120, 140, 160, or 180 °C) and constant air flow exposure for 3 min to 60 h [53]. The authors noticed the complete disappearance of TPC at 98 °C and 180 °C in, respectively, 60 h and 6 h, which is a notable difference. Moreover, they estimated the half-value concentrations (respectively, 55, 45, 25, 17 and 6 min) and concluded that the kinetic constant of the process at 180 °C is about 20 times higher than at 98 °C [53]. Importantly, the same group well correlated with the kinetics of the decay of TPC to that of the oil total antioxidant capacity, especially at 180 °C [54]. Unfortunately, the authors did not follow the decay of PC individually. More recently, Kishimoto et al. used a commercial tester (OxiTester) to compare TPC loss during heating and microwaving [55]. Although this method lacks the accurate validation of TPC quantification [56], the paper is discussed as it reported the accurate detection of temperature kinetics in both heating processes (see above, Section 4.2.1).

A work used MRM tandem mass spectrometry assisted by isotope dilution to test the kinetics of the thermal decomposition of individual PC, i.e., labelled Tyrosol (Ty), Hydroxytyrosol (Hy), *p*-HPEA-EDA and 3,4-DHPEA-EDA, in OO samples treated at 90, 170 or 220 °C for 30, 90 or 150 min [57]. With respect to Hy and Ty, the authors reported the complete disappearance of the former and higher resistance for the latter, as more than 50% of the initial amount lasted till the end of the treatment after 150 min at 220 °C, possibly due to the absence of the catechol moiety [57]. With respect to secoiridoids, while losses after 30 min at 90 °C were neglectable, heating at 220 °C for 150 min caused "complete decomposition" without particular differences between Ty and Hy derivatives. However, it cannot be excluded that differences were not appreciated by the authors because the samples probably had very different compositions, as demonstrated by the huge differences reported for intermediate times and temperature treatments [57].

Accordingly, later, Carrasco-Pancorbo used separative techniques (HPCL and CE) to characterize the deterioration of EVOO at 180 °C (100 mL for 30, 60, 90, 120, 150 and 180 min), demonstrating faster degradation for Hy and its derivatives (3,4-DHPEA-EDA and 3,4-DHPEA-EA) compared to Hy-Ac, *p*-HPEA-EA and lignans [58]. Similar results were obtained by Daskalaki et al. who confirmed that incubation at 180 °C for 60 min induced about 38% reduction of the Ty derivatives and total loss of the Hy ones and reported that lignans are also stable at such high temperatures. Moreover, they demonstrated that heating at a lower temperature of 100 °C only minimally affected Ty and Hy derivatives (20% loss) [59].

Going back in time to one of the first publications that used HPLC to study the fate of PC in EVOO during heating in the absence of foodstuff, the work from Brenes et al. [60] deserves special attention. Authors checked PC in two monocultivar EVOOs (Picual and Arbequina) heated at 180 °C for 1.5 to 25 h. Although this publication is highly cited, it has some limitations that to our knowledge have never been noticed and thus are worth mentioning. Firstly, as data for shorter incubations of oil was reported incompletely, one cannot exclude that the "strange behaviour" of one of the two oils (authors reported a surprising early increase in *p*-HPEA-EDA concentration during the first 5 h of heating) could happen also in the other one. Secondly, authors concluded that the "disappearance rate depended on each individual compound and the olive cultivar" and correlated lower retention of PC to oil chemical stability, i.e., lower content in polyunsaturated fatty acids [60]. However, a deeper observation of the authors' figures on Ty and Hy and their derivatives reveals that the two EVOOs analysed probably had very different amounts of TPC (that were actually not quantified). Such a difference regarding the content in polyunsaturated fatty acids (hypothesized by authors) could actually account for different losses in PC.

Accordingly, Allouche et al. demonstrated that TPC decrease in EVOOs from the same two cultivars (Arbequina with 82 mg/kg and Picual with 406 mg/kg TPC) was higher in proportion in the PC richer EVOO [61]. For example, at both extremes of the analysis (2 and 36 h at 180 °C), residual PC contents were 92% vs. 80% and 53% vs. 36% for Arbequina vs. Picual (data at 2 h was extrapolated from figures of the publication). This is in agreement with the idea that PC are the most effective antioxidants of OO [62] and thus play a role in protecting other components from degradation, i.e., tocopherol, as already suggested 20 years ago [63]. One more interesting point that can be noticed in the data produced by Allouche et al. is the different fate of the EA secoiridoids compared to the EDA ones, especially at 2 h of heating (see Figure 3, partially extrapolated from author publication). Unfortunately, no shorter (cooking-compatible) times were investigated.

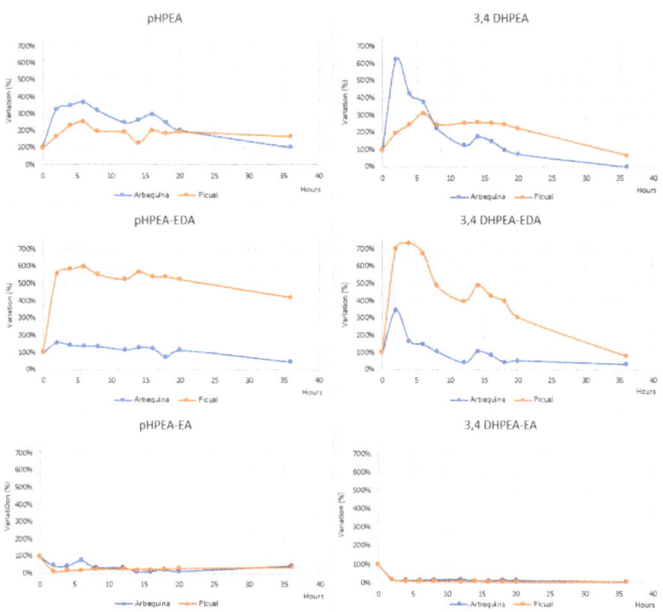

Figure 3. Effects of heating on EVOO phenolic compounds in absence of foodstuff. HPLC relative quantification (with respect of raw amounts) of Ty (**7** in Figure 1), Hy (**8**), 3,4-DHPEA-EA (**18**), pHPEA-EA (**19**), 3,4-DHPEA-EDA (**20**), *p*-HPEA-EDA (**21**) in two oils, with different initial amounts of TPC (82 and 406 mg/kg, respectively, for Arbequina and Picual), subjected to heating at 180 °C for 2–36 h in absence of vegetables. Mean changes were extrapolated from Allouche et al. [61].

Noteworthy, the data support the fact that the determination of the TPC can be insufficient and misleading for the nutritional evaluation of oils. In fact, while, similarly with Ty and Hy, the EDA secoiridoids (their derivatives linked to the dialdehydic forms of elenolic acid) hugely increased their concentrations, the EA ones (those linked to the mono-aldehydic forms of elenolic acid) were reduced by 60–80%. Both the increase in the EDA secoiridoids and the decrease in EA ones were actually already reported by Brenes [60], who hypothesized the coelution of oxidized forms for the increases; however, such a demonstration is still missing and the authors did not actually consider that the increase in aglycons concentration could simply be a consequence of the breaking of phenol-sugar glycosidic rising phenol concentration [64]. Finally, the differences between Ty and Hy (and their derivates species) showed similar lowering trends, but the lowering of the latter derivatives (the *o*-diphenolic aglycons) at prolonged heating was more marked, which is consistent with their higher antioxidant capacities [65].

With respect to other OO nutritionally relevant molecules, pinoresinol and 1-acetoxypinoresinol showed decays with different extents depending on EVOO content, with faster degradation in the phenol-richer EVOO [61]. Conversely, both luteolin and apigenin had opposite behaviour, with faster and higher degradation in the poorer EVOO, even if apigenin showed higher stability [61], accordingly with the lower number of phenolic hydroxyls [66] (Figure 1).

Later, Esposto and co-workers used a refined OO, void of hydrophilic PC, spiked with precise amounts of an already purified and characterized phenolic extract [67], to evaluate the overall and individual fates of PC after 30 min, 1, 2, 4, 6, 8, 10 and 12 h of heating at 180 °C [68]. They found that, even if oils containing higher amounts of PC showed higher resistance to the oxidation of oleuropein derivates and also, notably, the formation of negative volatile compounds and the loss of α-tocopherol, heating induced a huge degradation in PC, proportional with the initial amount [68]. In particular, they found that heating low-phenol EVOO induced the loss of more than 90% of original PC (expressed as the sum of 3,4-DHPEA and 3,4-DHPEA-EDA) already after 30 min, while a similar extent of loss was obtained only after prolonged heating or richer oils, 4 and 8 h, respectively, for oils with 200 and 400 mg/kg of PC [68], which is actually the opposite to what was observed by heating in the presence of water (see Section 4.2.3) or vegetables (see Section 4.3). Notably, they reported time-dependent formation of the oxidized dialdehydic form of decarboxymethyl elenolic acid (ox-EDA), by hydrolyzation of oxidized 3,4-DHPEA-EDA [68], consistent with the lack of accumulation of oxidized and polymerized products of Hy, as previously reported by the same group [69]. Importantly, based on the authors' data, the preservation of the phenolic fractions and the accumulation of ox-EDA in an EVOO rich in PC (more than 1200 mg/kg) during heating was delayed, compared to that of a refined OO enriched with a similar TPC, but void in *p*-HPEA-EDA, 3,4-DHPEA-EA and *p*-HPEA-EA and lignans, suggesting the higher protecting properties of such molecules and questioning the reliability of the results obtained using refined oils with only individual spiked PC. The authors hypothesize that higher protection could reside on *p*-HPEA-EDA (oleocanthal), actually absent in their phenolic extract, according to the observation that its concentration in EVOO decreased by only 29% after one hour of heating, lasting up to 12 h of heating, compared to that of the double hydroxyl group substances (3,4-DHPEA, 3,4-DHPEA-EA and 3,4-DHPEA-EDA) that were reduced in a time-dependent way and by 80% after 12 h of heating the EVOO. As long as *p*-HPEA-EA is completely degraded after one hour of heating, contribution to ligstroside derivates is supposed to depend exclusively on *p*-HPEA-EDA. Nonetheless, *p*-HPEA-EA had a starting concentration nearly ten-times lower than other secoiridoids [68]. Unfortunately, no data was reported for 3,4-DHPEA-EA, although the authors hypothesize that it may have converted into 3,4-DHPEA-EDA by the opening and decarboxylation of the elenolic ring, a process that could compensate for 3,4-DHPEA-EDA oxidation [68].

4.2.1. Microwaving

As reported above (Section 4.2), Goulas et al. concluded that microwaving (50 mL for 5 min at 500 W) was the only heating procedure, compared to boiling, frying and baking, that did not induce the loss of TPC (estimated both by Folin–Ciocateau and by fluorimetric assay) [52]. Deeper analysis showed that a concentration-dependent loss of kinetics also occurs by microwaving, but to lower extents. Cerretani et al. [70] evaluated changes occurring to three phenolic fractions (simple PC, secoiridoids and lignans) in different oils and at different times (1.5, 3, 6, 9, 12 and 15 min) of microwaving (90 mL in duplicate, at 720W). Based on their data, in EVOO, the loss starts immediately, while in non-EVOO containing lower concentrations of PC (13 vs. 93 mg/kg), it starts after 6 min of microwaving. However, at the end of treatment (15 min), PC were still present in EVOO (−83%), while in OO they disappeared completely. With respect to single PC, Ty showed higher resistance, which is consistent with what was reported above using heating. Within secoiridoids, 3,4-DHPEA-EDA showed higher resistance, followed by *p*-HPEA-EA and 3,4-DHPEA-EA. Lignans only showed slight decreases, even at the longest times, accordingly with previous results from Brenes [60], however in this case, shorter processes were tested, lower power was used (15 g of oil heated at 0.5 kW for 5 or 10 min) and minor changes were reported for Hy derivatives (20–30% loss) and increases for Ty ones [60].

On the other hand, Kishimoto et al. recently reported that microwaving (500 W) induced higher oxidative degradation of EVOO compared to normal heating (200 °C) [55]. As mentioned above, the method used by Kishimoto and co-workers for TPC quantification is poorly validated. Nonetheless, looking deeper at Kishimoto's data, which used only one oil and valuably examined temperature kinetics, one can see that after 5 min, the loss of PC was double in microwaving (around 29%) than in heating (around 14%), even if microwaved samples reached a lower temperature (140 °C vs. 200 °C). Nonetheless, different starting amounts were used (respectively, 50 and 200 g for microwaving and heating). Thus, weight, time and microwave power are crucial aspects that can deeply affect the final temperature and possibly the fate of PC. In fact, based on data from Cerretani et al. [70], microwaved samples can attain temperatures (315 °C) at the end of the treatments (15 min) higher than those normally obtained with other methods.

4.2.2. Pan-Frying

Very recently, Lozano-Castellón and co-workers studied the effects on PC by mimicking the home kitchen process of pan-frying [71]. For such a purpose, they heated a small amount of oil (20 g at 120 or 170 °C for 15 to 60 min) in a pan without controlling oxygen or light, and by means of MS-coupled UPLC they analysed changes occurring in EVOO PC [71]. Unfortunately, they did not repeat the same conditions using a deep-fryer or other oil-cooking techniques, thus their results are not comparable with other methods. In any case, they found that TPC decreased by 40% and 75%, respectively, at 120 °C and 170 °C, and that individual PC had different degradation rates [71]. With respect to longer cooking times, time did not affect TPC loss, which is consistent with previous results showing a similar decrease for 1 h or for 5 h [52]. Notably, they reported that the nutritionally relevant molecule *p*-HPEA-EDA (oleocanthal) was the more resistant compound, similarly with the higher resistance of Ty derivatives compared to Hy ones following heating, as reported by Silva et al. [72] and mentioned above (Section 4.2).

4.2.3. Boiling

Brenes et al. [60] were the first to investigate oil heating in the presence of water (60 g of oil in 2.5 L of water in a pressure cooker for at 109 °C 30 min). They reported "complete hydrolysis" of secoiridoids and "moving" of Ty and Hy to the water phase. However, as data for the oily phase were not reported, leakage could have been overestimated. Accordingly, less important losses for oil heated in presence of water were reported later [72]. Specifically, heating oil in boiling water (60 g of oil in 400 mL of water for 15 or 60 min) was shown to induce higher losses of PC, in proportion to EVOO (27 and 53% loss, respectively, at 15 and 60 min) than from non-EVOO oil (7 and 14%) [72]. Higher resistance of non-EVOO oil was associated with lower losses of secoiridoids, which were almost null except for a 16% decrease for 3,4-DHPEA-EA, even after 60 min of boiling [72], according to other indications of lower stability [61]. Interestingly, based on authors data, even if PC loss was quantitatively higher for EVOO, PC in EVOO showed a higher time-resistance of leaking to the water phase [72]. In any case, between 60 to 80% of secoiridoids leaked to the water phase, increasing from 3,4-DHPEA-EDA > 3,4-DHPEA-EA > *p*-HPEA-EDA > *p*-HPEA-EA. Higher losses observed by Brenes could rely on pressure treatment, in which higher temperature was reached (109 °C).

4.2.4. Air Baking

Goulas and co-workers [52] reported that hot-air oven baked oil (50 mL at 180 °C for 45 or 90 min) reduced TPC (calculated both by Foline–Ciocateau and by fluorimetric assay) significantly less than heated oil (50 mL at 180 °C for 1 or 5 h). Lower losses by baked (−11%) vs. heated oil (−75%) were not discussed by authors and are actually difficult to explain. However, as poor details were reported on the temperature of samples during processing, the role of the temperature kinetics cannot be excluded.

Apart from PC with stronger heat resistance (*p*-HPEA-EDA and *p*-HPEA-EA) or simpler PC (Hy and Ty) that remained unchanged, oil roasting (heating in air oven at 180 °C for 1 h, thus actually air baking) was reported more recently to induce a 16% loss of TPC and to significantly alter the phenolic composition [73]. Notably, the addition of vegetables (carrots, potato or/and onion in the 1:1 proportion) was shown to increase PC loss from oil by almost 80% of the initial molecules [73], possibly due to a migration to vegetable that was, unfortunately, not checked by authors (the role of vegetables is discussed below, Section 4.3). Nonetheless, significant alterations, even in the simpler PC Hy and Ty during roasting, in the absence of vegetables, were also reported by other authors [60].

4.3. Effects of Oil Heating with Vegetables
4.3.1. The Fate of Phenolic Compounds of EVOO during Cooking

As mentioned, vegetables can deeply modify OO PC during heating, and this process is highly influenced by the type of cooking process. One of the main factors affecting cooking performance in terms of protection of PC of oil, but also those of the food matrices, resides on whether the process is carried out in polar or apolar media. In fact, while hydrothermal processes deeply affect hydrosoluble antioxidants, such as PC (see Section 4.2.3), the use of apolar deep or surface frying in theory prevents losses in phenol concentrations of the vegetable matrix.

One of the first works to analyse the fate of PC in the presence of vegetables reported higher susceptibility of dihydroxyphenol components (3,4-DHPEA, 3,4-DHPEA-EDA and 3,4-DHPEA-EA), compared to Ty (*p*-HPEA) and its derivatives (*p*-HPEA-EDA and *p*-HPEA-EA) [74]. The authors deep-fried fresh sliced potato slices (200 g) at 180 °C for 10 min repeatedly with the same EVOO (2 L), twice a day for 6 days, allowing the oil to cool to less than 50 °C between operations, for an overall frying time of 2 h. The authors observed that 3,4-DHPEA and its secoiridoid derivatives rapidly diminished with the number of frying operations (log relationship), with components decreasing by 40–50% of their original concentration already after the first process and by more of 90% after six processes. Even more rapid loss was reported for lignans pinoresinol and 1-acetoxypinoresinol that fell drastically already in the first frying operation. On the other hand, Ty and its secoiridoid derivates showed slower reduction, with a linear relationship with the number of fryings). Higher susceptibility of dihydroxyphenols was attributed to the known higher antioxidant activity of Hy [65,75,76] and, we believe, erroneously (as stated in Section 4.2), with higher degradation in linoleic-richer oil compared to oleic-richer one [60].

Later, Casal and co-workers reported that domestic deep frying of fresh potatoes at 170 °C for 3 h induced a loss of almost 50% of TPC of EVOO oil, measured by the Folin–Ciocalteu method [46]. This is actually not surprising, if we consider that heating can result in the loss of original PC in EVOO already after 30 min at 180 °C, as reported above (Section 4.2), rather it seems that the vegetable protects oil PC. However, as the goal of the authors was to determine the degradation effects of long time frying up to 27 h (i.e., till the amount of total polar compounds reached 25%, in accordance to a Portuguese law), shorter times, more typical in home cooking, were unfortunately not tested [46]. Another work studied the relationships between olive ripening and long lasting EVOO stability during up to 40 cycles, 8 min each, of potato-frying [77]. Thanks to the processing of olives from the same cultivar and olive season, although from increasing ripening indexes, the authors tested three EVOOs with decreasing PC content (394, 289, and 78 ppm) and found that greater PC were associated with lower polar compound content and a longer Rancimat induction period [77]. Unfortunately, the negative correlation was reported in an incomplete fashion for the three oils and without detailing its timing throughout the 40 cycles and, above all, was very similar to that for tocopherols.

As mentioned above (Section 4.2.3), Silva et al. [72] reported that boiling in water induced PC loss from non-EVOO oil and, to a greater extent, from EVOO oil. Moreover, they qualified the loss of PC as "dramatic" if heating was performed in the presence of vegetables, i.e., potato, carrot, or onion, in both types of OO, even after only 15 min and for all PC tested, except in the presence of onion, especially from non-EVOO oil [72]. However, based on the authors' data, as long as the longer incubation time (60 vs. 15 min) results in higher recovered amounts of Hy, one cannot exclude the contribution of the vegetable itself to the molecule, that apparently was not analysed (authors reported quantification only of few molecules, i.e., quercetin-3,4-diglycoside, quercetin-40-monoglycoside, *p*-hydroxyphenylbenzoic acid and chlorogenic acid). On the other hand, authors suggest that increased losses of boiling with vegetables are due to the presence of elements, such as potassium or calcium and also in metals such as iron and copper, which can reduce PC very quickly and therefore contribute to their degradation, as previously reported, especially for Hy [78]. However, as onions can have important amounts of iron and copper (as reported by authors themselves), the higher yield of Hy, specifically and exclusively in samples boiled with onion, appears contradictory.

The work from Ramírez-Anaya et al. is frequently cited and thus deserves particular attention [79]. The authors studied the fate of individual PC (HPLC) and of TPC (Folin–Ciocalteu) in EVOO following the cooking of vegetables (potato, pumpkin, tomato or eggplant) for 10 min by deep frying at 180 °C (120 g of vegetable in 600 mL of oil) or by sautéing at 80–100 °C (120 g in 60 mL) [79]. They found that both cooking procedures reduced TPC in oil to a greater extent for sautéing, independently on the foodstuff cooked, with the only exception being deep frying in presence of tomato (see Figure 4, built using mean concentrations from [79]). However, as evidenced by differences in TPC at Time 0, the authors probably used different oils; this could explain some of the reported differences. Interestingly, the authors also investigated the fate of TPC in EVOO following boiling in water/oil (W/O, 10 min of heating of 120 g of vegetable at 100 °C in presence of, respectively, 540 mL and 60 mL of water and oil), finding that water increased TPC losses for all vegetable experiments (excluded eggplant, for which losses did not overcome those of sautéing). A limitation of the work is that authors did not perform tests without vegetables in order to assess their contribution to losses of the intrinsic oxidation of PC by heating, or of role of the hydrophilic medium in the migration of EVOO PC.

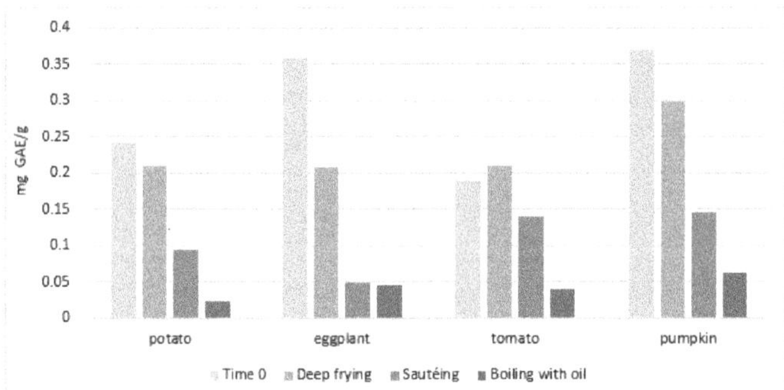

Figure 4. Fate of EVOO TPC following different heating processes in presence of different vegetables. Folin–Ciocalteu absolute quantification of TPC (mg GAE/g of oil) in oil following cooking of vegetables (120 g of potato, pumpkin, tomato or eggplant) for 10 min by deep frying at 180 °C in 600 mL of oil, by sautéing at 80–100 °C with 60 mL or by boiling at 100 °C in 540 mL of water and 60 mL of oil. Mean concentrations were extrapolated from Ramirez-Anaya et al. [79].

With respect to the effect of different techniques on individual PC, authors reported that major losses in PC were induced when vegetables were sautéed or boiled in W/O, down to complete elimination for Hy and Ty by sautéing of tomato and boiling in W/O of potato or pumpkin [79]. With respect to the effect of specific vegetables on EVOO PC, authors reported a net increase in the amount of pinoresinol and Hy in EVOO used to deep-fry tomato, which is different to that reported above for tomato by Gómez-Alonso [74].

However, looking deeper at the authors' data, the idea that the authors probably used different oils for all vegetable tested appears even more evident, especially for tomato (see Figure 5, that was built with μg/g mean concentrations from data from the publication, considering only molecules detected in all "time 0" oils (i.e., pinoresinol, oleuropein, *o*-vanillin, apigenin, Ty, Hy and *o*-coumaric, syringic and caffeic acids), and this renders the comparison of the effects of vegetables complicated, if not impossible.

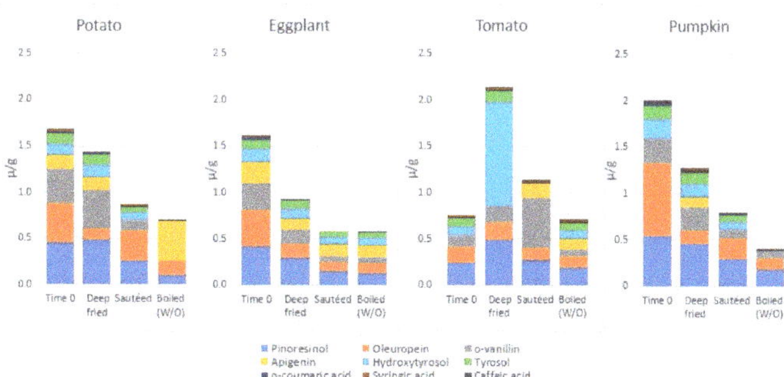

Figure 5. Effects of different heating processes on EVOO phenolic compounds of in presence of different vegetables. HPLC absolute quantification of pinoresinol (**22** in Figure 1), oleuropein (**17**), *o*-vanillin (**6**), apigenin (**14**), Ty **7**), Hy (**8**) and *o*-coumaric (**10**), syringic (**5**) and caffeic (**12**) acids (μg/g of oil) in oil following cooking of vegetables (120 g of potato, pumpkin, tomato or eggplant) for 10 min by deep frying at 180 °C in 600 mL of oil, by sautéing at 80–100 °C with 60 mL or by boiling at 100 °C in 540 mL of water and 60 mL of oil. Mean concentrations were extrapolated from Ramirez-Anaya et al. [79].

Nonetheless, looking at the authors' data on boiling W/O experiments, it appears that the type of vegetable used can significantly affect (based on the statistical analysis provided) the concentration of exclusive EVOO PC in the cooking water (i.e., oleuropein and pinoresinol, thus excluding those that can arise following the hydrolysis of other EVOO molecules, such as Hy and Ty) (Figure 6). However, as authors apparently used different oils, containing possibly different amphiphilic compounds, one cannot exclude that differences in PC migration between the two phases could be due to their interactions in water-oil interfaces [80], rather than on the vegetable cooked. Accordingly, retention of oleuropein following potato frying depends on oil characteristics, as demonstrated by the use of different oils enriched with the same amounts of TPC, as reported in Section 4.1. With respect to lignans, the frying of potatoes induced a strong loss of these molecules, which were instead reported stable in EVOO heated without food (Section 4.2).

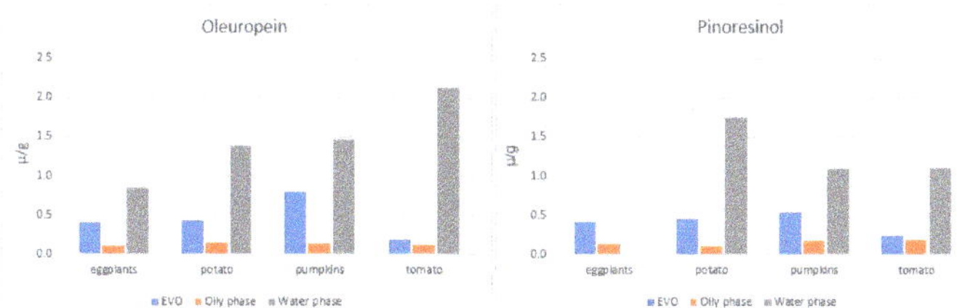

Figure 6. Oleuropein and pinoresinol concentration changes following water/oil cooking of different vegetables. HPLC absolute quantification of oleuropein and pinoresinol (respectively, **17** and **22** in Figure 1) (μg/g) in oily and water phases following cooking of vegetables (120 g of potato, pumpkin, tomato or eggplant) by boiling at 100 °C in 540 mL of water and 60 mL of oil for 10 min. Mean concentrations were extrapolated from Ramirez-Anaya et al. [79].

The use of spiked PC comes in handy again. As mentioned in Section 4.1, Chiou et al. used an olive leaf extract to enrich olive, sunflower and soybean oils to study the fate of spiked oleuropein during repeated potato frying [49]. The results unequivocally demonstrated that oleuropein is able to migrate to vegetable tissues. Specifically, the authors found that oleuropein content in the oil absorbed by potatoes was found more than ten-times enriched compared to that present in the frying oil, as if the water-rich food tissue protected the migrating relatively polar PC [49] and, with respect of the type of oils, the increase was seventeen, sixteen, and thirteen times higher in average for sunflower, olive and soybean oils, respectively [49].

De Alvarenga and co-workers recently studied the migration of PC from EVOO and vegetables during the preparation of the highly widespread Mediterranean cooking practice of tomato sauce consisting in two steps, 1 min frying of onion (400 g) and garlic (40 g) in 100 g of EVOO sofrito, followed by the addition of 460 g of tomato and further heating for 30 min at 100 °C [81]. Specifically, authors found that, among the eight phenolic molecules already present and detectable in raw EVOO by means of UPLC-MS (apigenin, elenolic acid, ferulic acid, ligstroside, luteolin, *p*-coumaric, oleuropein and pinoresinol), all were considerably reduced in the oily phase of the sofrito, apart from one that remained unchanged (pinoresinol) and one, i.e., ferulic acid, that was considerably enriched (by more than 15 times), possibly coming from garlic and tomato [81]. The migration of PC from EVOO to vegetables was indeed demonstrated by the quantification of individual EVOO PC (oleuropein, Ty, Hy, pinoresinol, *p*-coumaric and *p*-hydroxybenzoic acids) in vegetables following cooking with different techniques, as shown in Figure 7, extrapolated from Ramírez-Anaya's data [79]. Notably, the authors were able to demonstrate the enrichment of the vegetable's profile with specific EVOO PC, also by the W/O boiling technique.

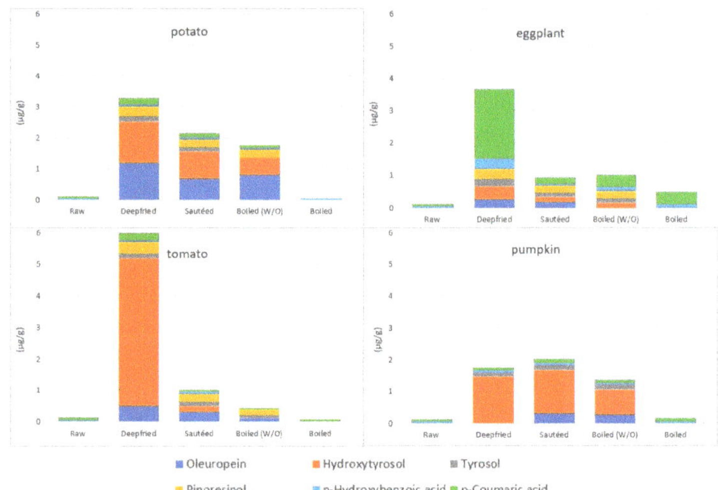

Figure 7. Incorporation of EVOO individual phenols in vegetables following cooking. Folin–Ciocalteu absolute quantification of oleuropein (**17** in Figure 1), Ty (**7**), Hy (**8**), pinoresinol (**22**), *p*-hydroxybenzoic (**1**) and *p*-coumaric (**11**) acids (µg/g) in vegetables (120 gr of potato, tomato, eggplant or pumpkin) before and after cooking for 10 min in 600 g of EVOO at 180 °C by deep-frying or with 60 g of EVOO at 80–100 °C by sautéing, or in 60 g oil plus 540 gr of water by water/oil (W/O) mixture boiling at 100 °C or in 600 g of water at 100 °C. Mean concentrations were extrapolated from Ramirez-Anaya et al. [79].

The fate of PC during sauce preparation was also previously studied by the group of Servili [82], whose work we have already mentioned [68] in the paragraph dealing with heating without vegetables (Section 4.2). Similarly, with their previous work, the authors spiked an extract of PC (containing Hy, Ty, 3,4-DHPEA-EDA and verbascoside) in 100 g of refined oil to be used for a sofrito preparation, obtained by heating at 100 °C chopped fresh celery, onion and carrot (20 g each) for 10 min, followed by the addition of tomato passata (800 g) and further boiling at 100 °C for 20 min. As expected for the presence in the extract of 3,4-DHPEA-EDA, the entire process induced an increase in Hy (but not Ty) concentration consistent with the loss of its derivative, as shown in Figure 8, which was built summing the absolute mean amounts of the 4 phenols from concentrations reported in the publication and obtained from two different amounts of spiked PC. Notably, based on authors data on the concentrations of individual PC in the oily and non-oily phases, Hy increase took place during sautéing but apparently migrated from the oily to the non-oily phase only after the addition of tomato, while the decrease in 3,4-DHPEA-EDA was not associated with the equivalent migration (Figure 9). Moreover, by spiking two concentrations of PC, the authors demonstrated that the loss of PC is not linear with the initial concentration (spiking with 60 mg induced lower-in-proportion loss compared to spiking with 40 mg), both after sautéing (8.6 vs. 22.2% loss, respectively) and after tomato cooking (26.2 vs. 39.9%) [82], which is opposite to what was observed by heating without vegetables (Section 4.2). Notably, the protective concentration-dependent effect of EVOO PC was also observed on specific PC in vegetable ingredients (quercetin-3-*O*-rutinoside, apigenin and luteolin), and on their carotenoids and α- tocopherol [82]. In particular, based on the authors' data, it appears that the more EVOO contains TPC, the less they will be lost by sautéing and tomato cooking.

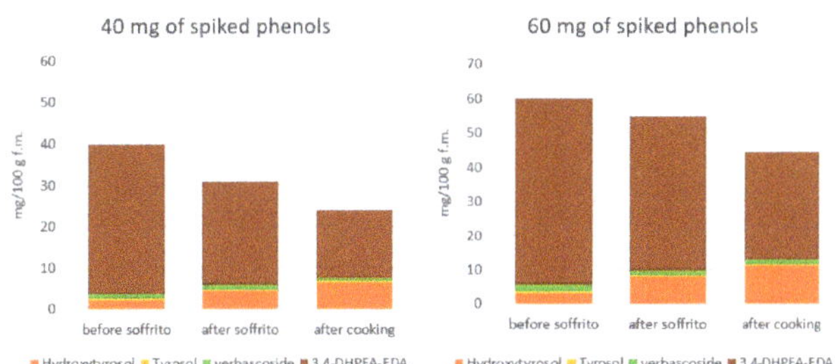

Figure 8. Fate of spiked phenols during tomato sauce preparation with refined oil. Figure 8 HPLC absolute quantification of Hy (**8** in Figure 1), Ty (**7**), 3,4-DHPEA-EDA (**20**) and verbascoside (**23**) (mg/100 g f.w.) following sofrito (celery, onion and carrot) and tomato passata cooking with a refined OO spiked with an extract containing indicated phenols at two different concentrations. Mean concentrations were extrapolated from Taticchi et al. [82].

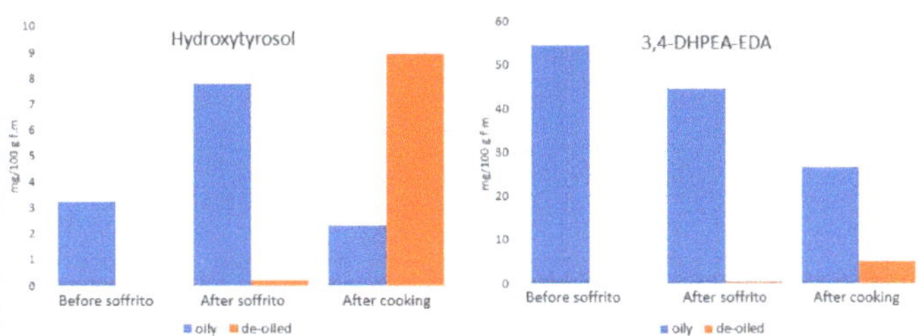

Figure 9. Migration of spiked Hy and 3,4-DHPEA-EDA during tomato sauce preparation with refined oil. HPLC absolute quantification of spiked Hy and 3,4-DHPEA-EDA (respectively, **8** and **20** in Figure 1) following sofrito (celery, onion, and carrot) and tomato passata cooking with a refined OO spiked with an extract containing indicated phenols at two different concentrations. Mean concentrations were extrapolated from Taticchi et al. [82].

4.3.2. The Incorporation of Phenolic Compounds of Vegetables in EVOO during Cooking

In the above-mentioned work on tomato sauce by de Alvarenga et al., even if the net increase in TPC in EVOO was not quantified, the authors demonstrate that cooking can enrich EVOO in individual vegetable-specific PC [81]. In fact, the oily fraction of the sofrito was enriched by molecules not present in raw EVOO, namely caffeic acid, caffeic acid hexoside, chlorogenic acid, naringenin, 3,4-dihydroxybenzoic acid, and quercetin, possibly thanks to the cooking-induced hydrolysis of progenitor phenolic glycosides yielding free hydroxylphenols groups with increased solubility in oil [81]. Notably, at the end of the process, the oily fraction of sofrito contained more PC than the water one, especially naringenin, whose concentration was similar to that present in raw tomato [81]. Authors ascribe the enrichment of sofrito in PC mainly to the step of frying onions, possibly yielding naringenin and ferulic at amounts consistent with increased absorption from tomato sauce cooked with OO, compared to tomato sauce cooked without OO (see below the work from Lamuela-Raventos' group, in the bioavailability Section 4.4.2). Unfortunately, the authors did not perform a control sofrito, skipping the onion frying step without omitting EVOO, which would have corroborated that enrichment was to be ascribed to EVOO and not to cooking.

Ramírez-Anaya and co-workers, already cited for their results on the fate of PC in EVOO following heating (Section 4.3.1), followed also the incorporation of individual PC of vegetables in EVOO, following deep frying, sautéing and boiling in water [79]. Although they did not quantify PC in vegetables, they reported that out of 14 vegetable molecules, five appeared and one increased their concentration in EVOO following cooking, compared to unused oils, i.e., 3,4-dihydroxybenzoic, gallic, *p*-hydroxyphenyl acetic and vanillic acids and luteolin and apigenin (see Figure 10 showing absolute mean concentrations from authors publication). Notably, the authors show that when using a heat matrix, a mixture of water and oil (boiled (W/O), PC migrated more and mainly towards the water phase.

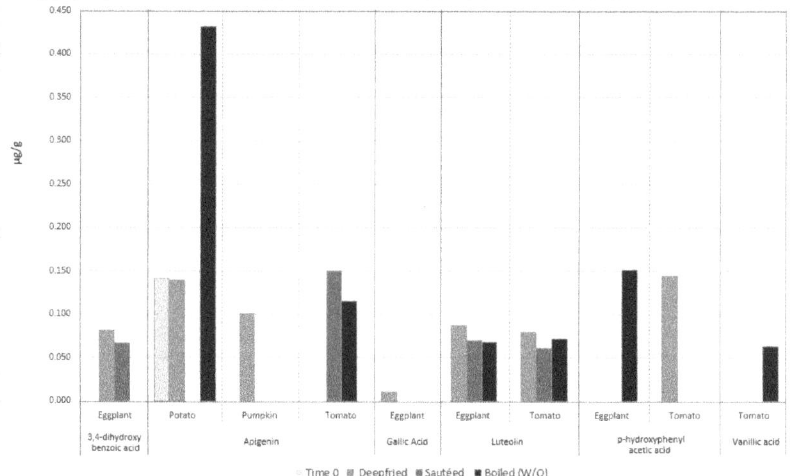

Figure 10. Incorporation of individual vegetable phenols in EVOO following deep frying, sautéing and boiling in water. HPLC absolute quantification of apigenin (**14** in Figure 1), luteolin (**15**) and 3,4-dihydroxybenzoic (**2**), gallic (**3**), *p*-hydroxyphenyl acetic (**9**), and vanillic (**4**) acids (µg/g of oil) in oil before and after cooking of vegetables (120 gr of potato, pumpkin, tomato or eggplant) for 10 min by deep frying at 180 °C in 600 mL of oil, by sautéing at 80–100 °C with 60 mL or by boiling at 100 °C in 540 mL of water and 60 mL of oil. Mean concentrations were extrapolated from Ramirez-Anaya et al. [79].

Such studies are relevant as they can help to define the rules for the predictability of PC migration during different cooking procedures. However, from a nutritional point of view, as no one would heat the residual frying oil (unless in the obvious case of sautéing), the study of the migration of PC to the opposite direction, i.e., from EVOO to vegetables, is trivially more relevant.

4.3.3. The Incorporation of EVOO Phenolic Compounds in Vegetables during Cooking

Reports investigating the migration of EVOO PC in vegetables following EVOO-cooking yielded contradictory results. In fact, in the work from Kalogeropoulos et al., a net transfer of PC from EVOO to vegetables (potato, green pepper, zucchini and eggplant) was reported by sautéing, despite the fact that both EVOO and vegetables had lost between 30 to 75% of their own TPC [50]. Specifically, based on the authors' data, chlorogenic acid increased its concentration by six times in green pepper (from 0.1 to 0.6 mg/100 g), accounting largely, according to the authors, for the "retention" of EVOO PC in the vegetable. However, as the molecule had a significantly lower concentration in fresh EVOO (0.02 mg/100 g), it looks unlikely that chlorogenic acid found in fried green pepper originated from EVOO. An increment in the TPC content of vegetables following pan-frying (and, to some extent, following sautéing) with EVOO was reported later, also by Ramírez-Anaya and co-workers [79]. Specifically, authors quantified PC in 120 g of diced (1 cm^3) potato, tomato, eggplant or pumpkin, cooked for 10 min in 600 g of EVOO at 180 °C by deep-frying, or in 60 g of EVOO at 80–100 °C by sautéing, or in 60 g oil plus 540 g of water by W/O mixture boiling at 100 °C (Figure 11). According to the authors' data, deep frying significantly increased TPC concentration in all four vegetables. As shown in Figure 11 extrapolated from the authors' data, the highest increase in TPC concentration following deep frying was for tomatoes (25 times more), followed by eggplant (10 times), pumpkin (almost five times) and potatoes (57% more). As no indication was reported on the amount and types of PC present in oil used, one cannot exclude a migration effect from oil. However, the quantification of TPC in W/O-boiled vegetables indicates the poor contribution of EVOO to vegetable enrichment of TPC (Figure 11). In fact, boiling vegetables in the W/O mixture led to a slight loss of PC from all vegetables except eggplant, whose increase was, however, oil-independent, as it was also found in water-boiled vegetables (120 g of each vegetable in 600 g of water at 100 °C). A possible explanation comes from erroneous quantification of TPC by the Folin–Ciocalteu method, or more specifically to the observation of deep changes in fat (increased) and moisture (reduced), correlating with TPC changes [83]. These observations indicate that the correct evaluation of the chemical transformations, possibly occurring in vegetables following sautéing or frying, requires parallel evaluation, not only of the effects of heating on oil itself (oil heating alone, without vegetables), but also of those on vegetables using similar conditions but without oil, for example in an oven. In fact, such controls could have indicated a change in the accessibility of PC at the end of the heating procedure. Nonetheless, as reported above, the authors, in spite of this, were able to demonstrate some migration of PC from EVOO to vegetables (see Section 4.3.1 and Figure 7) [83].

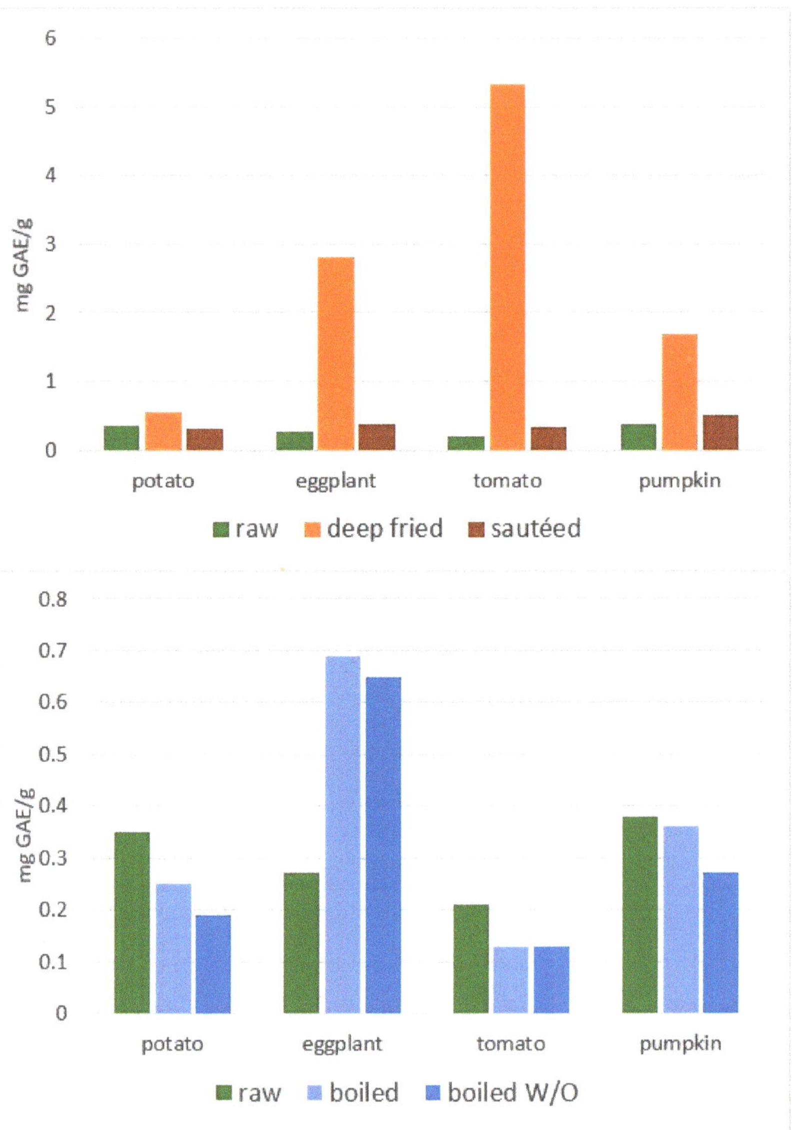

Figure 11. Incorporation of EVOO TPC in vegetables following cooking. Figure 11 Folin–Ciocalteu absolute quantification of TPC (mg GAE/g) in vegetables (120 g of potato, tomato, eggplant or pumpkin) before and after cooking for 10 min in 600 g of EVOO at 180 °C by deep-frying or with 60 g of EVOO at 80–100 °C by sautéing, or in 60 g oil plus 540 g of water by water/oil (W/O) mixture boiling at 100 °C or in 600 g of water at 100 °C. Mean concentrations were extrapolated from Ramirez-Anaya et al. [79].

4.3.4. The Fate of Phenolic Compounds of Vegetables during EVOO-Cooking

Recent reports indicate that oil cooking techniques induce higher losses in the PC of vegetables compared to other oil-free cooking techniques. For example, while microwaving and boiling induced negligible TPC losses, significant losses were reported in red cabbage for stir-frying [84] and in purple-fleshed potatoes for frying, stir-frying and air-frying [85].

Specifically, Tian et al. showed that the decrease in TPC was the highest for stir-fried potatoes (72.44% reduction). However, the comparison is worth little, as stir-frying was performed with potatoes cut eight-times thinner compared to other frying techniques. In this regard, the authors also tested the new technique of air-frying, using three-times less fat compared to frying (10 mL of soybean oil for 300 g of potatoes instead of 3 L for the same quantity of potatoes), finding a higher reduction of TPC (32.52% reduction) compared to frying (14.08%), possibly because air-frying lasted longer (18 min at 180 °C instead of 2 min at 191 °C for frying). Before concluding that air-frying is a healthier frying technique, based only on the fact that less oil is used, one should also consider that it induces more than doubled the reduction of TPC [85]. Similarly, even if PC were not quantified in the white coconut oil used, Gunathilake et al. reported that frying induced, in six different edible leaves, higher losses of TPC and flavonoids compared to boiling and steaming, and the extent of the losses depended on the vegetable species [86]. On the other hand, Mashiane and co-workers reported that within Cucurbitaceae stir-frying determined lower losses than microwaving [87]. However, authors stir-fried pumpkin leaves using an EVOO (10% volume compared to the weight of leaves), the PC of which were not quantified.

With respect to a single PC, in their already repeatedly mentioned article, Silva et al. reported that frying for 15 min in EVOO conserved most of onion and carrot quercetin and *p*-hydroxyphenylbenzoic, respectively, while potatoes kept only small amounts of their chlorogenic acid [72], accordingly with previous reports, showing that frying did not affect quercetin concentration in onion, while boiling induced 30% loss of glycosides [88]. Similarly, Jung et al. [89] performed an analysis of the fate in the roots of sweet potato cooked with six different home-processing techniques (boiling, deep frying, microwaving, oven baking, sautéing, and steaming) of six chlorogenic acids, demonstrating that, in general, deep frying showed the greatest reductions overall. Notably, authors from the same group subsequently used the fate of these PC to set up a functional mathematical index for predicting the effects of food processing [90]. Some works indicate apparent opposite effects. Martini and co-workers reported a significant increase in eggplant total hydroxycinnamic acids concentration following all tested cooking techniques (baking, boiling, frying and grilling), and a higher increase for sunflower vs. OO [91]. Similarly, for onion, green pepper or cardoon, identical frying conditions almost doubled the concentrations of chlorogenic acids and flavonoids in cooked vegetables, with respect to raw ingredients, using sunflower but not olive oil. However, even if higher temperatures were used (150 instead of 115 °C), increases in chlorogenic acids were even higher for griddling without oil [92]. Managa et al. reported that stir frying (at 125 to 140 °C) Chinese cabbage leaves with EVOO (10 mL) in nightshade (120 g) increased their concentration of kaempferol and quercetin derivatives and caffeoylmalic and chlorogenic acids, especially the latter [93]. However, similar to Martini's results, concentrations also increased (doubled with respect to raw) with other cooking techniques tested, with overall similar yields for microwaving, boiling and steaming, but lower than stir frying (quadrupled), supporting the general role of heating in also increasing individual PC accessibility in some specific vegetables (accessibility should not be confused with bioaccessibility, see Section 4.4.1). Accordingly, changes in vegetable matrices or enzymes were associated with a steaming or boiling-dependent increase in TPC content in artichoke leaves [94]. Similarly, a comparison of the effects of cooking treatments (boiling, steaming and microwave-cooking) of several Mediterranean wild edible species without employing oil, reported that any kind of heating increased the accessibility/availability of chlorogenic acid and rutin, and this could possibly also contribute to the net increase in TPC associated with any cooking procedure [95].

4.4. Changes in Bioaccessibility and Bioavailability of Vegetable PC Following Cooking with Oil

A terminological explanation is needed. Normally, bioaccessibility is related to digestion and absorption efficiencies of a nutrient, or any bioactive compound orally administered. It is normally expressed as the ratio between the amount of the constituent released and absorbed and its total amount ingested, regardless of whether the body is then able to use it in its metabolic processes. Basically, bioaccessibility depends on events driven by (1) digestion (that makes the bioactive molecules potentially bioaccessible matter); (2) absorption through epithelial tissue; and (3) pre-systemic metabolism. The term bioavailability is wider as it also includes the nutritional efficiency of the bioactive compound, i.e., the capability of the body to really "use" and take advantage of a particular component. Thus, bioavailability depends firstly on the systemic distribution of the compound and secondly on its potential interaction with target tissues. It is expressed as the ratio between the amount of the constituent actually utilized in metabolic functions (or stored to be subsequently used in metabolic functions) and the total amount introduced with the diet.

Several studies have been published to describe the bioaccessibility and bioavailability of PC in different food matrices. The review published by Shahidi and Peng cited 356 references and concluded that the bioaccessibility and bioavailability of different PC strictly depends on the effective dose taken, the general physical condition of the individual, and on the activity and efficiency of various internal mechanisms (i.e., digestion, absorption, transport, metabolic processes, excretion and microbiota activities) [96]. In fact, no more than 10% of the total intake of dietary PC are directly absorbed and almost all are metabolized in the gut microbiome to more easily absorb metabolites [97]. However, some recent indications support the possibility that an increase in the absorption of PC could be mor easily improved by their incorporation into OO.

Obviously, the thermal treatment of any cooking process may also modify the structure of the bioactive molecules present in the food, which may consequently influence bioaccessibility and bioavailability. In general, it can be stated that the cooking process and/or the heat treatment exerts a positive effect on bioaccessibility and bioavailability due to the softening of the cells [98]. Moreover, several cooking procedures involve the addition of other ingredients that can positively or negatively influence the bioaccessibility and bioavailability of all the components. A typical example is the culinary use of OO, which is known to improve the bioavailability of lipophilic compounds (such as carotenoids) by acting as food excipients and enhancing their extraction [99].

Unfortunately, only few papers among those dealing with the effects of thermal treatment of PC of OO, have measured changes in bioaccessibility and bioavailability. Moreover, most are in vitro studies or were performed on a very small number of subjects. For that reason, very little is known about that matter, and even less is known about the impact of a vegetable foodstuff-OO combination during processing on PC bioaccessibility and bioavailability.

4.4.1. Bioaccessibility

Very few and inconclusive studies have been published on the influence of cooking on the bioaccessibility of PC, in part due to inaccurate terminology. In fact, as mentioned in the above paragraph, studies on bioaccessibility should at least be performed though in vitro simulated gastrointestinal digestion, in this case defining bioaccessibility as the ratio between PC contents after and before in vitro digestion. Thus, higher than 100% values of bioaccessibility are not expected and, if reported, probably reflect incorrect extraction or quantification, due to, for example, the formation of water-soluble Maillard reaction products during cooking [100]. This is the case for the already reported work in Section 4.3.4 from Martini and colleagues, which compared the effects of different cooking techniques on the stability of dark purple eggplants PC. In particular, the authors reported higher than 100% values, possibly because they calculated bioaccessibility using PC concentration of a methanolic extract instead of that of the starting tissue. In any case, in relative terms, compared to other cooking techniques, they found the lowest bioaccessibility for frying [91]. Juàniz and colleagues measured the effect of griddling cardoon or frying it in olive or sunflower oil. The author reported that only 2% of PC were bioaccessible in raw cardoon, whereas in cooked samples, up to between 60 and 67% of TPC remained bioaccessible after gastrointestinal digestion, with griddling showing the best performance [101].

4.4.2. Bioavailability

Excluding studies performed on cereals, the heat treatment for which is carried out without the use of a fat matrix, papers available in the scientific literature dealing with PC bioavailability mainly refers to the preparation of the tomato sauces and sofrito prepared with tomatoes, onion and VOO or EVOO, classic condiments for the preparation of typical dishes of the Mediterranean diet. Worth a mention is the fact that, regardless of the effects of the presence of oil, the results about the increase in bioavailability of tomato PC following cooking were already known as early as 2004, for example for naringenin and chlorogenic acid [102]. Among them are three randomized controlled cross-over studies performed by the work led by Lamuela-Raventós. In the first pilot randomized controlled cross-over study, scholars studied the effect of the addition of an oil matrix during tomato sauce processing, on the accessibility/extractability and bioavailability of 11 PC of tomato [103]. The authors found that cooking increased the accessibility or extractability of 9 out of 11 compounds (in particular naringenin, rutin and ferulic acid); however, that was never improved by the presence of 5% oil, which is somehow unexpected, considering similar previous work showing increased yield when using 10% instead of 5% EVOO [104]. Indeed, oil presence decreased accessibility or extractability for about half of them compared to the sauce cooked without oil. Nevertheless, the presence of OO (EVOO or refined) in the preparation of tomato sauces increased plasma concentrations in naringenin glucuronides, but not in a statistically significant way, possibly due to high individual variability among very few subjects ($n = 5$). Notably, ingestion of the oil-enriched tomato sauces was associated with re-absorption events of PC, possibly induced by a lipid matrix-stimulated enterohepatic circulation [103]. Interestingly, based on the authors' data, pharmacokinetics differences were also noticed for naringenin between EVOO and refined oils. Unfortunately, the authors did not characterize PC nor fat present in the two oils, as protecting or influencing tomato PC, which could have at least partially explained the observed differences. Three years later, the group investigated the same molecules in another randomized controlled study (carried out on eight subjects), using only a refined oil void of any phenol and characterized for fat content, obtaining different results for oil presence during tomato cooking, i.e., statistically not significant increased accessibility or extractability for some PC (especially naringenin and caffeic acid hexose) but no effects on bioavailability [105]. The increase in naringenin could depend on its ability to be released from the cuticle of tomato where it is trapped. In their last controlled, randomized crossover feeding trial, performed with a significantly higher number of subjects ($n = 40$), the group demonstrated the higher bioavailability of naringenin glucuronide and quercetin, even in presence of statistically not significant increases in PC concentration, following oil addition in tomato cooking [106]. Again, the authors did not characterize the refined oil for PC content nor fat. The use of an EVOO with known amounts of PC or, better, of a refined oil with defined amounts of added PC, could have added more information on PC ability to protect tomato PC and increase their bioavailability.

Some other experiments studied the vascular protective [107] or insulin sensitivity properties [108] of the tomato-based sofrito in obese rats. However, even if PC were quantified in experimental food administered to animals, the nutrients of the tomato-based sofrito were divided into specific diets such as to allow to discern the effects of oil and/or PC. Unpublished data from the review by Garcimartín and co-workers support different effects of fried oils in rat jejunum, i.e., increased antioxidant enzymes following administration of discontinuously used EVO compared to sunflower [10]. However, to our knowledge, no results have been made available so far. Nevertheless, some data indicate that the beneficial effects of sofrito on rats could be attributed to PC [107,108]. Unfortunately, these molecules were quantified in experimental food administered to animals only in one publication [107].

5. Conclusions

The analysis of the literature of TPC fate during oil cooking appears at times conflicting. Some contradictory results can be explained by the poorly pertinent usage of terminology on cooking procedures, different protocols adopted in the different studies aimed at the same objective, different types and quality of oils used in the experiments, and different quality and quantity of PC present in the used oils and in the studied vegetables. For example, despite there being several experimental indications (also in food-less studied) that PC prevent tocopherol degradation during frying, more experiments are needed in order even to conclude to what extent PC, and not tocopherols, are relevant for maintaining fried oil shelf life. Obviously, the type of cooking (mainly its temperature and duration), deeply modifies the status and the composition of the food as a whole, beyond the sole presence of PC, and this concept should be taken into consideration when selecting a procedure to cook our foods. With respect simply to the PC fate, it is clear that the strategy adopted to minimize the loss of PC should account for a low cooking temperature and short cooking time (less than 15 and 6 min, respectively, for standard and microwaving heating time processing, to prevent the loss of the healthy but less stable OO secoiridoids 3,4-DHPEA-EDA and 3,4-DHPEA-EDA). Regarding the frying techniques, results are hardly comparable as a result of the different incubation times, which were typically shorter for frying. Regarding the type of oil, EVOO is basically better than OO, except in frying and apparently in microwaving, as higher TPC content does not confer increased PC stability. With regards to water/oil cooking, the choice should depend on the raw ingredient to be cooked. In general, when the PC content of the raw vegetable is high, cooking in the presence of water will not significantly affect PC concentration. However, the consumption also of the cooking water would benefit their uptake. On the other hand, when content is low, sautéing and frying with EVOO can compensate for the foodstuffs' weaknesses or enrich them in PC.

Overall, if one wants to accurately compare the effects of a particular cooking procedure in the presence of a certain vegetable (or water or other polar/apolar media), on the biological activity of (to simplify) one PC of EVOO, he should (1) use the same oil (or oils for oil-oil comparisons) and oil/food ratios for all heating treatments; (2) independently subject the oil and the vegetable to similar heating (temperature, time, pressure, evaporation) and processing (dicing, slicing) treatments; (3) run parallel experiments with other cooking techniques using at least one identical physical condition (for example the same temperature for deep- and air frying, or, for microwaving, thoroughly following the heating kinetics of the apparatus); (4) use parallel oil samples with quantitatively precise spiked amounts of the purified PC and a spiked refined oil with no PC at all; (5) preliminarily assess the recoverability of spiked PC in raw conditions; (6) analyse the kinetic of losses/accumulation of both individual PC and their derivatives (for example oleuropein and the four secoiridoids, their oxidized/hydrolysed forms, and Ty and Hy, by HPLC or MS) and TPC (Folin–Ciocalteu or another validated method) in both oil and vegetable heated together and separately (especially for sautéing), starting with short incubation times (minutes) and up to home-compatible usages (up to one hour) or industrial (days with/without cycle replenishments); (7) in order to assess the role of oil tocopherols, tocotrienols, pigments, minerals or fats or other molecules, analyse their content; and (8) do not confuse accessibility with bioaccessibility and bioavailability.

From a nutritional point of view, it is worth mentioning that, thanks to its light frying conditions in EVOO, the sofrito obtained sautéing tomatoes, carrots and onions at low temperature and for a short time length, is actually the base for many Mediterranean dishes and recipes and has been included in a validated questionnaire used to evaluate adherence to the Mediterranean diet [109]. As indicated by the transfer from oil to vegetables of molecules that are exclusively found in EVOO (for example oleuropein derivatives, pinoresinol, Ty and Hy), frying could also be viewed as a potential nutritional solution to increase the uptake of PC. The question remains as to how relevant this enrichment could be from a healthy point of view. In terms of uptake, experiments concerning the controlled enrichment of EVOO indicate that normal amounts of PC in EVOO are still enough concentrated to theoretically provide bioactive molecules at amounts claimed to lower blood LDL oxidation and thus the risk of cardiovascular diseases [110], i.e., at the minimum level of supplementation of 5 mg per day [82]. However, as analyses of bioavailability are very limited, controlled clinical trials are crucial for confirmation. Moreover, experimental data strongly suggest that the amount of PC transferred in food is proportional with the total amount of oil being absorbed by food. Defining a nutritional goal by just increasing the PC intake without considering the composition of the ingested food, as a whole, is nutritionally meaningless. One should consider: (1) the uptake of fat, (2) whether such fried EVOO is healthy, (3) whether there other unwanted compounds, and (4) if this procedure economically advantageous. Accordingly, the guidelines of all western countries concerning healthy nutrition advise against the frequent use of deep frying, in order to limit the ingestion of oxidized fatty acids and/or undesirable compounds (i.e., heterocyclic amines, acrylamide, acrolein, hexanal etc). Thus, the authors that speculated that frying with EVOO could improve the dietary fatty acid profile of ingested food thanks to the presence of PC [10] should consider that the dietary fatty acid profile would also equally improve with the use of less violent and more healthy cooking procedures or, even better, by using raw EVOO in our dishes.

Funding: This research received no external funding.

Institutional Review Board Statement: Not applicable.

Informed Consent Statement: Not applicable.

Conflicts of Interest: The authors declare no conflict of interest.

References

1. Bastida, S.; Sánchez-Muniz, F.J. Chapter 21—Frying: A Cultural Way of Cooking in the Mediterranean Diet. In *The Mediterranean Diet*; Preedy, V.R., Watson, R.R., Eds.; Academic Press: San Diego, CA, USA, 2015; pp. 217–234. ISBN 978-0-12-407849-9.
2. Mesias, M.; Delgado-Andrade, C.; Morales, F.J. Evaluation of domestic frying habits and consumer's preferences in Spanish households. *SDRP J. Food Sci. Technol.* **2018**, *3*, 450–459. [CrossRef]
3. Hosseini, H.; Ghorbani, M.; Meshginfar, N.; Mahoonak, A.S. A Review on Frying: Procedure, Fat, Deterioration Progress and Health Hazards. *J. Am. Oil Chem. Soc.* **2016**, *93*, 445–466. [CrossRef]
4. Ng, C.-Y.; Leong, X.-F.; Masbah, N.; Adam, S.K.; Kamisah, Y.; Jaarin, K. Heated vegetable oils and cardiovascular disease risk factors. *Vascul. Pharmacol.* **2014**, *61*, 1–9. [CrossRef] [PubMed]
5. Cinquanta, L.; Esti, M.; Di Matteo, M. Oxidative stability of virgin olive oils. *J. Am. Oil Chem. Soc.* **2001**, *78*, 1197. [CrossRef]
6. Napolitano, A.; Morales, F.; Sacchi, R.; Fogliano, V. Relationship between Virgin Olive Oil Phenolic Compounds and Acrylamide Formation in Fried Crisps. *J. Agric. Food Chem.* **2008**, *56*, 2034–2040. [CrossRef]
7. Murador, D.; Braga, A.R.; Da Cunha, D.; De Rosso, V. Alterations in phenolic compound levels and antioxidant activity in response to cooking technique effects: A meta-analytic investigation. *Crit. Rev. Food Sci. Nutr.* **2018**, *58*, 169–177. [CrossRef] [PubMed]
8. Chiou, A.; Kalogeropoulos, N. Virgin Olive Oil as Frying Oil. *Compr. Rev. Food Sci. Food Saf.* **2017**, *16*, 632–646. [CrossRef]
9. Varela, G.; Ruiz-Rosso, B. Frying process in the relation fat/degenerative diseases. *Grasas Aceites* **1998**, *49*, 359–365. [CrossRef]
10. Garcimartín, A.; Macho-González, A.; Caso, G.; Benedí, J.; Bastida, S.; Sánchez-Muniz, F.J. Chapter 19—Frying a cultural way of cooking in the Mediterranean diet and how to obtain improved fried foods. In *The Mediterranean Diet*, 2nd ed.; Preedy, V.R., Watson, R.R., Eds.; Academic Press: San Diego, CA, USA, 2020; pp. 191–207, ISBN 978-0-12-818649-7.
11. de la Rosa, L.A.; Moreno-Escamilla, J.O.; Rodrigo-García, J.; Alvarez-Parrilla, E. Chapter 12—Phenolic Compounds. In *Postharvest Physiology and Biochemistry of Fruits and Vegetables*; Yahia, E.M., Carrillo-Lopez, A., Eds.; Woodhead Publishing: Cambridge, England, 2019; pp. 253–271, ISBN 978-0-12-813278-4.
12. Vuolo, M.M.; Lima, V.S.; Maróstica Junior, M.R. Chapter 2—Phenolic Compounds: Structure, Classification, and Antioxidant Power. In *Bioactive Compounds*; Campos, M.R.S., Ed.; Woodhead Publishing: Cambridge, England, 2019; pp. 33–50, ISBN 978-0-12-814774-0.
13. Swain, T.; Bate-Smith, E.C. Flavonoid compounds. *Comp. Biochem.* **1962**, *Volume III*, 755–809.
14. Basli, A.; Belkacem, N.; Amrani, I. Health Benefits of Phenolic Compounds Against Cancers. In *Phenolic Compounds-Biological Activity*; IntechOpen: Rijeka, Croatia, 2017.
15. Tsimogiannis, D.; Oreopoulou, V. Classification of Phenolic Compounds in Plants. In *Polyphenols in Plants*; Elsevier: Amsterdam, The Netherlands, 2019; pp. 263–284.
16. Romero, C.; Brenes, M. Analysis of Total Contents of Hydroxytyrosol and Tyrosol in Olive Oils. *J. Agric. Food Chem.* **2012**, *60*, 9017–9022. [CrossRef]
17. Birt, D.F.; Jeffery, E. Flavonoids. *Adv. Nutr.* **2013**, *4*, 576–577. [CrossRef]

18. López-Biedma, A.; Sánchez-Quesada, C.; Delgado-Rodríguez, M.; Gaforio, J.J. The biological activities of natural lignans from olives and virgin olive oils: A review. *J. Funct. Foods* **2016**, *26*, 36–47. [CrossRef]
19. Luque-Muñoz, A.; Tapia, R.; Haidour, A.; Justicia, J.; Cuerva, J.M. Direct determination of phenolic secoiridoids in olive oil by ultra-high performance liquid chromatography-triple quadruple mass spectrometry analysis. *Sci. Rep.* **2019**, *9*, 15545. [CrossRef]
20. Knaggs, A.R. The biosynthesis of shikimate metabolites. *Nat. Prod. Rep.* **2003**, *20*, 119–136. [CrossRef]
21. Fraser, C.M.; Chapple, C. The phenylpropanoid pathway in Arabidopsis. *Arab. B* **2011**, *9*, e0152. [CrossRef]
22. Sharma, A.; Shahzad, B.; Rehman, A.; Bhardwaj, R.; Landi, M.; Zheng, B. Response of Phenylpropanoid Pathway and the Role of Polyphenols in Plants under Abiotic Stress. *Molecules* **2019**, *24*, 2452. [CrossRef]
23. Ávila-Román, J.; Soliz-Rueda, J.R.; Bravo, F.I.; Aragonès, G.; Suárez, M.; Arola-Arnal, A.; Mulero, M.; Salvadó, M.-J.; Arola, L.; Torres-Fuentes, C.; et al. Phenolic compounds and biological rhythms: Who takes the lead? *Trends Food Sci. Technol.* **2021**, *113*, 77–85. [CrossRef]
24. Fernandez-Panchon, M.S.; Villano, D.; Troncoso, A.M.; Garcia-Parrilla, M.C. Antioxidant Activity of Phenolic Compounds: From In Vitro Results to In Vivo Evidence. *Crit. Rev. Food Sci. Nutr.* **2008**, *48*, 649–671. [CrossRef] [PubMed]
25. Aguilera, Y.; Martin-Cabrejas, M.A.; González de Mejia, E. Phenolic compounds in fruits and beverages consumed as part of the mediterranean diet: Their role in prevention of chronic diseases. *Phytochem. Rev.* **2016**, *15*, 405–423. [CrossRef]
26. Ambra, R.; Lucchetti, S.; Pastore, G. The health benefits of oleocanthal and other oil phenols. In *Handbook of olive Oil: Phenolic Compounds, Production and Helath Benefits*; Nova Science Publisher: New York, NY, USA, 2017; pp. 215–236, ISBN 978-1-53612-356-2.
27. Karković Marković, A.; Torić, J.; Barbarić, M.; Jakobušić Brala, C. Hydroxytyrosol, Tyrosol and Derivatives and Their Potential Effects on Human Health. *Molecules* **2019**, *24*, 2001. [CrossRef] [PubMed]
28. Anter, J.; Tasset, I.; Demyda-Peyrás, S.; Ranchal, I.; Moreno-Millán, M.; Romero-Jimenez, M.; Muntané, J.; Luque de Castro, M.D.; Muñoz-Serrano, A.; Alonso-Moraga, Á. Evaluation of potential antigenotoxic, cytotoxic and proapoptotic effects of the olive oil by-product "alperujo", hydroxytyrosol, tyrosol and verbascoside. *Mutat. Res. Toxicol. Environ. Mutagen.* **2014**, *772*, 25–33. [CrossRef]
29. Gheena, S.; Ezhilarasan, D. Syringic acid triggers reactive oxygen species–mediated cytotoxicity in HepG2 cells. *Hum. Exp. Toxicol.* **2019**, *38*, 694–702. [CrossRef] [PubMed]
30. Celano, M.; Maggisano, V.; Lepore, S.M.; Russo, D.; Bulotta, S. Secoiridoids of olive and derivatives as potential coadjuvant drugs in cancer: A critical analysis of experimental studies. *Pharmacol. Res.* **2019**, *142*, 77–86. [CrossRef] [PubMed]
31. Beauchamp, G.K.; Keast, R.S.J.; Morel, D.; Lin, J.; Pika, J.; Han, Q.; Lee, C.-H.; Smith, A.B.; Breslin, P.A.S. Ibuprofen-like activity in extra-virgin olive oil. *Nature* **2005**, *437*, 45–46. [CrossRef]
32. Sharma, S.H.; Rajamanickam, V.; Nagarajan, S. Antiproliferative effect of p-Coumaric acid targets UPR activation by downregulating Grp78 in colon cancer. *Chem. Biol. Interact.* **2018**, *291*, 16–28. [CrossRef] [PubMed]
33. Jung, H.W.; Mahesh, R.; Lee, J.G.; Lee, S.H.; Kim, Y.S.; Park, Y.-K. Pinoresinol from the fruits of Forsythia koreana inhibits inflammatory responses in LPS-activated microglia. *Neurosci. Lett.* **2010**, *480*, 215–220. [CrossRef]
34. Lee, J.Y.; Woo, E.-R.; Kang, K.W. Inhibition of lipopolysaccharide-inducible nitric oxide synthase expression by acteoside through blocking of AP-1 activation. *J. Ethnopharmacol.* **2005**, *97*, 561–566. [CrossRef] [PubMed]
35. Mirzaei, S.; Gholami, M.H.; Zabolian, A.; Saleki, H.; Farahani, M.V.; Hamzehlou, S.; Far, F.B.; Sharifzadeh, S.O.; Samarghandian, S.; Khan, H.; et al. Caffeic acid and its derivatives as potential modulators of oncogenic molecular pathways: New hope in the fight against cancer. *Pharmacol. Res.* **2021**, *171*, 105759. [CrossRef]
36. López-Biedma, A.; Sánchez-Quesada, C.; Beltrán, G.; Delgado-Rodríguez, M.; Gaforio, J.J. Phytoestrogen (+)-pinoresinol exerts antitumor activity in breast cancer cells with different oestrogen receptor statuses. *BMC Complement. Altern. Med.* **2016**, *16*, 350. [CrossRef]
37. Cheng, Y.; Han, X.; Mo, F.; Zeng, H.; Zhao, Y.; Wang, H.; Zheng, Y.; Ma, X. Apigenin inhibits the growth of colorectal cancer through down-regulation of E2F1/3 by miRNA-215-5p. *Phytomedicine* **2021**, *89*, 153603. [CrossRef]
38. El-Asfar, R.K.; El-Derany, M.O.; Sallam, A.-A.M.; Wahdan, S.A.; El-Demerdash, E.; Sayed, S.A.; El-Mesallamy, H.O. Luteolin mitigates tamoxifen-associated fatty liver and cognitive impairment in rats by modulating beta-catenin. *Eur. J. Pharmacol.* **2021**, *908*, 174337. [CrossRef] [PubMed]
39. Owumi, S.E.; Nwozo, S.O.; Arunsi, U.O.; Oyelere, A.K.; Odunola, O.A. Co-administration of Luteolin mitigated toxicity in rats' lungs associated with doxorubicin treatment. *Toxicol. Appl. Pharmacol.* **2021**, *411*, 115380. [CrossRef]
40. Mathew, S.; Abraham, T.E.; Zakaria, Z.A. Reactivity of phenolic compounds towards free radicals under in vitro conditions. *J. Food Sci. Technol.* **2015**, *52*, 5790–5798. [CrossRef] [PubMed]
41. Hu, R.; Wu, S.; Li, B.; Tan, J.; Yan, J.; Wang, Y.; Tang, Z.; Liu, M.; Fu, C.; Zhang, H.; et al. Dietary ferulic acid and vanillic acid on inflammation, gut barrier function and growth performance in lipopolysaccharide-challenged piglets. *Anim. Nutr.* **2021**, *8*, 144–152. [CrossRef] [PubMed]
42. Li, Y.; Zhang, L.; Wang, X.; Wu, W.; Qin, R. Effect of Syringic acid on antioxidant biomarkers and associated inflammatory markers in mice model of asthma. *Drug Dev. Res.* **2019**, *80*, 253–261. [CrossRef] [PubMed]
43. Lende, A.B.; Kshirsagar, A.D.; Deshpande, A.D.; Muley, M.M.; Patil, R.R.; Bafna, P.A.; Naik, S.R. Anti-inflammatory and analgesic activity of protocatechuic acid in rats and mice. *Inflammopharmacology* **2011**, *19*, 255. [CrossRef]
44. Sánchez-Muniz, F.J.; Bastida, S.; Márquez-Ruiz, G.; Dobarganes, C. Effect of heating and frying on oil and food fatty acids. In *Fatty Acids in Foods and Their Health Implications*; CRC Press: Boca Raton, FL, USA, 2007; pp. 525–558, ISBN 0429127553.

45. Nogueira-de-Almeida, C.A.; Castro, G.A. Effects of heat treatment by immersion in household conditions on olive oil as compared to other culinary oils: A descriptive study. *Int. J. Food Stud.* **2018**, *7*, 89–99. [CrossRef]

46. Casal, S.; Malheiro, R.; Sendas, A.; Oliveira, B.P.P.; Pereira, J.A. Olive oil stability under deep-frying conditions. *Food Chem. Toxicol.* **2010**, *48*, 2972–2979. [CrossRef]

47. Chiou, A.; Salta, F.N.; Kalogeropoulos, N.; Mylona, A.; Ntalla, I.; Andrikopoulos, N.K. Retention and Distribution of Polyphenols after Pan-Frying of French Fries in Oils Enriched with Olive Leaf Extract. *J. Food Sci.* **2007**, *72*, S574–S584. [CrossRef]

48. Chiou, A.; Kalogeropoulos, N.; Salta, F.N.; Efstathiou, P.; Andrikopoulos, N.K. Pan-frying of French fries in three different edible oils enriched with olive leaf extract: Oxidative stability and fate of microconstituents. *LWT—Food Sci. Technol.* **2009**, *42*, 1090–1097. [CrossRef]

49. Chiou, A.; Kalogeropoulos, N.; Efstathiou, P.; Papoutsi, M.; Andrikopoulos, N.K. French Fries oleuropein content during the successive deep frying in oils enriched with an olive leaf extract. *Int. J. Food Sci. Technol.* **2013**, *48*, 1165–1171. [CrossRef]

50. Kalogeropoulos, N.; Mylona, A.; Chiou, A.; Ioannou, M.S.; Andrikopoulos, N.K. Retention and distribution of natural antioxidants (α-tocopherol, polyphenols and terpenic acids) after shallow frying of vegetables in virgin olive oil. *LWT—Food Sci. Technol.* **2007**, *40*, 1008–1017. [CrossRef]

51. Santos, C.S.P.; Cunha, S.C.; Casal, S. Deep or air frying? A comparative study with different vegetable oils. *Eur. J. Lipid Sci. Technol.* **2017**, *119*, 1600375. [CrossRef]

52. Goulas, V.; Orphanides, A.; Pelava, E.; Gekas, V. Impact of Thermal Processing Methods on Polyphenols and Antioxidant Activity of Olive Oil Polar Fraction. *J. Food Process. Preserv.* **2015**, *39*, 1919–1924. [CrossRef]

53. Campanella, L.; Nuccilli, A.; Tomassetti, M.; Vecchio, S. Biosensor analysis for the kinetic study of polyphenols deterioration during the forced thermal oxidation of extra-virgin olive oil. *Talanta* **2008**, *74*, 1287–1298. [CrossRef] [PubMed]

54. Amati, L.; Campanella, L.; Dragone, R.; Nuccilli, A.; Tomassetti, M.; Vecchio, S. New Investigation of the Isothermal Oxidation of Extra Virgin Olive Oil: Determination of Free Radicals, Total Polyphenols, Total Antioxidant Capacity, and Kinetic Data. *J. Agric. Food Chem.* **2008**, *56*, 8287–8295. [CrossRef]

55. Kishimoto, N. Microwave heating induces oxidative degradation of extra virgin olive oil. *Food Sci. Technol. Res.* **2019**, *25*, 75–79. [CrossRef]

56. Kamvissis, V.N.; Barbounis, E.G.; Megoulas, N.C.; Koupparis, M.A. A Novel Photometric Method for Evaluation of the Oxidative Stability of Virgin Olive Oils. *J. AOAC Int.* **2008**, *91*, 794–801. [CrossRef]

57. Attya, M.; Benabdelkamel, H.; Perri, E.; Russo, A.; Sindona, G. Effects of Conventional Heating on the Stability of Major Olive Oil Phenolic Compounds by Tandem Mass Spectrometry and Isotope Dilution Assay. *Molecules* **2010**, *15*, 8734–8746. [CrossRef]

58. Carrasco-Pancorbo, A.; Cerretani, L.; Bendini, A.; Segura-Carretero, A.; Lercker, G.; Fernández-Gutiérrez, A. Evaluation of the Influence of Thermal Oxidation on the Phenolic Composition and on the Antioxidant Activity of Extra-Virgin Olive Oils. *J. Agric. Food Chem.* **2007**, *55*, 4771–4780. [CrossRef] [PubMed]

59. Daskalaki, D.; Kefi, G.; Kotsiou, K.; Tasioula-Margari, M. Evaluation of phenolic compounds degradation in virgin olive oil during storage and heating. *J. Food Nutr. Res.* **2009**, *48*, 31–41.

60. Brenes, M.; García, A.; Dobarganes, M.C.; Velasco, J.; Romero, C. Influence of Thermal Treatments Simulating Cooking Processes on the Polyphenol Content in Virgin Olive Oil. *J. Agric. Food Chem.* **2002**, *50*, 5962–5967. [CrossRef] [PubMed]

61. Allouche, Y.; Jiménez, A.; Gaforio, J.J.; Uceda, M.; Beltrán, G. How Heating Affects Extra Virgin Olive Oil Quality Indexes and Chemical Composition. *J. Agric. Food Chem.* **2007**, *55*, 9646–9654. [CrossRef]

62. Gutiérrez, F.; Arnaud, T.; Garrido, A. Contribution of polyphenols to the oxidative stability of virgin olive oil. *J. Sci. Food Agric.* **2001**, *81*, 1463–1470. [CrossRef]

63. Pellegrini, N.; Visioli, F.; Buratti, S.; Brighenti, F. Direct Analysis of Total Antioxidant Activity of Olive Oil and Studies on the Influence of Heating. *J. Agric. Food Chem.* **2001**, *49*, 2532–2538. [CrossRef]

64. Singleton, V.L.; Orthofer, R.; Lamuela-Raventós, R.M.B.T.-M. In E. Analysis of total phenols and other oxidation substrates and antioxidants by means of folin-ciocalteu reagent. In *Oxidants and Antioxidants Part A*; Academic Press: San Diego, CA, USA, 1999; Volume 299, pp. 152–178, ISBN 0076-6879.

65. Papadopoulos, G.; Boskou, D. Antioxidant effect of natural phenols on olive oil. *J. Am. Oil Chem. Soc.* **1991**, *68*, 669–671. [CrossRef]

66. Rice-Evans, C.A.; Miller, N.J.; Paganga, G. Structure-antioxidant activity relationships of flavonoids and phenolic acids. *Free Radic. Biol. Med.* **1996**, *20*, 933–956. [CrossRef]

67. Servili, M.; Esposto, S.; Veneziani, G.; Urbani, S.; Taticchi, A.; Di Maio, I.; Selvaggini, R.; Sordini, B.; Montedoro, G. Improvement of bioactive phenol content in virgin olive oil with an olive-vegetation water concentrate produced by membrane treatment. *Food Chem.* **2011**, *124*, 1308–1315. [CrossRef]

68. Esposto, S.; Taticchi, A.; Di Maio, I.; Urbani, S.; Veneziani, G.; Selvaggini, R.; Sordini, B.; Servili, M. Effect of an olive phenolic extract on the quality of vegetable oils during frying. *Food Chem.* **2015**, *176*, 184–192. [CrossRef]

69. Di Maio, I.; Esposto, S.; Taticchi, A.; Selvaggini, R.; Veneziani, G.; Urbani, S.; Servili, M. HPLC–ESI-MS investigation of tyrosol and hydroxytyrosol oxidation products in virgin olive oil. *Food Chem.* **2011**, *125*, 21–28. [CrossRef]

70. Cerretani, L.; Bendini, A.; Rodriguez-Estrada, M.T.; Vittadini, E.; Chiavaro, E. Microwave heating of different commercial categories of olive oil: Part I. Effect on chemical oxidative stability indices and phenolic compounds. *Food Chem.* **2009**, *115*, 1381–1388. [CrossRef]

71. Lozano-Castellón, J.; Vallverdú-Queralt, A.; Rinaldi de Alvarenga, J.F.; Illán, M.; Torrado-Prat, X.; Lamuela-Raventós, R.M. Domestic sautéing with EVOO: Change in the phenolic profile. *Antioxidants* **2020**, *9*, 77. [CrossRef]
72. Silva, L.; Garcia, B.; Paiva-Martins, F. Oxidative stability of olive oil and its polyphenolic compounds after boiling vegetable process. *LWT—Food Sci. Technol.* **2010**, *43*, 1336–1344. [CrossRef]
73. Brkić Bubola, K.; Klisović, D.; Lukić, I.; Novoselić, A. Vegetable species significantly affects the phenolic composition and oxidative stability of extra virgin olive oil used for roasting. *LWT* **2020**, *129*, 109628. [CrossRef]
74. Gómez-Alonso, S.; Fregapane, G.; Salvador, M.D.; Gordon, M.H. Changes in Phenolic Composition and Antioxidant Activity of Virgin Olive Oil during Frying. *J. Agric. Food Chem.* **2003**, *51*, 667–672. [CrossRef] [PubMed]
75. Montedoro, G.; Servili, M.; Baldioli, M.; Miniati, E. Simple and hydrolyzable phenolic compounds in virgin olive oil. 2. Initial characterization of the hydrolyzable fraction. *J. Agric. Food Chem.* **1992**, *40*, 1577–1580. [CrossRef]
76. Baldioli, M.; Servili, M.; Perretti, G.; Montedoro, G.F. Antioxidant activity of tocopherols and phenolic compounds of virgin olive oil. *J. Am. Oil Chem. Soc.* **1996**, *73*, 1589–1593. [CrossRef]
77. Olivero-David, R.; Mena, C.; Pérez-Jimenez, M.A.; Sastre, B.; Bastida, S.; Márquez-Ruiz, G.; Sánchez-Muniz, F.J. Influence of Picual Olive Ripening on Virgin Olive Oil Alteration and Stability during Potato Frying. *J. Agric. Food Chem.* **2014**, *62*, 11637–11646. [CrossRef]
78. Paiva-Martins, F.; Gordon, M.H. Interactions of Ferric Ions with Olive Oil Phenolic Compounds. *J. Agric. Food Chem.* **2005**, *53*, 2704–2709. [CrossRef] [PubMed]
79. Ramírez-Anaya, J.D.P.; Castañeda-Saucedo, M.C.; Olalla-Herrera, M.; Villalón-Mir, M.; de la Serrana, H.L.-G.; Samaniego-Sánchez, C. Changes in the Antioxidant Properties of Extra Virgin Olive Oil after Cooking Typical Mediterranean Vegetables. *Antioxidants* **2019**, *8*, 246. [CrossRef] [PubMed]
80. Zembyla, M.; Murray, B.S.; Radford, S.J.; Sarkar, A. Water-in-oil Pickering emulsions stabilized by an interfacial complex of water-insoluble polyphenol crystals and protein. *J. Colloid Interface Sci.* **2019**, *548*, 88–99. [CrossRef] [PubMed]
81. de Alvarenga, J.F.; Quifer-Rada, P.; Francetto Juliano, F.; Hurtado-Barroso, S.; Illan, M.; Torrado-Prat, X.; Lamuela-Raventós, R.M. Using Extra Virgin Olive Oil to Cook Vegetables Enhances Polyphenol and Carotenoid Extractability: A Study Applying the sofrito Technique. *Molecules* **2019**, *24*, 1555. [CrossRef] [PubMed]
82. Taticchi, A.; Esposto, S.; Urbani, S.; Veneziani, G.; Selvaggini, R.; Sordini, B.; Servili, M. Effect of an olive phenolic extract added to the oily phase of a tomato sauce, on the preservation of phenols and carotenoids during domestic cooking. *LWT* **2017**, *84*, 572–578. [CrossRef]
83. Samaniego-Sánchez, C.; Castañeda-Saucedo, M.C.; Villalón-Mir, M.; De La Serrana, H.L. Phenols and the antioxidant capacity of Mediterranean vegetables prepared with extra virgin olive oil using different domestic cooking techniques. *Food Chem.* **2015**, *188*, 430–438. [CrossRef]
84. Xu, F.; Zheng, Y.; Yang, Z.; Cao, S.; Shao, X.; Wang, H. Domestic cooking methods affect the nutritional quality of red cabbage. *Food Chem.* **2014**, *161*, 162–167. [CrossRef]
85. Tian, J.; Chen, J.; Lv, F.; Chen, S.; Chen, J.; Liu, D.; Ye, X. Domestic cooking methods affect the phytochemical composition and antioxidant activity of purple-fleshed potatoes. *Food Chem.* **2016**, *197*, 1264–1270. [CrossRef]
86. Gunathilake, K.D.P.P.; Ranaweera, K.K.D.S.; Rupasinghe, H.P.V. Effect of Different Cooking Methods on Polyphenols, Carotenoids and Antioxidant Activities of Selected Edible Leaves. *Antioxidants* **2018**, *7*, 117. [CrossRef]
87. Mashiane, P.; Mashitoa, F.M.; Slabbert, R.M.; Sivakumar, D. Impact of household cooking techniques on colour, antioxidant and sensory properties of African pumpkin and pumpkin leaves. *Int. J. Gastron. Food Sci.* **2021**, *23*, 100307. [CrossRef]
88. Ioku, K.; Aoyama, Y.; Tokuno, A.; Terao, J.; Nakatani, N.; Takei, Y. Various cooking methods and the flavonoid content in onion. *J. Nutr. Sci. Vitaminol.* **2001**, *47*, 78–83. [CrossRef]
89. Jung, J.-K.; Lee, S.-U.; Kozukue, N.; Levin, C.E.; Friedman, M. Distribution of phenolic compounds and antioxidative activities in parts of sweet potato (*Ipomoea batata* L.) plants and in home processed roots. *J. Food Compos. Anal.* **2011**, *24*, 29–37. [CrossRef]
90. Finotti, E.; Bersani, E.; Del Prete, E.; Friedman, M. A functional mathematical index for predicting effects of food processing on eight sweet potato (*Ipomoea batatas*) cultivars. *J. Food Compos. Anal.* **2012**, *27*, 81–86. [CrossRef]
91. Martini, S.; Conte, A.; Cattivelli, A.; Tagliazucchi, D. Domestic cooking methods affect the stability and bioaccessibility of dark purple eggplant (*Solanum melongena*) phenolic compounds. *Food Chem.* **2021**, *341*, 128298. [CrossRef]
92. Juániz, I.; Ludwig, I.A.; Huarte, E.; Pereira-Caro, G.; Moreno-Rojas, J.M.; Cid, C.; De Peña, M.-P. Influence of heat treatment on antioxidant capacity and (poly) phenolic compounds of selected vegetables. *Food Chem.* **2016**, *197*, 466–473. [CrossRef]
93. Managa, M.G.; Shai, J.; Phan, A.D.T.; Sultanbawa, Y.; Sivakumar, D. Impact of Household Cooking Techniques on African Nightshade and Chinese Cabbage on Phenolic Compounds, Antinutrients, in vitro Antioxidant, and β-Glucosidase Activity. *Front. Nutr.* **2020**, *7*, 292. [CrossRef]
94. Ferracane, R.; Pellegrini, N.; Visconti, A.; Graziani, G.; Chiavaro, E.; Miglio, C.; Fogliano, V. Effects of Different Cooking Methods on Antioxidant Profile, Antioxidant Capacity, and Physical Characteristics of Artichoke. *J. Agric. Food Chem.* **2008**, *56*, 8601–8608. [CrossRef]
95. Sergio, L.; Boari, F.; Pieralice, M.; Linsalata, V.; Cantore, V.; Di Venere, D. Bioactive phenolics and antioxidant capacity of some wild edible greens as affected by different cooking treatments. *Foods* **2020**, *9*, 1320. [CrossRef] [PubMed]
96. Shahidi, F.; Peng, H. Bioaccessibility and bioavailability of phenolic compounds. *J. Food Bioact.* **2018**, *4*, 11–68. [CrossRef]

97. Del Rio, D.; Rodriguez-Mateos, A.; Spencer, J.P.E.; Tognolini, M.; Borges, G.; Crozier, A. Dietary (poly)phenolics in human health: Structures, bioavailability, and evidence of protective effects against chronic diseases. *Antioxid. Redox Signal.* **2013**, *18*, 1818–1892. [CrossRef] [PubMed]

98. Palermo, M.; Pellegrini, N.; Fogliano, V. The effect of cooking on the phytochemical content of vegetables. *J. Sci. Food Agric.* **2014**, *94*, 1057–1070. [CrossRef]

99. Li, Q.; Li, T.; Liu, C.; Chen, J.; Zhang, R.; Zhang, Z.; Dai, T.; McClements, D.J. Potential physicochemical basis of Mediterranean diet effect: Ability of emulsified olive oil to increase carotenoid bioaccessibility in raw and cooked tomatoes. *Food Res. Int.* **2016**, *89*, 320–329. [CrossRef]

100. Nicoli, M.C.; Anese, M.; Parpinel, M. Influence of processing on the antioxidant properties of fruit and vegetables. *Trends Food Sci. Technol.* **1999**, *10*, 94–100. [CrossRef]

101. Juániz, I.; Ludwig, I.A.; Bresciani, L.; Dall'Asta, M.; Mena, P.; Del Rio, D.; Cid, C.; de Peña, M.P. Bioaccessibility of (poly)phenolic compounds of raw and cooked cardoon (*Cynara cardunculus* L.) after simulated gastrointestinal digestion and fermentation by human colonic microbiota. *J. Funct. Foods* **2017**, *32*, 195–207. [CrossRef]

102. Bugianesi, R.; Salucci, M.; Leonardi, C.; Ferracane, R.; Catasta, G.; Azzini, E.; Maiani, G. Effect of domestic cooking on human bioavailability of naringenin, chlorogenic acid, lycopene and β-carotene in cherry tomatoes. *Eur. J. Nutr.* **2004**, *43*, 360–366. [CrossRef]

103. Tulipani, S.; Martinez Huelamo, M.; Rotches Ribalta, M.; Estruch, R.; Ferrer, E.E.; Andres-Lacueva, C.; Illan, M.; Lamuela-Raventós, R.M. Oil matrix effects on plasma exposure and urinary excretion of phenolic compounds from tomato sauces: Evidence from a human pilot study. *Food Chem.* **2012**, *130*, 581–590. [CrossRef]

104. Vallverdú-Queralt, A.; Regueiro, J.; Rinaldi De Alvarenga, J.F.; Torrado, X.; Lamuela-Raventos, R.M. Home cooking and phenolics: Effect of thermal treatment and addition of extra virgin olive oil on the phenolic profile of tomato sauces. *J. Agric. Food Chem.* **2014**, *62*, 3314–3320. [CrossRef] [PubMed]

105. Martínez-Huélamo, M.; Tulipani, S.; Estruch, R.; Escribano, E.; Illán, M.; Corella, D.; Lamuela-Raventós, R.M. The tomato sauce making process affects the bioaccessibility and bioavailability of tomato phenolics: A pharmacokinetic study. *Food Chem.* **2015**, *173*, 864–872. [CrossRef]

106. Martínez-Huélamo, M.; Vallverdú-Queralt, A.; Di Lecce, G.; Valderas-Martínez, P.; Tulipani, S.; Jáuregui, O.; Escribano-Ferrer, E.; Estruch, R.; Illan, M.; Lamuela-Raventós, R.M. Bioavailability of tomato polyphenols is enhanced by processing and fat addition: Evidence from a randomized feeding trial. *Mol. Nutr. Food Res.* **2016**, *60*, 1578–1589. [CrossRef] [PubMed]

107. Rodriguez-Rodriguez, R.; Jiménez-Altayó, F.; Alsina, L.; Onetti, Y.; Rinaldi de Alvarenga, J.F.; Claro, C.; Ogalla, E.; Casals, N.; Lamuela-Raventos, R.M. Mediterranean tomato-based sofrito protects against vascular alterations in obese Zucker rats by preserving NO bioavailability. *Mol. Nutr. Food Res.* **2017**, *61*, 1601010. [CrossRef] [PubMed]

108. Sandoval, V.; Rodríguez-Rodríguez, R.; Martínez-Garza, Ú.; Rosell-Cardona, C.; Lamuela-Raventós, R.M.; Marrero, P.F.; Haro, D.; Relat, J. Mediterranean Tomato-Based Sofrito Sauce Improves Fibroblast Growth Factor 21 (FGF21) Signaling in White Adipose Tissue of Obese ZUCKER Rats. *Mol. Nutr. Food Res.* **2018**, *62*, 1700606. [CrossRef] [PubMed]

109. Martínez-González, M.A.; García-Arellano, A.; Toledo, E.; Salas-Salvadó, J.; Buil-Cosiales, P.; Corella, D.; Covas, M.I.; Schröder, H.; Arós, F.; Gómez-Gracia, E.; et al. A 14-Item Mediterranean Diet Assessment Tool and Obesity Indexes among High-Risk Subjects: The PREDIMED Trial. *PLoS ONE* **2012**, *7*, e43134. [CrossRef]

110. EFSA Panel on Dietetic Products, N. and A. (NDA) Scientific Opinion on the substantiation of a health claim related to 3 g/day plant sterols/stanols and lowering blood LDL-cholesterol and reduced risk of (coronary) heart disease pursuant to Article 19 of Regulation (EC) No 1924/2006. *EFSA J.* **2012**, *10*, 2693. [CrossRef]

Article

Boiling Technique-Based Food Processing Effects on the Bioactive and Antimicrobial Properties of Basil and Rosemary

Ahmad Mohammad Salamatullah, Khizar Hayat *, Shaista Arzoo, Abdulhakeem Alzahrani, Mohammed Asif Ahmed, Hany M. Yehia, Tawfiq Alsulami, Nawal Al-Badr, Bandar Ali M Al-Zaied and Mohammed Musaad Althbiti

Department of Food Science & Nutrition, College of Food and Agricultural Sciences, King Saud University, Riyadh 11451, Saudi Arabia; asalamh@ksu.edu.sa (A.M.S.); sarzoo@ksu.edu.sa (S.A.); aabdulhakeem@ksu.edu.sa (A.A.); masifa@ksu.edu.sa (M.A.A.); hanyehia@ksu.edu.sa (H.M.Y.); talsulami@ksu.edu.sa (T.A.); nalbader@ksu.edu.sa (N.A.-B.); 437102343@student.ksu.edu.sa (B.A.M.A.-Z.); 437101927@student.ksu.edu.sa (M.M.A.)
* Correspondence: khayat@ksu.edu.sa

Citation: Salamatullah, A.M.; Hayat, K.; Arzoo, S.; Alzahrani, A.; Ahmed, M.A.; Yehia, H.M.; Alsulami, T.; Al-Badr, N.; Al-Zaied, B.A.M.; Althbiti, M.M. Boiling Technique-Based Food Processing Effects on the Bioactive and Antimicrobial Properties of Basil and Rosemary. *Molecules* **2021**, *26*, 7373. https://doi.org/10.3390/molecules26237373

Academic Editors: Mirella Nardini and Gabriele Rocchetti

Received: 23 October 2021
Accepted: 1 December 2021
Published: 4 December 2021

Publisher's Note: MDPI stays neutral with regard to jurisdictional claims in published maps and institutional affiliations.

Abstract: Rosemary (*Rosmarinus officinalis*) and basil (*Ocimum sanctum* Linn) are mostly used as herbal teas, made by steeping whole or ground herbs in boiling water. Hence, it is important to know the effect of boiling time on the bioactivity of these herbs. The effect of different boiling times (5, 10, and 15 min) on the antioxidant and antimicrobial properties, and some selected phenolic compounds of these herbs was examined in this study. Experimental results revealed that basil displayed the highest total polyphenol content (TPC), total flavonoid content (TFC), and antioxidant activity when it was boiled for 5 min, and the lowest TPC was obtained when it was boiled for 15 min. On the other hand, rosemary had the highest TPC, TFC, and antioxidant potential after being boiled for 15 min, while it had the lowest after being boiled for 5 min. There was no growth inhibition of rosemary extracts against gram-negative bacteria, whereas higher growth inhibition was observed against gram-positive bacteria. The MIC and MBC of rosemary ethanolic extract against *Listeria monocytogenes* were 5 and 5 mg/mL and against *B. subtilis* were 10 and 10 mg/mL, respectively. While MIC and MBC of methanolic extract against *L. monocytogenes* were 5 and 5 mg/mL and against *Bacillus subtilis* were and 5 and 5 mg/mL, respectively. Salicylic acid was the most abundant (324.7 mg/100 g dry weight (dw)) phenolic compound in the rosemary sample boiled for 5 min, and acetyl salicylic acid was the most abundant (122.61 mg/10 g dw) phenolic compound in the basil sample boiled for 15 min.

Keywords: food processing; herbal tea; boiling; antioxidant activity; antimicrobial activity

1. Introduction

The reactive oxygen species (ROS) play a significant role in numerous cellular activities such as signaling transduction, gene transcription, and immune response [1]. ROS, either oxygen ions or oxygen-containing radicals are usually present at little concentrations [2]. The excess production of ROS either from the external sources or due to the endogenous metabolic processes in the human body causes oxidative damage to biomolecules resulting in several diseases such as neurodegenerative diseases, diabetes, cancers, chronic inflammatory diseases, and atherosclerosis [3–5]. The human body is equipped with antioxidants that counterbalance the harmful effects of oxidants as they are capable of scavenging the ROS and reducing the oxidation of cellular molecules [6].

Antioxidant-rich herbs serve as a great source of antioxidants in foods that strengthen the body's ability to fight free radical damage and thus decreasing the risk of many diseases [7,8]. In addition to their content of antioxidants, herbs are well-known for their antimicrobial, antiseptic, diuretic, anti-inflammatory, analgesic, anthelmintic, and carminative properties [9,10]. Therefore, the lamiaceae family members including mint, thyme, basil, rosemary, sage, savory, and oregano are traditionally added to foods as flavors or used as medicines such as basil (*Ocimum sanctum* L.) and rosemary (*Rosmarinus officinalis* L.) [11].

Basil (*Ocimum sanctum* L.) is an aromatic herb and includes over 150 species, native to tropical areas of Africa, Asia, Central America, and South America [12]. It possesses stimulant and expectorant properties, anti-diabetic, anti-carcinogenic, anti-inflammatory, and antimicrobial characteristics. The properties of basil make it traditionally used to manage the multiple medical conditions such as chest complications, cough, bronchitis, stress, anxiety, gastritis, dysentery, skin diseases, asthma, diarrhea, fever, arthritis, and eye diseases [13–21]. Basil also guards against the lethal effects of industrial chemicals and many pharmaceutical drugs [22]. Basil herbs are rich source of vitamin A, C, and minerals including calcium, iron, and zinc [23]. The leaf volatile oil contains eugenol, eugenol, ursolic acid, carvacrol, methyl carvicol, linalool, and sitosterol [19,24]. The presence of these components depends on the basil species, weather, location and growing conditions, and growth period/level during harvest [25–28]. Similarly, rosemary (*Rosmarinus officinalis* L.) is also one of the most popular perennial culinary herbs of the lamiaceae family. This plant comes from the Mediterranean region and is cultivated all over the world. In folk medicine, rosemary is used to control numerous diseases such as headache, stomachache, dysmenorrhea, epilepsy, depression, nervous agitation, rheumatic pain, fatigue, spasms, and improvement of memory. In addition, it possesses antioxidant, antimicrobial, anti-inflammatory, anti-apoptotic, anti-tumorigenic, antinociceptive, and neuroprotective properties [29,30]. Rosemary provides protein, vitamins, minerals, and fiber which are known to have disease preventing properties [31]. It has many phytochemicals which constitute natural compounds as phenolic diterpenes, flavonoids, and phenolic acids. The main constituents of the rosemary essential oil are 1,8-cineole, camphor, α-pinene, camphene, borneol, β-pinene, and limonene [32–34]. Tawfeeq et al., [35] and Jiang et al., [36] stated 1,8-cineole to be the main component of rosemary essential oil while Bendeddouche et al., [37] reported camphor followed by 1,8-cineole to be the main components of rosemary essential oil. This difference may be due to the vegetative stage and bioclimatic conditions [33]. The most common polyphenols, the secondary metabolites of rosemary are apigenin, homoplantaginin, diosmin, gallocatechin, luteolin, and genkwanin. Apart from this, it also contains various phenolic acids such as chlorogenic, rosmarinic, and caffeic acid [38–40]. In a study on antioxidant and antibacterial properties of some fresh and dried Labiatae herbs, the fresh and commercial rosemary had the highest antioxidant activity and phenolic content, but oven-dried rosemary ranked third [41]. The reduction in antioxidant values following thermal treatments has been credited to the enzymatic degradation of phenolic compounds, thermal degradation of phytochemicals, and loss of antioxidant enzyme activities [42]. The strong antioxidant properties of the commercial brand of rosemary may be due to freeze-drying in which heat is not involved [41]. In contrast to this, other studies have observed an increase in antioxidant activity and phenolic content following the thermal treatment and suggested that phenolic compounds may be released through the breakdown of cellular constituents and formed new compounds with increased antioxidant capacity [43,44].

In several Asian nations, basil and rosemary as therapeutic herbs are prepared in the traditional form of herbal teas by extricating whole or ground herbs in bubbling water. Despite the availability of numerous studies examining the antioxidant properties of basil and rosemary, hardly any research has been published evaluating the effect of boiling conditions on their antioxidant contents. Consequently, the purpose of this study was to determine the possible effects of boiling on the total polyphenol and flavonoid contents, antioxidant activity, and antimicrobial proprieties of basil and rosemary leaves at various times.

2. Results and Discussion

2.1. Effect of Boiling on Total Polyphenol Content of Basil and Rosemary

The effect of boiling time on the total polyphenol content (TPC) of basil and rosemary is provided in Table 1. It clearly shows that the boiling time significantly affected the TPC of both samples. The highest TPC of basil was achieved by boiling it for 5 min while the lowest TPC was obtained when the boiling time was 15 min. It also showed that the boiling time had an adverse effect on the phenolic compounds of basil. In a recent study, it was noticed that the boiling process caused a reduction in the total phenol content and the antioxidant activity of the celery roots [45]. However, the TPC content of the rosemary showed an increasing trend with the increase in the boiling time. For example, the TPC of the rosemary sample boiled for 5 and 15 min was 122.84 and 140.43 mg, respectively, gallic acid equivalent (GAE) per gram dry weight (dw). This may be due to the heat during boiling, which may rupture the cell wall of the material causing the release of phenolic compounds in the solvent [46,47]. Our results are in line with a recent study where the boiling process of green and red rooibos (Aspalathus linearis) herbal tea delivered a higher TPC and antioxidant activity [48]. In brief, rosemary exhibited a significantly ($p < 0.05$) higher TPC than the basil in all analyzed samples, which can be attributed to the phenolic compounds content and antioxidant potential of the plants itself [49,50].

Table 1. Effect of boiling on the bioactive properties of basil and rosemary.

Sample	Process Time	Total Polyphenol Content (mg GAE/g dw)	Total Flavonoid Content (mg CE/g dw)	DPPH (IC_{50} mg/g)	Ferric Reducing Power (Absorbance 760 nm)
Basil	5 min	69.24 ± 1.03 [c]	39.66 ± 0.08 [d]	6.08 ± 0.15 [b]	0.815 ± 0.012 [c]
	10 min	66.22 ± 3.89 [c]	39.00 ± 0.63 [d,e]	6.39 ± 0.15 [a]	0.789 ± 0.102 [c,d]
	15 min	54.64 ± 6.13 [d]	36.79 ± 0.26 [e]	6.62 ± 0.27 [a]	0.712 ± 0.009 [d]
Rosemary	5 min	122.84 ± 5.79 [b]	78.36 ± 1.55 [c]	0.79 ± 0.01 [c]	1.426 ± 0.013 [b]
	10 min	119.24 ± 2.47 [b]	86.85 ± 2.80 [b]	0.82 ± 0.03 [c]	1.503 ± 0.040 [a,b]
	15 min	140.43 ± 4.44 [a]	109.73 ± 0.33 [a]	0.66 ± 0.01 [d]	1.526 ± 0.037 [a]

The values for each assay are expressed as mean ± standard deviation of three replicates. Different superscript letters in the same column represent the significant differences in data ($p < 0.05$).

2.2. Effect of Boiling on Total Flavonoid Content of Basil and Rosemary

As shown in Table 1, the total flavonoid content (TFC) of basil was decreased by increasing the boiling time. A total flavonoid content of 39.66 and 36.79 mg catechin equivalent (CE) per gram dry weight was obtained for the basil samples boiled for 5 and 15 min, respectively. An earlier study reported a loss of phenolic compounds and the antioxidant activity on boiling for 10 min of various green vegetables [51]. In contrary to the basil, the total flavonoid content of rosemary was increased significantly by increasing the boiling time. The highest flavonoid content (109.73 mg CE/g dw) was noted for the rosemary sample boiled for 15 min, while 5 min of boiling provided a flavonoid content of 122.84 mg CE/g dw. In a recent study, it was reported that the boiled edible leaves of *Sesbania grandiflora*, *Cassia auriculata*, *Centella asiatica*, and *Gymnema lactiferum* showed an increase in the total content of phenols and flavonoids compared with the fresh ones [49]. Similar to the TPC, the TFC of all the rosemary samples was significantly higher than those of the basil samples.

2.3. Effect of Boiling on DPPH Scavenging of Basil and Rosemary

The DPPH scavenging of basil and rosemary in terms of their 50% inhibitory concentration (IC_{50}) is provided in Table 1. The higher the IC_{50} value, the lower is the antioxidant potential of the sample. For basil samples, 5 min of boiling significantly resulted in the lowest IC_{50} value (6.08 mg) compared with the samples boiled for 10 min (6.39 mg) and

15 min (6.62 mg), respectively. For the rosemary sample, the lowest IC_{50} value, the highest DPPH scavenging potential, was achieved at 15 min of boiling. A study conducted by Gunathilake et al., [49] showed that the boiled edible leaves of C. auriculata, Passiflora edulis, and C. asiatica resulted in a higher DPPH scavenging compared with their fresh counterparts. However, Arias-Rico et al., [52] revealed that the DPPH scavenging and polyphenols of the leaves and stems of *Chenopodium nuttalliae* Safford, *Suaeda torreyana* S. Watson, *Portulaca oleracea* L., *Chenopodium album* L., and *Porophyllum ruderale* (Jacq.) were decreased by boiling either for 3 or 5 min. The rosemary showed significantly ($p < 0.05$) higher DPPH scavenging compared with the basil. These results echoed the TPC and TFC results showing that the DPPH potential of the samples was due to at least a part of their TPC and TFC contents.

2.4. Effect of Boiling on Ferric Reducing Power of Basil and Rosemary

The effect of boiling on the ferric reducing power of basil and rosemary is provided in Table 1. The reducing power of basil boiled for 5 min was significantly ($p < 0.05$) higher (0.815) compared with the samples boiled for 10 min (0.789) and 15 min (0.712). For the rosemary sample, 15 min of boiling provided the highest reducing power while 5 min of boiling exhibited the lowest reducing power. Nie et al. [53] assessed the effect of boiling time (10–120 min) on the nutritional value and the antioxidant capacity of a mushroom Lentinus edodes. They found that the antioxidant capacity was increased during the first 30 min of boiling and then kept stable or declined on prolongation of the boiling time. Overall, the rosemary showed higher reducing power than that of the basil extracts, which are in an agreement with the results of TPC and TFC.

2.5. Antimicrobial Activity

Both ethanolic and methanolic rosemary extracts had no effect on the gram-negative bacteria (*Escherichia coli* and *Salmonella typhimurium*), but they affected the gram-positive bacteria (*L. monocytogenes* and *B. subtilis*) (Figure 1A,B). Both extracts inhibited the growth of *L. monocytogenes* and provided a zone of inhibition ranging from 15 to 20 mm in diameter (Table 2). Similar zones of inhibition were observed with *B. subtilis* using ethanol and methanol extracts. The zones of inhibition exhibited a diameter of 14 and 15 mm with ethanol and methanol, respectively (Table 2). In comparison with rosemary, basil extracts in ethanol or methanol showed no effect on the tested pathogens except for *B. subtilis*, which was partially inhibited (with the zone of inhibition being less than 8 mm in diameter) (Figure 1C). This step was carried out using 50 µL of rosemary extract (100 mg/mL) to determine if there was an effect against the tested gram-positive bacteria. The effect of cold and hot aqueous extract of rosemary has shown a high inhibition rate against *Proteus* sp., *Klebsella* sp., *E. coli*, and *Pseudomonas* sp., because it contains a number of hydroxyl groups that act as a hydrogen donor making it very important and powerful [54].

The MIC of the ethanolic and methanolic rosemary extracts against *L. monocytogenes* was 5 and 5 mg/mL, respectively, and against *B. subtilis* were 10 and 5 mg/mL, respectively. Similarly, the MBC of the ethanolic and methanolic rosemary extracts against *L. monocytogenes* were 5 and 5 mg/mL respectively, and against *B. subtilis* were 10 and 5 mg/mL respectively. It was revealed that when methanol was used as a solvent to extract rosemary, there was no growth of both *L. monocytogenes* and *B. subtilis* on Mueller–Hinton agar, indicating that rosemary extraction in methanol has a bactericidal effect against both pathogenic bacteria [55]. On the other hand, the ethanolic extract of rosemary had bactericidal effect only against *L. monocytogenes*. In a study conducted by Gonelimali et al. [56], it was revealed that the ethanolic and water extract of selected plants (roselle, rosemary, clove, and thyme) efficiently suppressed the growth of food pathogens and spoilage microorganisms with variable potency. The ethanolic extract of rosemary exhibited an inhibitory effect against four of the pathogenic strains (*E. coli*, *Salmonella enteritidis*, *Bacillus cereus*, and *Salmonella aureus*), while aqueous extract of rosemary was effective against all strains except *S. enteritidis*.

Table 2. Zone of inhibition (mm), MIC and MBC of the of ethanolic and methanolic extracts of rosemary against *L. monocytogenes* and *B. subtilis.*

Plant Extract / Microorganisms	Rosemary Extract							
	Ethanolic Extract Rosemary				Methanolic Extract Rosemary			
	Zone of Inhibition (mm)	MIC (mg/mL)	MBC (mg/mL)	Growth on MH Agar	Zone of Inhibition (mm)	MIC (mg/mL)	MBC (mg/mL)	Growth on MH Agar
L. monocytogenes ATCC 19114	20	5	5	NG/Bactericide	15	5	5	NG/Bactericide
B. subtilis ATCC 6633	14	10	10	+++/Bacteriostatic	15	5	5	NG/Bactericide
S. typhimurium ATCC 14028	-	-	-	-	-	-	-	-
E. coli ATCC 10798	-	-	-	-	-	-	-	-

NG: no growth; -: no effect., +++: good growth; ATCC: American Type Culture Collection

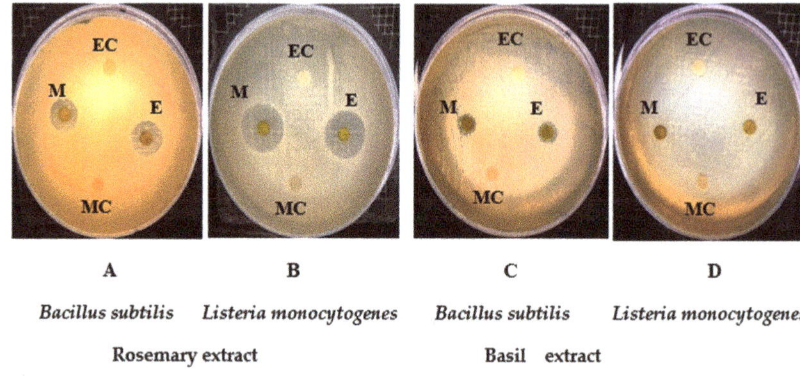

Figure 1. Zone of inhibition for both ethanolic and methanolic extracts of rosemary and basil against *B. subtilis* (**A,C**) and *L. monocytogenes* (**B,D**). E: Ethanolic extract of rosemary; M: Methanolic extract of rosemary; EC: Ethanol control; MC: Methanol control.

The antimicrobial activity of rosemary depends mainly on its phenolic and flavonoid content [57]. It cannot be dissolved in water, but in drought conditions it can reduce its resistance [58]. Leaves of rosemary, either fresh or dried, are aromatic and usually added in small quantities for cooking to improve taste. Due to its antioxidant content, rosemary extract is also used as a natural preservative to extend the shelf life of perishable foods according to the UK Food Standards Agency [59]. Thus, the evidence indicates that the antioxidant content of rosemary may help in fighting the bacterial infections and serious diseases such as cancer. The immune-boosting properties of rosemary have also been shown to help in defense against Helicobacter pylori, which causes gastric ulcers, and Staphylococcus infections, which causes a wide variety of clinical diseases [60]. A variety of factors can affect the antioxidant content of rosemary, such as the plant quality, the geographical origin, the date of harvest, the method of extraction, and even the climatic conditions where the plant was grown [61].

2.6. HPLC Analysis of Phenolic Compounds

Results of phenolic compounds analyzed by HPLC are reported in Table 3. Chlorogenic acid was higher in the rosemary sample boiled for 15 min (70.61 mg/100 g dw) compared with those boiled for 5 and 10 min. The rosemary sample boiled for 5 min, however, contained the highest concentration of salicylic acid (324.7 mg/100 g dw). The quercetin was only detected in the rosemary sample and its highest amount (243.64 mg/100 g dw) was found in the sample boiled for 15 min. The phenolic compounds content in the rosemary sample boiled for 10 min ranged from 3.78 to 137.10 mg/100 g dw. Moreover, vanillin (2.29 mg/100 g dw) was found in the lowest amount in the rosemary sample boiled for 5 min. The total phenolic compounds were higher (808.38 mg/100 g dw) in the rosemary sample boiled for 15 min, followed by the 5 min (746.12 mg/100 g dw) and 10 min boiled sample (529.03 mg/100 g dw).

In basil, the highest phenolic compound was acetyl salicylic acid (122.61 mg/100 g dw) then chlorogenic acid (20.67 mg/100 g dw) in the 15 min boiled sample. While, with respect to other compounds, 3,5 dinitro salicylic acid was the lowest (0.42 mg/100 g dw) phenolic compound in the basil sample. Total phenolic compounds were higher (163.06 mg/100 g dw) in the 15 min boiled basil sample, compared with the 5 min and 10 min boiled samples which contained 117.99 mg/100 g dw and 111.14 mg/100 g dw, respectively.

Table 3. Effect of boiling on phenolic compounds of rosemary and basil by HPLC (mg/100 g) dry weight (dw).

Sample	Process Time	Resorcinol	Chlorogenic Acid	Caffeic Acid	Vanillin	Acetyl Salicylic Acid	3,5-DNSA	Salicylic Acid	Quercetin	Total
Basil	5 min	ND	20.38 ± 0.15	2.03 ± 0.23	1.64 ± 0.04	63.58 ± 0.83	24.87 ± 1.48	5.49 ± 0.09	ND	117.99 ± 2.82
	10 min	ND	19.23 ± 0.56	4.36 ± 0.67	16.52 ± 0.23	48.18 ± 0.67	19.32 ± 0.67	3.53 ± 0.68	ND	111.14 ± 2.81
	15 min	ND	20.67 ± 0.69	2.72 ± 0.51	15.20 ± 0.35	122.61 ± 0.58	0.42 ± 0.35	1.44 ± 0.92	ND	163.06 ± 3.40
Rosemary	5 min	4.84 ± 0.26	25.56 ± 2.07	14.41 ± 1.38	2.28 ± 0.66	165.35 ± 1.21	42.99 ± 1.80	324.70 ± 2.50	165.99 ± 2.15	746.12 ± 12.03
	10 min	5.77 ± 0.28	5.76 ± 0.05	51.50 ± 1.32	3.78 ± 1.41	135.79 ± 3.62	87.17 ± 1.17	102.16 ± 1.11	137.10 ± 1.40	529.03 ± 10.36
	15 min	5.28 ± 0.08	70.61 ± 0.37	14.63 ± 0.44	2.35 ± 0.37	161.81 ± 1.17	143.15 ± 0.78	166.91 ± 3.68	243.64 ± 0.61	808.38 ± 7.50

ND: Not detected.

The HPLC results of rosemary were in accordance with the results provided in Table 1. Total polyphenol content and reducing power (Table 1) were increased at 15 min of boiling for the rosemary sample, similarly the phenolic compounds were high at 15 min of boiling (808.37 mg/100 g dw) as quantified by HPLC. While, for the basil sample, the total polyphenol content and reducing power decreased with the increase in the boiling time (Table 1) in contrast to the HPLC results where the phenolic compounds were increased at 15 min of boiling compared with 5 min of boiling of the basil sample. It showed that there might be some other compounds (except detected by HPLC in this study) contributing to the antioxidant activity of the basil sample. Elansary et al. [62] revealed that the caffeic acid was 27.6 mg/100 g dw in methanolic rosemary extract. Furthermore, Flanigan et al. [63] reported that caffeic acid was 19.9 mg/100 g dw in basil. Begum et al. [33] reported the phenolic acids (rosmarinic, chlorogenic and caffeic acid) were >3% in rosemary samples.

3. Materials and Methods

3.1. Raw Materials

The raw material basil (*Ocimum sanctum* L.) was collected from a vegetable garden in King Saud University, Saudi Arabia, while the rosemary (*Rosmarinus officinalis*) was acquired from the local market in Riyadh, Saudi Arabia. The samples were cleaned, washed with water, cut into small pieces, and dried overnight in an oven dryer at 40 °C. The samples were ground using a coffee grinder and sieved with stainless steel wire mesh (25 mm). Powdered samples were stored at −20 °C in an airtight container until used.

3.2. Preparation of Phenolic Extracts

Two grams (2.0 ± 0.05 g) of powdered samples were extracted with 20 mL of nano-pure water. The suspension was heated to the boiling point, and results were observed at 5, 10, and 15 min. The boiling time was selected based on the results of pre-trials. The pre-trials were conducted for different boiling times and data were obtained for TPC and the reducing power. The boiling times which were suitable for both the basil and the rosemary were selected for this study. Once the boiling was finished for their respective time, the mixture was cooled at room temperature. The obtained mixture was then centrifuged at room temperature for 10 min at $3000 \times g$ (HERMLE Labortechnik GmbH. Siemensstr. 25 D-78564 Wehingen, Germany) and filtered using Whatman filter paper number 2. The obtained extracts were stored at 4 °C and used for analyses.

3.3. Total Polyphenol Content (TPC)

The Folin–Ciocalteu (FC) method was followed to detect TPC [64]. Initially, 125 μL of undiluted FC reagent was added to 25 μL of extract. Subsequently, the mixture was shaken for 1 min at room temperature after the addition of 1.5 mL of nano pure water. After 1 min of shaking, 375 μL of 20% sodium carbonate and 475 μL of water were added to the mixture and the final volume was made to 2500 μL. TPC detection was achieved spectroscopically at 760 nm (Jasco, V-630 spectrophotometer, Easton, MD, USA) after 30 min of incubation at room temperature. The TPC was expressed as the gallic acid equivalent per gram dry weight of the sample (mg GAE/g dw).

3.4. Total Flavonoid Content (TFC)

In the current study, the TFC was measured according to the procedure suggested by Hayat [64]. A total of 1 mL of water was added to 250 μL of extract. Then, 75 μL of each 5% (*w/v*) sodium nitrite and 10% (*w/v*) aluminum chloride were added. The mixture was allowed to stand at room temperature for 5 min. After that, 0.5 mL of sodium hydroxide (1 M) and 0.6 mL of water were added to the mixture and then vortexed. Blank samples were prepared following the same steps but without extract. Total flavonoid detection was achieved spectroscopically at 510 nm (Jasco, V-630 spectrophotometer, USA). TFC was expressed as the mg catechin equivalent per gram dry weight of the sample (mg CE/g dw).

3.5. DPPH Scavenging

The free radical scavenging capacity of the basil and rosemary extract was analyzed using DPPH according to the method suggested by Gülçin et al. [65] with slight modifications. Firstly, 130 µL aliquot of extract was mixed with 0.1 mM DPPH and then incubated in the dark for 30 min. The absorbance was measured at 510 nm (Jasco, V-630 spectrophotometer, USA). Control samples were prepared following the same steps, however, ethanol, as a blank, was added to the control samples instead of extract. The scavenging percentage was calculated as Equation (1):

$$\text{DPPH scavenging \%} = A\text{control} - A\text{sample}/A\text{control} \times 100 \tag{1}$$

3.6. Reducing Power

The ferric reducing power of the sample was analyzed according to the steps used by Hayat et al., [66]. Initially, potassium ferricyanide (1.25 mL) was mixed with the extract (0.5 mL). Then, 1.25 mL of sodium phosphate buffer (0.2 M, pH 6.6) was added to the mixture and incubated for 20 min at 50 °C. Subsequently, 1.25 mL of trichloroacetic acid was added and then centrifuged at $3000 \times g$ for 10 min at room temperature. Lastly, 1.25 mL of water and 0.25 mL of ferric chloride were added to 1.25 mL of aliquot from the supernatant. Blank samples were prepared without extract and the absorbance was recorded at 700 nm (Jasco, V-630 spectrophotometer, USA).

3.7. Bacterial Strains

The standard reference strains of American Type Culture Collection (ATCC): *Listeria monocytogenes* ATCC 19114 (Microbiologics Inc., St. Cloud, MN, USA), *Bacillus subtilis* ATCC 6633 (Microbiologics Inc., St. Cloud, MN, USA), *Escherichia coli* ATCC 10798 (Microbiologics Inc., St. Cloud, MN, USA), and *Salmonella typhimurium* ATCC 14028 (Microbiologics Inc., St. Cloud, MN, USA), were used in this study.

3.8. Antibiotic Proprieties of Basil and Rosemary

An agar diffusion assay [67] was performed in order to assess the antimicrobial efficiency of the basil and rosemary extracts against the foodborne pathogens listed previously. Active bacterial strains were inoculated in Mueller–Hinton broth (Oxoid, CM0405) at a concentration of 10^6 CFU/mL and incubated overnight. From this broth, 100 µL was spread-plated onto Mueller–Hinton agar (Oxoid CM 0337). The basil and rosemary extracts were dissolved separately in 96% ethanol and methanol to provide a final concentration of 100 mg/mL and all samples were stored at 4 °C until further analysis. Three sterilized 6 mm Whatman filter paper disks were saturated as follows: the first disc was saturated with 50 µL of methanolic basil extract, while the second disc was saturated with 100 µL of the same substance. For a control, the disc was saturated with only methanol. The same procedure was followed with the methanolic rosemary extract, as well as the ethanolic extracts of both herbs (using ethanol on the control disc instead). The inhibition zone (mm) was determined for each tested pathogen with each extract after incubation at the suitable temperature.

3.9. Determination of Minimum Inhibitory and Minimum Bactericidal Concentrations

Each colony of the four tested bacteria were inoculated into Mueller–Hinton broth (Oxoid, CM0405) and incubated at 37 °C for 24 h. The bacterial culture was adjusted to ~0.5 optical density at 650 nm using a spectrophotometer (GloMax® Multi), corresponding to a concentration of 1×10^8 CFU/mL (0.5 McFarland standard). One hundred microliters of ethanolic and methanolic extracts of basil were separately dispersed into two a 96-well flat-bottom plate (Costar®) and mixed well with 100 µL of bacterial inoculum. The optical density was measured after a 24 h incubation period [68,69].

A microdilution assay was performed to assess the MIC of rosemary and basil extracts according to the procedure suggested by Andrew [69]. First, 100 µL of Mueller–Hinton broth (Oxoid, CM0405) was distributed onto the microtiter plate. Then, 100 µL of the test extracts (basil and rosemary at a concentration of 100 mg/mL) were dispensed into to the first well of the microtiter plate and serially diluted. A 100 µL aliquot of the bacterial inoculum ($\sim$5 $\times$ 10^5 CFU/mL) was mixed well with each extract dilution. The plates were incubated at 37 °C for 24 h with shaking before determining the optical density of the bacterial growth. The MIC was recorded as the lowest concentration of the test agent that resulted in bacterial growth inhibition [69].

The MBC of the tested substances was derived from the positive density recorded from the MIC results. Therefore, 10 µL was pipetted and sub-cultured onto the Mueller–Hinton agar and incubated for 24 h at 37 °C [70]. The MBC was determined to be the concentration, at which no visible growth of bacteria appeared on the cultured plate, i.e., the lowest concentration of substance required to achieve the bactericidal killing [68].

3.10. HPLC Analysis of Phenolic Compounds

Phenolic compounds (chlorogenic acid, acetyl acetic acid, tannic acid, quercetin, resorcinol, caffeic acid, 1,2-dihydroxybenzene, vanillin, 3,5-dinitrosalicylic acid (3,5-DNSA), and salicylic acid) in basil and rosemary samples were analyzed with HPLC according to the method of He et al. [71] with some modification. The Shimadzu (prominence) HPLC system (Kyoto, Japan) equipped with binary pump LC-20 AB, UV detector SPD-10 A was used in this work. Phenolic compounds were separated on the column Zorbax SB-C_{18} (250 $\times$ 4.6 mm, 5µm) with mobile phase, (A) 1% acetic acid and (B) 100% methanol, and detected at wavelength 280 nm. The binary gradient program for the mobile phase with 1 mL/min flow rate was 0–10 min for 15–30% (B); 10–20 min for 30–40% (B); 20–30 min for 40–50% (B); 30–41 min for 50–60% (B), and 41–45 min for 15% (B). Phenolic compounds in basil and rosemary samples were compared with the retention time of standards. All samples were analyzed in duplicate and the arithmetical mean $\pm$ standard error was reported.

3.11. Statistical Analysis

The statistical analysis was performed using SAS (v9.2, 2000–2008; SAS Institute Inc., Cary, NC, USA) for data analysis. All the tests were performed in triplicates and the results were presented as mean $\pm$ standard deviation (SD). A one-way analysis of variance (ANOVA, $p \leq 0.05$) was used to analyze the differences among the treatment groups and a post-hoc analysis using Duncan's multiple range tests was performed if any significant differences were found.

4. Conclusions

The duration of boiling time has a significant influence on the medicinal properties of traditional herbs. Boiling for 5 min showed the higher TPC, TFC, and antioxidant activity of the basil samples, while 15 min of boiling exhibited better results for rosemary samples. Overall, the rosemary extracts were effectively better than the basil extracts with respect to TPC, TFC, and antioxidant capacity. The MIC and MIB of rosemary ethanolic extract against *L. monocytogenes* were 5 and 5 mg/mL, respectively, and against *B. subtilis* were 10 and 10 mg/mL, respectively. Similar zones of inhibition were noted against *B. subtilis* and were 14 and 15 mm in diameter using ethanolic and methanolic extracts, respectively. Basil extracts in either ethanol or methanol had no effect on the same tested microorganisms in comparison with rosemary except against *B. subtilis*. While the MIC and MBC of methanolic extract for *L. monocytogenes* and *B. subtilis*, were 5 mg/mL, respectively. According to the HPLC results, among the tested compounds, acetyl salicylic acid was obtained in the highest phenolic compound in the basil sample boiled for 15 min, while, salicylic acid was the highest phenolic compound in the rosemary sample boiled for 5 min. The results of this study might be helpful to reap the maximum health benefits of the teas produced from these herbs. In addition, as the response surface methodology (RSM) is one of the most commonly used experimental designs for the optimization of process conditions, so based on this study, the RSM approach in the near future is also suggested to determine the optimum conditions for the extraction process.

Author Contributions: Conceptualization, A.M.S., K.H.; methodology, K.H., M.A.A.; formal analysis, K.H., B.A.M.A.-Z., M.M.A.; investigation, K.H.; resources, A.M.S.; data curation, K.H., S.A., A.A., M.A.A., H.M.Y.; writing—original draft preparation, A.M.S., K.H., S.A., M.A.A.; writing—review and editing, A.M.S., N.A.-B., A.A., H.M.Y., T.A.; visualization, A.M.S., A.A.; supervision, A.M.S., K.H.; project administration, A.M.S., K.H., A.A.; funding acquisition, A.M.S. All authors have read and agreed to the published version of the manuscript.

Funding: The authors extend their appreciation to the Deanship of Scientific Research at King Saud University for funding this work through research group no (RG-1441-360).

Institutional Review Board Statement: Not applicable.

Informed Consent Statement: Not applicable.

Data Availability Statement: The data presented in this study are available on request from the corresponding author.

Conflicts of Interest: The authors declare no conflict of interest. The funders had no role in the design of the study; in the collection, analyses, or interpretation of data; in the writing of the manuscript, or in the decision to publish the results.

Sample Availability: Samples of the compounds are not available from the authors.

References

1. Liu, Z.; Ren, Z.; Zhang, J.; Chuang, C.C.; Kandaswamy, E.; Zhou, T.; Zuo, L. Role of ROS and nutritional antioxidants in human diseases. *Front. Physiol.* **2018**, *9*, 477. [CrossRef]
2. Auten, R.L.; Davis, J.M. Oxygen toxicity and reactive oxygen species: The devil is in the details. *Pediatric Res.* **2009**, *66*, 121–127. [CrossRef]
3. Langseth, L. *Oxidants, Antioxidants, and Disease Prevention*; ILSI Europe: Brussels's, Belgium, 1995.
4. Gutteridge, J.M.; Halliwell, B. Free radicals and antioxidants in the year 2000. A historical look to the future. *Ann. N Y. Acad. Sci.* **2000**, *899*, 136–147. [CrossRef]
5. McCord, J.M. The evolution of free radicals and oxidative stress. *Am. J. Med.* **2000**, *108*, 652–659. [CrossRef]
6. Gilgun-Sherki, Y.; Melamed, E.; Offen, D. Oxidative stress induced-neurodegenerative diseases: The need for antioxidants that penetrate the blood brain barrier. *Neuropharmacology* **2001**, *40*, 959–975. [CrossRef]
7. Embuscado, M.E. Herbs and spices as antioxidants for food preservation. In *Handbook of Antioxidants for Food Preservation*; Elsevier Inc: Amsterdam, The Netherlands, 2015; pp. 251–283.
8. Kaefer, C.M.; Milner, J.A. The role of herbs and spices in cancer prevention. *J. Nutr. Biochem.* **2008**, *19*, 347–361. [CrossRef]
9. Kheroda Devi, M.; Thangjam, I.; Singh, W. Phytochemical screening of selected twelve medicinal plants commonly used as spices and condiments in manipur, north-east India. *Int. J. Curr. Res. Life Sci.* **2019**, *8*, 2945–2947.
10. Firenzuoli, F.; Gori, L. Herbal medicine today: Clinical and research issues. *Evid. Based Complement. Altern. Med.* **2007**, *4*, 37–40. [CrossRef]
11. Park, J.B. Identification and quantification of a major anti-oxidant and anti-inflammatory phenolic compound found in basil, lemon thyme, mint, oregano, rosemary, sage, and thyme. *Int. J. Food Sci. Nutr.* **2011**, *62*, 577–584. [CrossRef] [PubMed]
12. Simon, J.E.; Morales, M.R.; Phippen, W.B.; Vieira, R.F.; Hao, Z. Basil: A source of aroma compounds and a popular culinary and ornamental herb. *Perspect. New Crop. New Uses* **1999**, *16*, 499–505.
13. Mohan, L.; Amberkar, M.; Kumari, M. *Ocimum sanctum linn.* (TULSI)-an overview. *Int. J. Pharm. Sci. Rev. Res.* **2011**, *7*, 51–53.
14. Rahman, S.; Islam, R.; Kamruzzaman, M.; Alam, K.; Jamal, A. *Ocimum sanctum* L.: A review of phytochemical and pharmacological profile. *Am. J. Drug Discov. Develop.* **2011**, *1*, 1–15.
15. Pattanayak, P.; Behera, P.; Das, D.; Panda, S.K. *Ocimum sanctum* Linn. A reservoir plant for therapeutic applications: An overview. *Pharmacogn. Rev.* **2010**, *4*, 95–105. [CrossRef] [PubMed]
16. Mondal, S.; Mirdha, B.R.; Mahapatra, S.C. The science behind sacredness of Tulsi (*Ocimum sanctum* Linn.). *Indian. J. Physiol. Pharmacol.* **2009**, *53*, 291–306.
17. Narendhirakannan, R.; Subramanian, S.; Kandaswamy, M. Biochemical evaluation of antidiabetogenic properties of some commonly used Indian plants on streptozotocin-induced diabetes in experimental rats. *Clin. Exp. Pharmacol. Physiol.* **2006**, *33*, 1150–1157. [CrossRef]
18. Goel, R.K.; Sairam, K.; Dorababu, M.; Prabha, T.; Rao Ch, V. Effect of standardized extract of *Ocimum sanctum* Linn. on gastric mucosal offensive and defensive factors. *Indian J. Exp. Biol.* **2005**, *43*, 715–721.
19. Kelm, M.A.; Nair, M.G.; Strasburg, G.M.; DeWitt, D.L. Antioxidant and cyclooxygenase inhibitory phenolic compounds from *Ocimum sanctum* Linn. *Phytomedicine* **2000**, *7*, 7–13. [CrossRef]

20. Prakash, J.; Gupta, S.K. Chemopreventive activity of *Ocimum sanctum* seed oil. *J. Ethnopharmacol.* **2000**, *72*, 29–34. [CrossRef]

21. Vasudevan, P.; Kashyap, S.; Sharma, S. Bioactive Botanicals from Basil (*Ocimum* sp.). *J. Sci. Res.* **1999**, *58*, 332–338.

22. Cohen, M.M. Tulsi-Ocimum sanctum: A herb for all reasons. *J. Ayurved. Integrat. Med.* **2014**, *5*, 251. [CrossRef]

23. Anbarasu, K.; Vijayalakshmi, G. Improved shelf life of protein-rich tofu using *Ocimum sanctum* (tulsi) extracts to benefit Indian rural population. *J. Food Sci.* **2007**, *72*, M300–M305. [CrossRef] [PubMed]

24. Shishodia, S.; Majumdar, S.; Banerjee, S.; Aggarwal, B.B. Ursolic acid inhibits nuclear factor-kappaB activation induced by carcinogenic agents through suppression of IkappaBalpha kinase and p65 phosphorylation: Correlation with down-regulation of cyclooxygenase 2, matrix metalloproteinase 9, and cyclin D1. *Cancer Res.* **2003**, *63*, 4375–4383.

25. Scagel, C.F.; Lee, J. Phenolic composition of basil plants is differentially altered by plant nutrient status and inoculation with mycorrhizal fungi. *HortScience* **2012**, *47*, 660–671. [CrossRef]

26. Nurzynska-Wierdak, R. Sweet basil (*Ocimum basilicum* L.) flowering affected by foliar nitrogen application. *Acta Agrobotan.* **2011**, *64*, 57–64. [CrossRef]

27. Kruma, Z.; Andjelkovic, M.; Verhe, R.; Kreicbergs, V.; Karklina, D.; Venskutonis, P. Phenolic compounds in basil, oregano and thyme. *Foodbalt* **2008**, *5*, 99–103.

28. Hussain, A.I.; Anwar, F.; Hussain Sherazi, S.T.; Przybylski, R. Chemical composition, antioxidant and antimicrobial activities of basil (*Ocimum basilicum*) essential oils depends on seasonal variations. *Food Chem.* **2008**, *108*, 986–995. [CrossRef]

29. Duke, J.A. *Handbook of Medicinal Herbs*; CRC Press: Boca Raton, NY, USA, 2000; pp. 630–632.

30. Ghasemzadeh, R.M.; Hosseinzadeh, H. Therapeutic effects of rosemary (*Rosmarinus officinalis* L.) and its active constituents on nervous system disorders. *Iran J. Basic Med. Sci.* **2020**, *23*, 1100–1112.

31. United States Department of Agriculture, Agricultural Research Services USDA. National Nutrient Database for Standard Reference, Release 25. NutrientData Laboratory. 2015. Available online: http://ndb.nal.usda.gov/ndb/foods/show/256 (accessed on 30 November 2021).

32. Touafek, O.; Nacer, A.; Kabouche, A.; Kabouche, Z.; Bruneau, C. Chemical composition of the essential oil of *Rosmarinus officinalis* cultivated in the Algerian Sahara. *Chem. Nat. Comp.* **2004**, *40*, 28–29. [CrossRef]

33. Begum, A.; Sandhya, S.; Shaffath Ali, S.; Vinod, K.R.; Reddy, S.; Banji, D. An in-depth review on the medicinal flora *Rosmarinus officinalis* (Lamiaceae). *Acta Sci. Pol. Technol. Aliment.* **2013**, *12*, 61–73. [PubMed]

34. Atti-Santos, A.C.; Rossato, M.; Pauletti, G.F.; Rota, L.D.; Rech, J.C.; Pansera, M.R.; Agostini, F.; Serafini, L.A.; Moyne, P. Physico-chemical evaluation of *Rosmarinus officinalis* L. essential oils. *Braz. Arch. Biol. Technol.* **2005**, *48*, 1035–1039. [CrossRef]

35. Tawfeeq, A.A.; Mahdi, M.F.; Abaas, I.S.; Alwan, A.H. Phytochemical and antibacterial studies of leaves of *Rosmarinus officinalis* cultivated in Karbala, Iraq. *Al-Mustansiriyah J. Pharmaceut. Sci.* **2018**, *17*, 9.

36. Jiang, Y.; Wu, N.; Fu, Y.J.; Wang, W.; Luo, M.; Zhao, C.J.; Zu, Y.G.; Liu, X.L. Chemical composition and antimicrobial activity of the essential oil of rosemary. *Environ. Toxicol. Pharmacol.* **2011**, *32*, 63–68. [CrossRef]

37. Bendeddouche, M.S.; Benhassaini, H.; Hazem, Z.; Romane, A. Essential oil analysis and antibacterial activity of *Rosmarinus tournefortii* from Algeria. *Nat. Prod. Commun.* **2011**, *6*, 1511–1514. [CrossRef]

38. Bai, N.; He, K.; Roller, M.; Lai, C.S.; Shao, X.; Pan, M.H.; Ho, C.T. Flavonoids and phenolic compounds from *Rosmarinus officinalis*. *J. Agric. Food. Chem.* **2010**, *58*, 5363–5367. [CrossRef] [PubMed]

39. del Bano, M.J.; Lorente, J.; Castillo, J.; Benavente-Garcia, O.; Marin, M.P.; Del Rio, J.A.; Ortuno, A.; Ibarra, I. Flavonoid distribution during the development of leaves, flowers, stems, and roots of *Rosmarinus officinalis*: Postulation of a biosynthetic pathway. *J. Agric. Food Chem.* **2004**, *52*, 4987–4992. [CrossRef] [PubMed]

40. Samuelsson, G.; Bohlin, L. *Drugs of Natural Origin. A Treatise of Pharmacognosy*; CRC Press Inc.: Boca Raton, FL, USA; London, UK, 2017.

41. Chan, E.W.C.; Kong, L.Q.; Yee, K.Y.; Chua, W.Y.; Loo, T.Y. Antioxidant and antibacterial properties of some fresh and dried Labiatae herbs. *Free Radic. Antiox.* **2012**, *2*, 20–27. [CrossRef]

42. Larrauri, J.A.; Ruperez, P.; Saura-Calixto, F. Effect of drying temperature on the stability of polyphenols and antioxidant activity of red grape pomace peels. *J. Agric. Food Chem.* **1997**, *45*, 1390–1393. [CrossRef]

43. Dewanto, V.; Wu, X.Z.; Adom, K.K.; Liu, R.H. Thermal processing enhances the nutritional value of tomatoes by increasing total antioxidant activity. *J. Agric. Food Chem.* **2002**, *50*, 3010–3014. [CrossRef]

44. Kim, S.Y.; Jeong, S.M.; Kim, S.J.; Jeon, K.I.; Park, E.; Park, H.R.; Lee, S.C. Effects of heat treatment on the antioxidative and antigenotoxic activity of extracts from persimmon (Diospyros kaki L.) peel. *Biosci. Biotechnol. Biochem.* **2006**, *70*, 999–1002. [CrossRef] [PubMed]

45. Salamatullah, A.M.; Özcan, M.M.; Alkaltham, M.S.; Uslu, N.; Hayat, K. Influence of boiling on total phenol, antioxidant activity, and phenolic compounds of celery (*Apium graveolens* L) root. *J. Food Proc. Preserv.* **2021**, *45*, e15171. [CrossRef]

46. Hayat, K.; Abbas, S.; Hussain, S.; Shahzad, S.A.; Tahir, M.U. Effect of microwave and conventional oven heating on phenolic constituents, fatty acids, minerals and antioxidant potential of fennel seed. *Ind. Crop. Prod.* **2019**, *140*, 111610. [CrossRef]

47. Karrar, E.; Sheth, S.; Wei, W.; Wang, X.G. Effect of microwave heating on lipid composition, oxidative stability, color value, chemical properties, and antioxidant activity of gurum (*Citrullus lanatus* var. Colocynthoide) seed oil. *Biocatal. Agric. Biotechnol.* **2020**, *23*, 101504. [CrossRef]

48. Elisabetta, D.; Patricia, C.; Gabriele, R.; Biancamaria, S.; Luca, T.; Elizabeth, J.; Dalene, D.B.; Luigi, L. Impact of cold versus hot brewing on the phenolic profile and antioxidant capacity of Rooibos (*Aspalathus linearis*) herbal tea. *Antioxidants* **2019**, *8*, 499.

49. Gunathilake, K.; Ranaweera, K.; Rupasinghe, H.P.V. Effect of different cooking methods on polyphenols, carotenoids and antioxidant activities of selected edible leaves. *Antioxidants* **2018**, *7*, 117. [CrossRef]

50. Salamatullah, A.M.; Uslu, N.; Özcan, M.M.; Alkaltham, M.S.; Hayat, K. The effect of oven drying on bioactive compounds, antioxidant activity, and phenolic compounds of white and red-skinned onion slices. *J. Food Proc. Preserv.* **2021**, *45*, e15173. [CrossRef]

51. Vinha, A.F.; Alves, R.C.; Barreira, S.V.; Costa, A.S.; Oliveira, M.B. Impact of boiling on phytochemicals and antioxidant activity of green vegetables consumed in the Mediterranean diet. *Food Funct.* **2015**, *6*, 1157–1163. [CrossRef]

52. Arias-Rico, J.; Macias-Leon, F.J.; Alanis-Garcia, E.; Cruz-Cansino, N.D.S.; Jaramillo-Morales, O.A.; Barrera-Galvez, R.; Ramirez-Moreno, E. Study of edible plants: Effects of boiling on nutritional, antioxidant, and physicochemical properties. *Foods* **2020**, *9*, 599. [CrossRef]

53. Nie, Y.; Yu, M.; Zhou, H.; Zhang, P.; Yang, W.; Li, B. Effect of boiling time on nutritional characteristics and antioxidant activities of *Lentinus edodes* and its broth. *CyTA J. Food* **2020**, *18*, 543–550. [CrossRef]

54. Mohammed, J.; Hateem, S.M.; Sattar, O.D.A. Effect of aqueous, alcoholic and acidic extract of rosemary leaves *Rosmarinus officinalis* in inhibiting the effect of free radicals manufactured and inhibitory effect in some microorganisms and detection of some active compounds. *J. Phys. Conf. Ser.* **2020**, *1664*, 012079. [CrossRef]

55. Pankey, G.A.; Sabath, L.D. Clinical relevance of bacteriostatic versus bactericidal mechanisms of action in the treatment of Gram-positive bacterial infections. *Clin. Infect. Dis.* **2004**, *38*, 864–870. [CrossRef] [PubMed]

56. Gonelimali, F.D.; Lin, J.; Miao, W.; Xuan, J.; Charles, F.; Chen, M.; Hatab, S.R. Antimicrobial properties and mechanism of action of some plant extracts against food pathogens and spoilage microorganisms. *Front. Microbiol.* **2018**, *9*, 1639. [CrossRef]

57. Peng, Y.; Yuan, J.; Liu, F.; Ye, J. Determination of active components in rosemary by capillary electrophoresis with electrochemical detection. *J. Pharm. Biomed. Anal.* **2005**, *39*, 431–437. [CrossRef] [PubMed]

58. Kh, M. Antibacterial Effect of *Allium sativum* and *Rosmarinus officinalis* essential oil on major mastitis pathogens in dairy cattle. *J. Cell Tissue* **2014**, *5*, 79–88.

59. Food Standards Agency, Approved additives and E numbers. Available online: https://www.food.gov.uk/business-guidance/approved-additives-and-e-numbers (accessed on 15 November 2021).

60. Habtemariam, S. The therapeutic potential of rosemary (*Rosmarinus officinalis*) diterpenes for Alzheimer's disease. *Evid. Based Complement. Alternat. Med.* **2016**, *2016*, 2680409. [CrossRef]

61. Andrade, J.M.; Faustino, C.; Garcia, C.; Ladeiras, D.; Reis, C.P.; Rijo, P. *Rosmarinus officinalis* L.: An update review of its phytochemistry and biological activity. *Future Sci. OA* **2018**, *4*, FSO283. [CrossRef]

62. Elansary, H.O.; Szopa, A.; Kubica, P.; Ekiert, H.; El-Ansary, D.O.; Al-Mana, F.A.; Mahmoud, E.A. Saudi Rosmarinus officinalis and *Ocimum basilicum* L. polyphenols and biological activities. *Processes* **2020**, *8*, 446. [CrossRef]

63. Flanigan, P.M.; Niemeyer, E.D. Effect of cultivar on phenolic levels, anthocyanin composition, and antioxidant properties in purple basil (*Ocimum basilicum* L.). *Food Chem.* **2014**, *164*, 518–526. [CrossRef] [PubMed]

64. Hayat, K. Impact of drying methods on the functional properties of peppermint (*Mentha piperita* L.) leaves. *Sci. Lett* **2020**, *8*, 36–42.

65. Gülçin, İ.; Kìreçì, E.; Akkemìk, E.; Topal, F.; Hisar, O. Antioxidant and antimicrobial activities of an aquatic plant: Duckweed (*Lemna minor* L.). *Turk. J. Biol.* **2010**, *34*, 175–188.

66. Hayat, K.; Zhang, X.; Chen, H.; Xia, S.; Jia, C.; Zhong, F. Liberation and separation of phenolic compounds from citrus mandarin peels by microwave heating and its effect on antioxidant activity. *Sep. Purif. Technol.* **2010**, *73*, 371–376. [CrossRef]

67. Vineetha, N.; Vignesh, R.; Sridhar, D. Preparation, standardization of antibiotic discs and study of resistance pattern for first-line antibiotics in isolates from clinical samples. *Int. J. Appl. Res.* **2015**, *1*, 624–631.

68. CLSI. *Reference Method for Broth Dilution Antifungal Susceptibility Testing of Yeasts*; Clinical and Laboratory Standards Institute: Wayne, PA, USA, 2008; pp. 1–28.

69. Andrews, J.M. Determination of minimum inhibitory concentrations. *J. Antimicrob. Chemother.* **2001**, *48*, 5–16. [CrossRef]

70. de Castro, R.D.; Lima, E.O. Anti-candida activity and chemical composition of *Cinnamomum zeylanicum* blume essential oil. *Braz. Arch. Biol. Technol.* **2013**, *56*, 749–755. [CrossRef]

71. He, J.; Yin, T.; Chen, Y.; Cai, L.; Tai, Z.; Li, Z.; Liu, C.; Wang, Y.; Ding, Y. Phenolic compounds and antioxidant activities of edible flowers of *Pyrus pashia*. *J. Funct. Foods* **2015**, *17*, 371–379. [CrossRef]

Article

In Vitro and In Silico Interaction Studies with Red Wine Polyphenols against Different Proteins from Human Serum [†]

Raja Mohamed Beema Shafreen [1], Selvaraj Alagu Lakshmi [2], Shunmugiah Karutha Pandian [2], Young-Mo Kim [3], Joseph Deutsch [4], Elena Katrich [4] and Shela Gorinstein [4,*]

[1] Department of Biotechnology, Dr. Umayal Ramanathan College for Women, Algappapuram, Karaikudi 630003, India; beema.shafreen@gmail.com

[2] Department of Biotechnology, Alagappa University, Science Campus, Karaikudi 630003, India; lakshmivinay.317@gmail.com (S.A.L.); sk_pandian@rediffmail.com (S.K.P.)

[3] Industry Academic Collaboration Foundation, Kwangju Women's University, Gwangju 62396, Korea; bliss0816@kwu.ac.kr

[4] Institute for Drug Research, School of Pharmacy, Faculty of Medicine, The Hebrew University of Jerusalem, Jerusalem 9112001, Israel; josephd@ekmd.huji.ac.il (J.D.); ekatrich@gmail.com (E.K.)

* Correspondence: shela.gorin@mail.huji.ac.il; Tel.: +972-2-6758690

[†] This article is dedicated to the memory of my dear brother Prof. Simon Trakhtenberg, who died on 20 November 2011, exactly 10 years ago. Simon encouraged our research group during all his life, especially in vivo studies on alcoholic and non-alcoholic beverages. He was a special friend, outstanding scientist and longtime cooperator. He will always remain in our hearts.

Abstract: Previous reports have shown that consumption of wine has several health benefits; however, there are different types of wine. In the present study, red wines were investigated for their compositions of active ingredients. The interaction of each component in terms of its binding mode with different serum proteins was unraveled, and the components were implicated as drug candidates in clinical settings. Overall, the study indicates that red wines have a composition of flavonoids, non-flavonoids, and phenolic acids that can interact with the key regions of proteins to enhance their biological activity. Among them, rutin, resveratrol, and tannic acid have shown good binding affinity and possess beneficial properties that can enhance their role in clinical applications.

Keywords: beverages; health properties; antioxidant activities; fibrinogen; albumin; rutin; tannic acid; resveratrol; binding properties

Citation: Shafreen, R.M.B.; Lakshmi, S.A.; Pandian, S.K.; Kim, Y.-M.; Deutsch, J.; Katrich, E.; Gorinstein, S. In Vitro and In Silico Interaction Studies with Red Wine Polyphenols against Different Proteins from Human Serum. *Molecules* **2021**, *26*, 6686. https://doi.org/10.3390/molecules26216686

Academic Editor: Mirella Nardini

Received: 1 October 2021
Accepted: 1 November 2021
Published: 5 November 2021

Publisher's Note: MDPI stays neutral with regard to jurisdictional claims in published maps and institutional affiliations.

1. Introduction

Phenolic compounds are an essential part of the human diet and are of considerable interest due to their antioxidant properties and potential beneficial health effects. Their effects on human health depend on the amount consumed and on their bioavailability. Many studies have demonstrated that polyphenols also have good effects on the vascular system by lowering blood pressure, improving endothelial function, increasing antioxidant defenses, inhibiting platelet aggregation and low-density lipoprotein oxidation, and reducing inflammatory responses [1,2]; they are found in fruits [3], cereals, vegetables, legumes, chocolate, and beverages such as coffee, tea, beer, and wine. Polyphenols have been reported to inhibit platelet aggregation both in vitro and in vivo. The analysis of the results indicates a promising role for food polyphenols in preventing thrombosis and cardiovascular diseases, but, at the same time, suggests caution when transferring the in vitro findings in vivo [4–6]. It was reported in a number of recent studies [7–10] that the consumption of beer, wine, and fruits has already been associated with a multitude of beneficial effects due to their high polyphenolic content. Beverages have to be consumed moderately in order to obtain their positive influence on health. Only the high amount of bioactive compounds—and not the alcohol content—positively influences the human metabolism. The pro-oxidant effect of red wine polyphenols promotes an adaptive stress response in human erythrocytes, which enhances their antioxidant defense [11]. Phenolic compounds in wine, such as tannins, phenolic acids, and anthocyanins, are some of the determinants of its quality. It was shown that phenolic compounds participate not only in the appearance and sensory characteristics of wine, but also in its healthy properties [12]. As an example, research was conducted on 110 Italian red wines from a single vintage in order to determine the standard composition, color, and phenolic characteristics based on the parameters used in the wine industry [13]. The elevation of C-reactive protein (CRP) levels in blood was recognized as a cardiac disease risk factor. An increased nitration of fibrinogen has been reported in cardiovascular diseases. Consumption of wine is shown to reduce the risk of heart disease and improve longevity [14]. The defense mechanisms against nitrative modifications are crucial for the complex hemostasis process. Flavonoids have antioxidative properties and could protect biomolecules against the action of peroxynitrite [15]. There are reports that also show the results of phenolic compounds with the main human serum proteins [16]. Recently, research on wine has been widely cited, but in spite of a number of reports [17,18] and evidence that wine exerts beneficial effects on human health when it is consumed with moderation, how the main phenolics of red wine contribute to the quenching properties of the main human serum proteins has not been investigated. For this reason, red wine samples were investigated for their antioxidant activities, bioactive compounds, and interactions of wine polyphenols with the main serum proteins, with an emphasis on the influence of polyphenols and not the alcohol content on the health properties of wine. Fluorescence measurements, antioxidant assays, and docking analysis were applied for the first time to wine samples, and correlations were made among the results obtained in these studies.

2. Results

2.1. Bioactive Properties of the Investigated Wines

The results of the phenolic compounds are presented in Table 1.

Table 1. Bioactive substances and antioxidant activities of red wines x liter.

Indices	CSCarmel1	CSCarmel2	CSYarden1	CSYarden2
Polyph, mgGAE	2190.83 ± 9.43 [a]	2230.73 ± 8.72 [a]	1560.33 ± 6.32 [b]	1610.42 ± 6.21 [b]
Flavan, mgCE	241.84 ± 3.62 [b]	253.94 ± 2.92 [ab]	272.51 ± 4.33 [a]	283.63 ± 3.73 [a]
Flavon, mgCE	408.63 ± 3.63 [a]	418.63 ± 5.11 [a]	292.42 ± 2.54 [b]	302.62 ± 5.24 [ab]
Tannins, mgCE	152.54 ± 1.82 [a]	156.24 ± 1.42 [a]	51.33 ± 0.92 [b]	52.43 ± 0.73 [b]
Anthoc, mgCGE	137.53 ± 2.24 [a]	140.23 ± 2.93 [a]	97.91 ± 1.83 [b]	101.22 ± 2.93 [b]
ABTS, mMTE	19.84 ± 0.34 [a]	20.45 ± 1.12 [a]	13.94 ± 1.11 [b]	14.85 ± 1.23 [b]
FRAP, mMTE	5.84 ± 0.54 [a]	6.18 ± 0.61 [a]	4.12 ± 0.34 [b]	4.46 ± 0.25 [ab]
CUPRAC, mMTE	27.11 ± 1.14 [a]	28.33 ± 1.65 [a]	19.64 ± 1.65 [b]	20.47 ± 1.76 [b]
DPPH, mMTE	9.65 ± 0.87 [a]	10.52 ± 1.12 [a]	7.14 ± 0.65 [b]	7.36 ± 0.73 [b]
Rutin, mg	8.63 ± 0.54 [a]	9.25 ± 0.87 [a]	6.81 ± 0.56 [b]	6.53 ± 0.55 [b]
Resveratro, mg	2.15 ± 0.18 [ab]	2.98 ± 0.12 [a]	1.71 ± 0.17 [b]	1.91 ± 0.17 [ab]
Quercetin, mg	7.32 ± 0.41 [ab]	8.24 ± 0.61 [a]	5.74 ± 0.43 [c]	6.49 ± 0.43 [b]
Caffeic acid, mg	10.15 ± 0.97 [a]	11.24 ± 1.12 [a]	8.64 ± 0.76 [b]	9.45 ± 0.75 [ab]
Catechin, mg	40.21 ± 0.37 [a]	42.17 ± 0.46 [a]	31.18 ± 0.23 [b]	34.15 ± 0.22 [ab]
Epicatechin, mg	26.14 ± 2.33 [a]	28.65 ± 2.43 [a]	21.94 ± 2.09 [b]	23.18 ± 1.89 [ab]

Values are means ± SD of five measurements; $n = 5$ samples per vintage, each subsampled and analyzed five times. Means within rows with the different superscripts are statistically different ($p < 0.05$; Student's t-test). Abbreviations: 1, 1-diphenyl-2-picrylhydrazyl (DPPH); Polyph, polyphenols; Flavon, flavonoids; Flavan, flavanols; GAE, gallic acid equivalent; CE, catechin equivalent; TE, trolox equivalent; Anthoc, anthocyanins; CGE, cyanidin-3-glucoside equivalent; ABTS, 2, 2-Azino-bis (3-ethyl-benzothiazoline-6-sulfonic acid) diammonium salt; FRAP, Ferric-reducing/antioxidant power; CUPRAC, Cupric reducing antioxidant capacity; CSCarmel1, CSCarmel2, Cabernet Sauvignon Carmel Selected, vintages 2017, 2019; CSYarden1 and CSYarden2, Cabernet Sauvignon Golan Heights Winery Yarden, vintages 2017 and 2019.

Our results for the total polyphenols were in the ranges found for Italian red wines at 1065 mg GAE/L, with the average values ranging between 1945 and 2033 and 2841–3578 mg GAE/L; the total tannins were the lowest at 533 mg CE/L, with averages of 1341 and 2043 and a maximum of 2965 mg CE/L [13]. The polyphenols, flavonoids, and antioxidant activities were in the range of organic wines rather than the range of conventional ones. The results obtained were comparable to those of conventional red wines with regard to the total polyphenols, flavonoid content, phenolic profile, and antioxidant activity [5]. The total phenolic content and total flavanols of the wines were found to vary from 1439.8 mg to 2966.0 mg GAE/L and from 439.4 to 1367.7 mg CE/L, respectively, which was in line with the presented data (Table 1). The phenolic content values observed in the present study were similar to those reported by other authors for Cabernet Sauvignon wines, showing values from 1300 to 2900 mg GAE/L [19]. The final flavonoid content was 555 mg CE/L. These results were lower than those reported previously (1390 mg CE/L) [19]. The anthocyanin content was also lower than previously reported for Cabernet Sauvignon wines (between 300 and 320 mg C3G/L). The results of FRAP 5.08 mmol TE/L, ABTS 10.68 mmol TE/L, and DPPH 3.71 mmol TE/L were lower than those reported by other authors for red wines, as well as in comparison with the presented data (Table 1). The polyphenol content was 1268 mg GAE/L; the tannin content varied from 133.9 to 318.3 mg CE/L, and the anthocyanin content was observed to be 104.08 mg cyanidin 3-glucoside equivalent per liter (mg C3G/L) [20–22]. Our results were in line with others [23], where twenty seven wines—among them, twenty four were red wines and three were white wines—were analyzed using liquid chromatography and spectrophotometric analysis, and the amounts of total polyphenols of the red wines varied from the lowest sample of 1220 mg GAE/L to the highest one of 2413 mg GAE/L. The amounts of anthocyanins in the same samples [23] varied from the lowest of 79 to the highest of 232 mg malvidin-3, 5-diglucoside/L. For the red wines, the antioxidant capacities varied between 13.4 and 20 mmol/L [24,25].

2.2. Binding Properties of Wines and Some Phenolic Compounds with the Main Human Proteins

The interactions of fibrinogen and human serum albumin with wine samples, tannic acid, quercetin, and caffeic acid in their cross-images are presented in Figures 1 and 2.

The binding properties of wine samples and some individual phenolic compounds and standards were compared in terms of their interactions with native human serum albumin and plasma-circulating fibrinogen. Tannic and caffeic acids and quercetin were also compared with the natural fibrinogen.

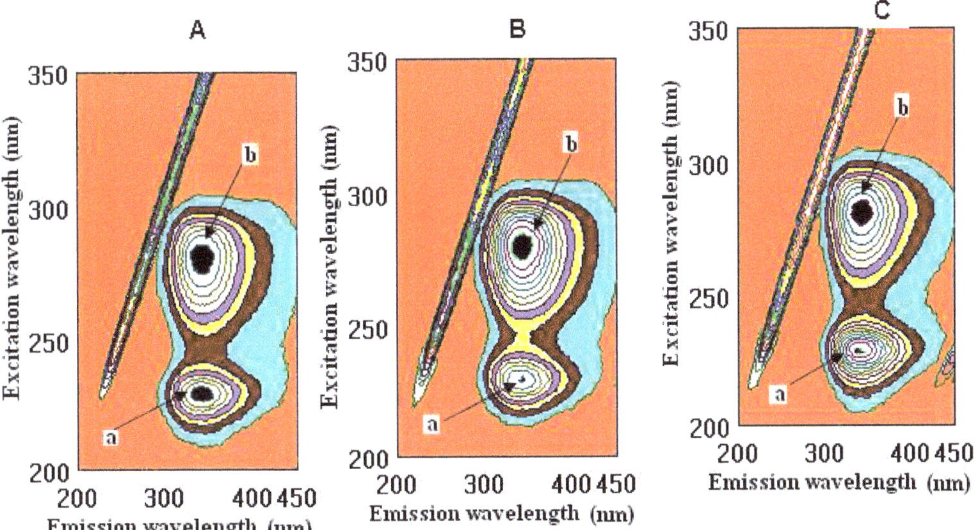

Figure 1. *Cont.*

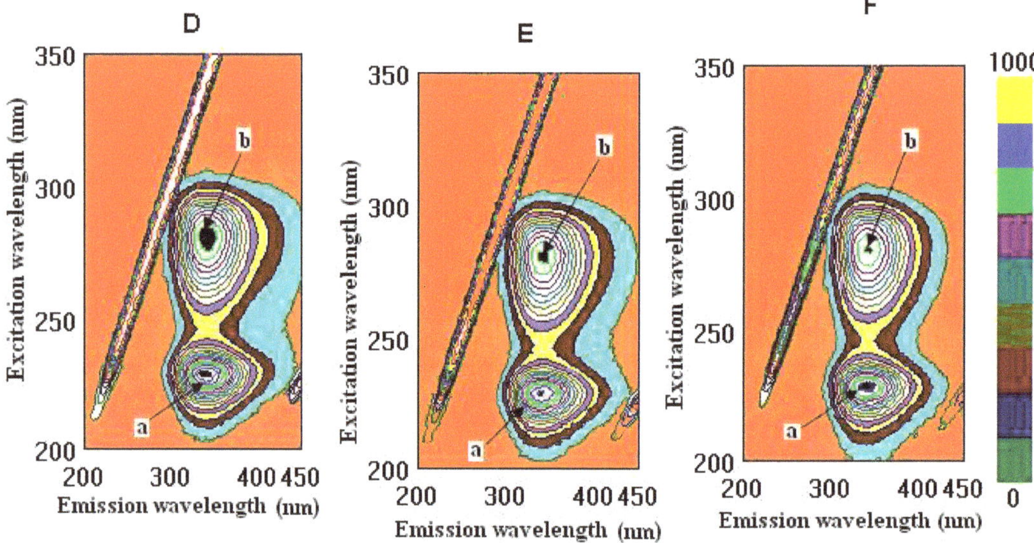

Figure 1. Fluorometric measurements in a three-dimensional fluorescence analysis (3D-FL) of red wine samples and standards after interaction with fibrinogen Cross-images of the results obtained from the 3D-FL of the investigated samples after interaction with fibrinogen are shown in the following order: (**A**) CSCarmel2; (**B**) CSYarden2; (**C**) tannic acid; (**D**) quercetin; (**E**) caffeic acid; (**F**) fibrinogen in solution. Abbreviations: CSCarmel2, Cabernet Sauvignon Carmel Selected, vintage 2019; CSYarden2, Cabernet Sauvignon Golan Heights Winery Yarden, vintage 2019. The locations of peaks **a** and **b** are shown in the figure and in Table 2 (for interpretation of the references to color in this figure legend, the reader is referred to the web version of this article).

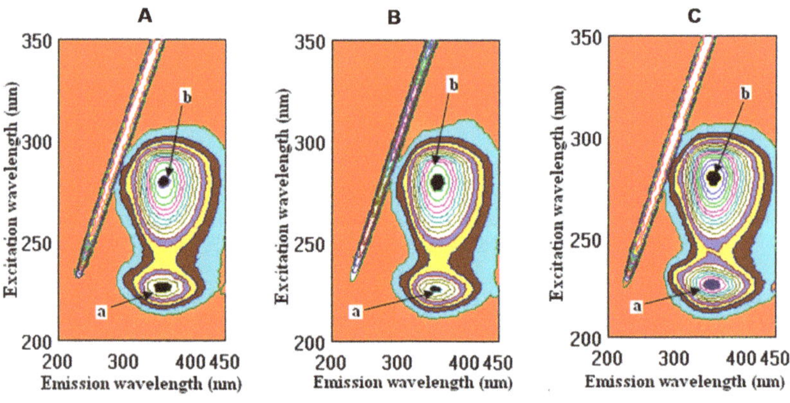

Figure 2. *Cont.*

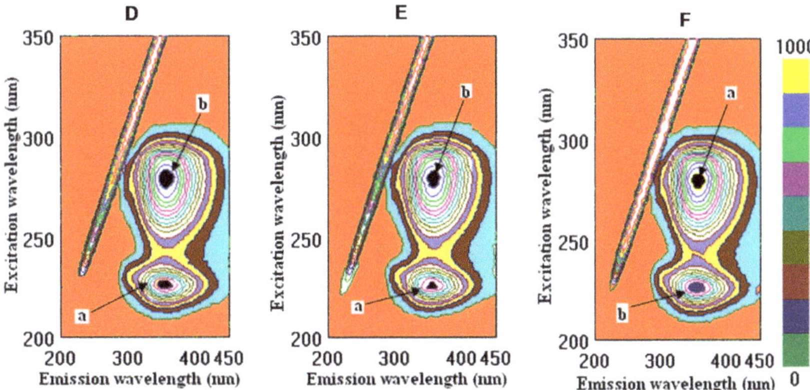

Figure 2. Fluorometric measurements in a three-dimensional fluorescence analysis (3D-FL) of red wine samples and standards after interactions with human serum albumin (HSA). Cross-images of the results obtained from the 3D-FL of the investigated samples after their interaction with HSA are shown in the following order: (**A**) CSCarmel2; (**B**) CSYarden2; (**C**) tannic acid; (**D**) quercetin; (**E**) caffeic acid; (**F**) HSA in solution. Abbreviations: CSCarmel2, Cabernet Sauvignon Carmel Selected, vintage 2019; CSYarden2, Cabernet Sauvignon Golan Heights Winery Yarden, vintage 2019. The locations of peaks **a** and **b** are shown in the figure and in Table 2 (for interpretation of the references to color in this figure legend, the reader is referred to the web version of this article).

Few studies of the roles played by different phenolic fractions in wine astringency using methods involving fluorescence spectra are available [26]. However, the polymeric proanthocyanidins were found to possess higher activity than the other fractions. This conclusion was corroborated by the findings in the spectral measurements of the fluorescence, which showed that the polymeric fractions exhibited a significantly stronger quenching of the fluorescence of bovine serum albumin (BSA). The fluorescence intensity of BSA in the absence of samples at the emission maximum (335 nm) was approximately 541.68. It was observed that, for BSA, there was a noticeable decrease in the fluorescence intensity upon binding to polymeric fractions. In the present study, at the emission maximum (228 nm), HSA was approximately 643.0. As was shown, CSYarden2 exhibited a significantly stronger function of binding for the fluorescence of HSA than in the other samples, except for tannic acid, based on the calculations of the intensity of peak **b** (Figure 2, Table 2). Among these investigated samples, the intensity decreased for peak **a** in the order ethanol > caffeic acid > quercetin > tannic acid > CSYarden2 > CSCarmel2 > CSYarden1 > CSCarmel1, which means that CSCarmel1 possessed a higher affinity for HSA than the other samples did. For peak **b**, a slight change in the order appeared: ethanol > quercetin > caffeic acid > CSCarmel2 > CSCarmel1 > tannic acid > CSYarden2 > CSYarden1. These results are in agreement with the data on BSA and its interactions with polymeric fractions that were also found in wines [27]. Recently, the interactions of flavonoid–metal complexes with serum albumin (SA) have been widely studied because the complexation has a significant impact on biological activities [28]. The strength of binding might be increased in the presence of Cu (II), as evidenced by the calculation of the binding constant. However, the drug binding sites in BSA and HSA are not altered during the complexation process. Both HSA and BSA exhibit similar kinds of fluorescence emissions based on excitation at 295 nm due to the presence of tryptophan residue (Trp 214 in HSA and Trp 134/Trp 213 in BSA). The fluorescence of BSA/HSA was observed to be quenched in the presence of both quercetin and its Cu (II) complex [28]. The fluorescence intensity of BSA decreased with an increase in flavonoid concentration in the model wine with or without ultrasonic irradiation, which is in line with the data shown in Table 2. A characteristic fluorescence emission spectrum of BSA is displayed with a maximum value at the wavelength of 336 nm, which is mainly attributed to the tryptophan (Trp) residues [17]. The data on the fluorescence measurements of the interaction of tannic and gallic acids with HSA (Table 2) are comparable with the data on the interactions of a typical gallotannin 1,2,3,4,6-penta-O-galloyl-β-D-glucopyranose (PGG) and two simple phenolic compounds, ellagic acid (EA) and gallic acid (GA), with BSA. Fluorescence experiments showed that PGG and EA could strongly interact with BSA. The binding constants showed a pH-dependent binding of phenolic acids by BSA. PGG has a greater impact on the secondary structure of BSA when compared with small molecular compounds GA and EA [29]—which is also typical during the interaction of wine polyphenols with tannic acid—in comparison with gallic acid. The interaction between polyphenols and HSA/BSA is widely used not only for pharmaceutical applications, but also as an index of the quality of red wines. The skin and seed extracts of grapes were analyzed in terms of their fluorescence spectra. The fluorescence emission spectra of BSA with three kinds of grape varieties—Cabernet Sauvignon, Merlot, and Cabernet Gernischet—were obtained upon the addition of the skin and seed extracts; the fluorescence quenching of BSA in the presence of extracts from grape skins and seeds was evaluated by using fluorescence spectrometry based on simulated conditions of human physiological conditions (pH = 7.4) in order to study the phenolics. A decrease in the fluorescence intensity caused by the quenching of the skin and seed extracts was observed, which is in line with the action of red wine polyphenols [30]. The presence of phenolic compounds can make a great contribution to the perception of astringency in red wines based on their interactions with proteins. Human salivary protein and bovine serum albumin were used in this study to investigate the relationship between astringency and polyphenol composition. The results indicated that a positive correlation existed between the percentage of polymeric proanthocyanidins and the total phenols. In comparison with other fractions, the polymeric fractions exhibited the highest affinity for proteins and, thus, the highest astringency [27]. The results of previous reports revealed that both original and lyophilized wines exercise statistically significant beneficial lipidemic and antioxidant effects by reducing total cholesterol (TC), low-density lipoprotein cholesterol, triglycerides, and lipid peroxides, as well as by elevating the high-density lipoprotein cholesterol–TC ratio. There were no statistically significant differences in the results between groups that were fed a basic diet (BD) supplemented with original wine versus groups fed BD supplemented with lyophilized wine. Therefore, it can be concluded that the biologically active compound of this beverage is the dry matter containing, inter alia, polyphenols in relatively high concentrations [4,31,32]. This conclusion is also in agreement with the present results, which show very low values of binding with ethanol (Table 2).

2.3. Docking Studies

As a complex mixture, red wine has been shown to have several health benefits in the improvement of cardiovascular disease and cancer, as well as in its antioxidant and anti-inflammatory properties. Red wine as such has been defined according to its antioxidant role; however, the functional role of each molecule has not been explored. Hence, in the present study, interactions of flavonoids (epicatechin, epigallocatechin, quercetin, rutin, and myricetin), non-flavonoids (resveratrol), and phenolic acids (caffeic, gallic, and tannic acids) were investigated with respect to different proteins from human serum (CRP, fibrinogen, GPX3, and HSA). From the interaction study, it was observed that rutin, a flavonoid, had the highest binding affinity with the serum proteins (Table 3).

Table 2. Spectral data of the investigated samples and some standards.

Indices	CSCarmel 1	CSCarmel 2	CSYarden 1	CSYarden 2	Tannic Acid	Quercetin	Caffeic Acid	Ethanol	Fib/HSA
λ_{em}/ex, nm, peak **a** Fib	230/341	230/341	230/340	230/341	229/343	228/340	229/340	228/341	229/342
FI, A.U., peak **a** Fib	468.8 ± 7.2 [c]	478.4 ± 8.2 [c]	484.6 ± 5.6 [c]	499.6 ± 7.3 [c]	657.4 ± 5.9 [b]	861.0 ± 8.5 [a]	836.1 ± 8.9 [a]	865.5 ± 7.9 [a]	883.6 ± 7.9 [a]
BP, %, peak **a** Fib	46.9 ± 3.8[a]	45.9 ± 6.9 [a]	45.2 ± 3.4 [a]	43.5 ± 7.1 [a]	26.6 ± 3.9 [b]	2.6 ± 0.7 [d]	5.4 ± 0.7 [c]	2.1 ± 0.3 [d]	-
λ_{em}/ex, nm, peak **b** Fib	280/345	281/345	278/344	278/345	280/347	280/343	278/341	281/342	282/341
FI, A.U., peak **b** Fib	524.6 ± 9.3 [c]	535.3 ± 6.9 [c]	679.3 ± 5.3 [bc]	702.3 ± 8.3 [b]	595.9 ± 5.4 [c]	706.9 ± 8.2 [b]	768.7 ± 7.1 [ab]	797.1 ± 7.8 [ab]	811.7 ± 8.4 [a]
BP, %, peak **b** Fib	35.4 ± 4.4 [a]	34.1 ± 2.8 [a]	16.3 ± 1.4 [bc]	13.7 ± 1.2 [bc]	26.6 ± 5.1 [b]	12.9 ± 1.1 [c]	5.3 ± 0.7 [d]	1.8 ± 0.1 [e]	-
λ_{em}/ex, nm, peak **a** HSA	227/356	228/357	226/354	226/355	226/358	229/356	225/358	228/355	228/353
FI, A.U., peak **a** HSA	406.2 ± 7.3 [c]	414.5 ± 7.2 [c]	411.5 ± 4.3 [c]	424.9 ± 3.1[c]	579.6 ± 4.7 [b]	582.8 ± 5.1 [b]	626.9 ± 7.4 [a]	633.0 ± 9.1 [a]	643.0 ± 6.3 [a]
BP, %, peak **a** HSA	36.9+3.8 [a]	35.5 ± 1.8 [a]	36.0 ± 3.2 [a]	34.0 ± 3.2 [a]	9.9 ± 0.9 [b]	9.4 ± 0.7 [b]	2.5 ± 0.4 [c]	1.6 ± 0.4 [c]	-
λ_{em}/ex, nm, peak **b** HSA	278/359	278/359	279/360	280/360	280/359	280/356	279/360	280/356	280/357
FI, A.U., peak **b** HSA	843.0 ± 6.4 [ab]	860.2 ± 7.4 [ab]	771.0 ± 7.4 [b]	794.8 ± 5.2 [b]	818.9 ± 9.9 [ab]	867.4 ± 8.8 [ab]	865.5 ± 7.9 [ab]	910.9 ± 8.3 [a]	920.1 ± 10.3 [a]
BP, %, peak **b** HSA	8.4 ± 0.7 [bc]	6.5 ± 0.8 [c]	16.2 ± 1.5 [a]	13.6 ± 1.9 [ab]	11.0 ± 1.0 [b]	5.7 ± 0.4 [c]	5.9 ± 0.6 [c]	1.0 ± 0.9 [d]	-

The values are means ± SD of five measurements with $n = 5$ samples per vintage, and each was subsampled and analyzed five times. Means within rows with different superscripts are statistically different ($p < 0.05$; Student's *t*-test). Abbreviations; CSCarmel1, CSCarmel2, Cabernet Sauvignon Carmel Selected, vintages 2017 and 2019; CSYarden1 and CSYarden2, Cabernet Sauvignon Golan Heights Winery Yarden, vintages 2017 and 2019; maximum emission/excitation peak (λ_{em}/ex); fluorescence; intensity (FI); arbitral units (A.U.); fibrinogen (Fib); binding to HSA (%) and binding to Fib (%) are the percent decreases in fluorescence emissions of the fractions of the binding sites of the proteins occupied by the ligands.

Table 3. Binding affinity score of wine components docked with serum proteins.

Compound Name	PubChem ID	Binding Affinity (kcal/mol)			
		CRP	Fibrinogen	GPX3	HSA
Epicatechin	72276	−7.8	−5.1	−6.6	−8.9
Epigallocatechin	72277	−8.3	−6.3	−6.4	−8.6
Resveratrol	445154	−7.4	−6.1	−6.8	−9.1
Rutin	5280805	−8.7	−7.9	−7.4	−9.9
Quercetin	5280343	−8.7	−5.3	−6.8	−9.2
Gallic acid	370	−6.3	−5.7	−6.2	−6.2
Tannic acid	16129778	−7.7	−6.4	−7.3	−10.4
Myricetin	5281672	−8.4	−6.3	−6.8	−9
Caffeic acid	689043	−6.4	−5	−5.7	−7.2

However, resveratrol showed a similar binding affinity to that of flavonoids, though its presence in the red wine was observed to be very low compared to the other polyphenols. Among the phenolic acids, tannic acid exhibited the highest binding affinity with the target proteins in comparison with gallic and caffeic acids.

2.3.1. Interaction Analysis with C-Reactive Protein (CRP)

The molecular docking of rutin with CRP showed a binding affinity of −8.7 kcal/mol and interactions with Val86, Ala92, Pro93, Val94, Val111, Asp112, and Lys114 (Figure 3). Among the amino acids, Asp112 and Lys114 played critical roles in the formation of the C1q binding with CRP. C1q, as the first subcomponent of the classical pathway, initiates an anti-inflammatory response in association with CRP and the phagocytosis of apoptotic cells [33]. Rutin showed a covalent interaction with Asp112 and a hydrogen interaction with Lys114, which supports the stable interaction of the ligand with the receptor. Similarly, with a binding affinity of −7.4 kcal/mol, resveratrol showed interactions with Thr41, Trp67, Val86, Val94, and Val111. Resveratrol has been established to have a hydrogen bond interaction with Thr41. The regions of CRP covering from 35 to 47 amino acids are considered to be important residues for therapeutical and diagnostic studies. Thus, the interaction of resveratrol with Thr41 from CRP is considered significant for clinical studies. Tannic acid, a phenolic acid that was used for this study, showed a binding affinity of −7.7 kcal/mol and interactions with several critical residues: Asp112, Lys114, and Tyr 175 of CRP.

2.3.2. Interaction Analysis with Fibrinogen

The docking of fibrinogen with wine compounds revealed rutin as the top scorer with the highest binding affinity of −7.9 kcal/mol, followed by tannic acid with a binding affinity of −6.4 kcal/mol. The amino acids from the gamma chain involved in the interactions were Phe168, Gln176, Gln177, Phe178, Leu179, Arg197, Asp199, Gly200, Val202, Asp203, Phe204, Lys205, Glu210, Phe215, His217, Leu218, Glu225, Leu228, Lys232, Gly331, Asn345, Gly346, Tyr349, Val347, Tyr348, Gln350, Gly351, Thr353, Tyr354, Ser358, and Pro360 (Figure 4). The D-fragment of the fibrinogen gamma chain (γC domain) acted as a ligand and interacted with leukocyte integrin αmβ2 (CD11b/CD18, also known as Mac-1 or CR3). In particular, two peptides, P2 (γ377–395) and P1 (γ190–202), have been reported to be implicated in binding integrin αmβ2. Though P2 and P1 were demonstrated to be binding sites for integrin αmβ2, P2 was reported to exhibit high binding affinity to the αmI domain of integrin αmβ2. The binding of fibrinogen binding to leukocytes' integrin recruits the phagocyte during inflammatory response [34,35]. In addition, the pleiotropic role of fibrinogen in neurological diseases has been described [36,37]. Intriguingly, our docking analysis of fibrinogen with components of red wine unveiled P1 as the binding site of flavonoids and phenolic acids.

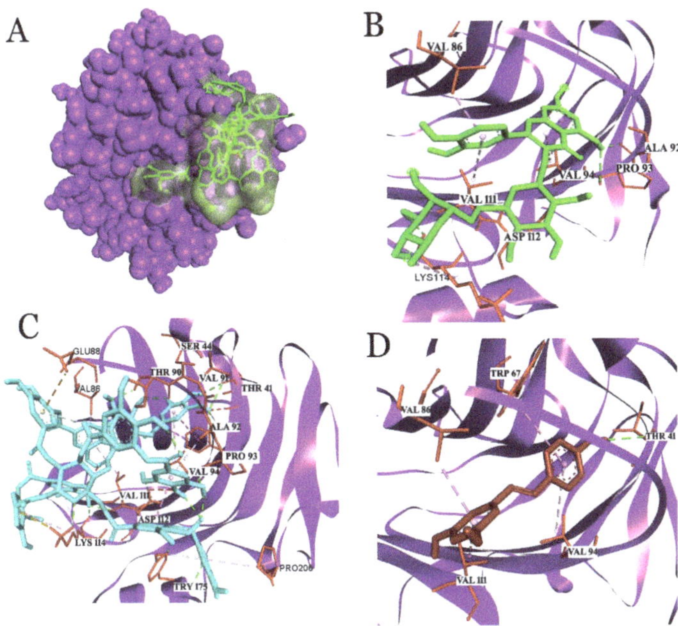

Figure 3. Interactions of ligands with the key residues of C-reactive protein (CRP). (**A**) The surface view represents the predicted binding pocket (green color) of the CRP; (**B**) the interaction of rutin with CRP; (**C**) the interaction of tannic acid with CRP; (**D**) the interaction of resveratrol with CRP.

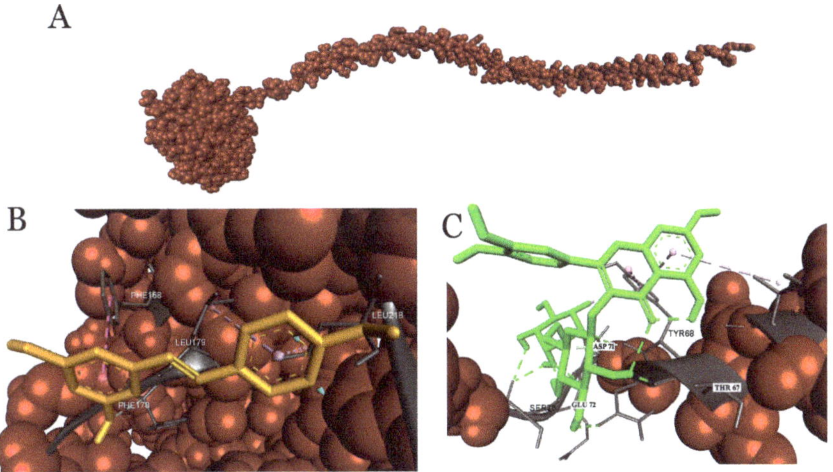

Figure 4. *Cont.*

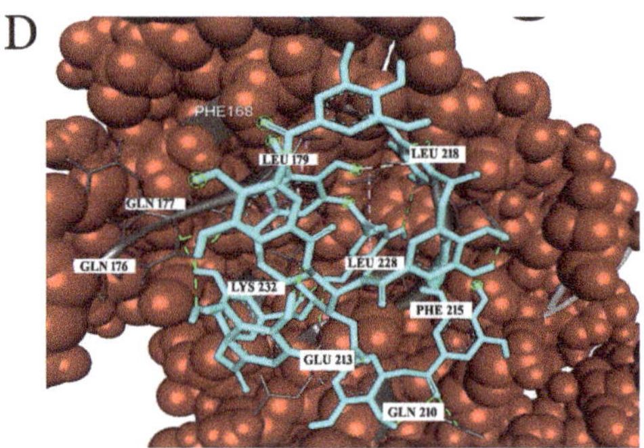

Figure 4. Molecular docking of ligands with fibrinogen. (**A**) Fibrinogen represented as a CPK model; expanded view of the ligands showing their interactions with the critical residues of fibrinogen; (**B**) rutin; (**C**) tannic acid; (**D**) resveratrol.

2.3.3. Interaction Analysis with Human Glutathione Peroxidase 3 (GPX3)

Similarly, in GPX3, rutin had the highest binding affinity of −7.4 kcal/mol, and the second top scorer, tannic acid, had a binding affinity of −7.3 kcal/mol; the amino acids that made non-covalent interactions were Thr20, Asp21, Tyr43, Gly44, Ala45, Leu46, Tyr53, His99, Phe135, Gln136, Lys137, Gly138, Asp139, Lys144, Glu145, Gln146, Lys147, Cys156, Pro157, Thr159, and Met196 (Figure 5). GPX3, an antioxidant enzyme in plasma proteins, contains Sec73 as an active site. Though the docking analysis revealed a high binding affinity for rutin and tannic acid, the non-flavonoid resveratrol showed an interaction at the active-site pocket residues of Sec73 [38,39]. Apart from this, residues from the pocket interface, such as Gly74, Leu75, Arg180, Trp181, His200, and Arg201, were implicated in the interactions.

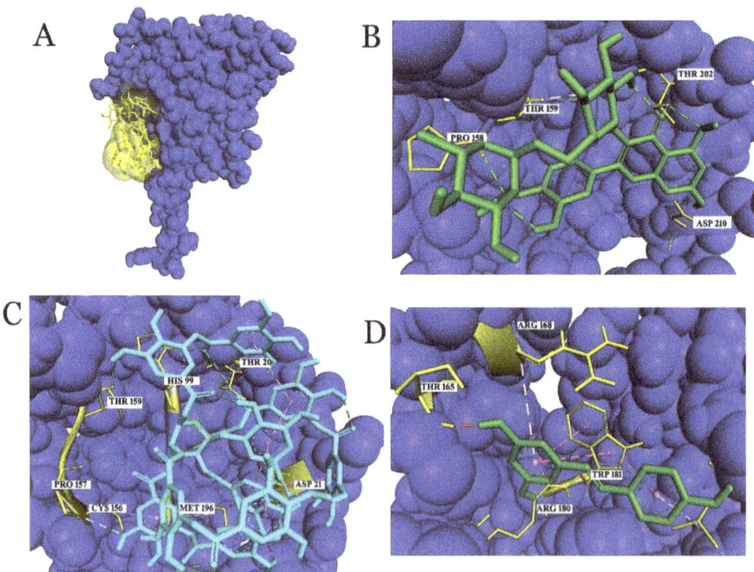

Figure 5. In silico docking of ligands into the binding pocket of human glutathione peroxidase 3 (GPX3). (**A**) CPK model of GPX3 protein, in which the yellow-colored surface represents the binding pocket; the expanded views of ligands interacting with the key amino acids of GPX3 (CPK model); (**B**) rutin; (**C**) tannic acid; (**D**) resveratrol.

2.3.4. Analysis of the Interaction with Human Serum Albumin (HSA)

In the case of HSA, tannic acid exhibited the highest binding affinity of -10.4 kcal/mol, followed by rutin with a binding affinity of -9.9 kcal/mol. The binding was driven by Asn109, Pro110, Asn111, Leu112, Arg114, Leu115, Arg117, Pro118, Met123, Ile142, Arg145, His146, Phe149, Leu154, Phe157, Tyr161, Leu182, Arg186, Gly189, Lys190, Val191, Ser193, Ala194, Arg197, and Leu 463 (Figure 6). Almost all of the compounds from wine used for the study were shown to have interactions with HSA. Interestingly, HSA is an important protein in plasma, which acts as a drug carrier and in the transportation of key enzymes, fatty acids, and biomolecules in the circulatory system. Thus, the interactions of the active ingredients in wine with HSA can be used in mainstream albumin fusion technology or albumin-based encapsulation methods for developing drug molecules for clinical applications.

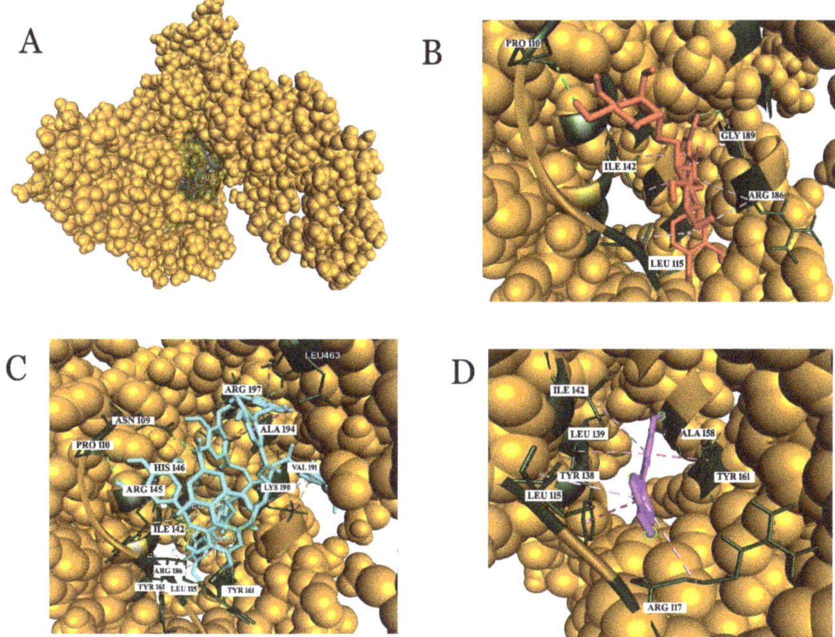

Figure 6. Analysis of interactions with human serum albumin (HSA). (**A**) The green-colored surface represents the binding pocket of the HSA protein (CPK model). (**B**) Interaction of rutin with the amino acids in the binding pocket, (**C**) tannic acid, and (**D**) resveratrol.

3. Discussion

Overall, tannic acid has been reported to interact with human plasma proteins, and our results are in good agreement with previous reports [40]. Hence, it is not surprising to observe the high binding affinity between tannic acids and plasma proteins, such as HSA, GPX3, and fibrinogen. Moreover, phenolic acids have been reported to interact with plasma proteins through non-covalent interactions, such as hydrogen bonds and electrostatic, van der Waals, or hydrophobic interactions [41]. Our docking analysis also showed the possible interactions, such as hydrogen bonds, van der Waals interactions, and electrostatic interactions, with the target proteins. Tannic acid has been reported to lower blood pressure in hypertensive rats [42], to have a protective role against acute doxorubicin-induced cardiotoxicity in Sprague–Dawley rats [43], to lower the malondialdehyde (MDA) level (a marker of oxidative stress), and to increase the antioxidant enzymes, such as superoxide dismutase, glutathione peroxidase, and catalase, thereby enhancing the antioxidant properties. In addition, anticancer [44], neuroprotective [45,46], anti-neuro-inflammatory [47], anti-inflammatory [48], antimicrobial, and antibiofilm activity [49,50] has been reported. On the contrary, tannic acids are mainly used in topical applications due to their bioavailability after ingestion, high molecular weight (>1000), low lipid solubility, and high affinity to bound to plasma proteins. In addition, their mode of elimination from the human system is still unclear. Rutin, a flavonoid, showed the highest binding affinity with all target proteins (CRP, fibrinogen, GPX3, and HSA) in our present study.

Rutin has also been investigated due to its beneficial effects on the strengthening of blood vessels and arteries [51], as well during the chronic condition of arthritis, as it can ease the flexibility of the veins [52]. Rutin has also been used for several pharmacological applications, including as a model drug, due to its antifungal, antiviral, and anti-inflammatory properties [53–55]. The cytotoxicity of heme and oxidative stress in endothelial cells were decreased in the BSA–diligand complexes relative to those of heme or BSA–heme complexes, and the co-presence of rutin played an important role. These results suggest the possibility and advantage of developing BSA-based carriers for the suppression of heme toxicity in biomedical applications [56]. Apart from this, rutin has also been applied in industry as a colorant, stabilizer, and preservative for food [57]. However, as an important common dietary flavonoid, studies of rutin have been focused on its formulation and bioavailability [58]. Thus rutin, was found to be a promising nutraceutical for the prevention and treatment of chronic human diseases according to hematological, biochemical, and histological evidence and due to the drug delivery [59,60]. However, rutin is the compound with the lowest molecular weight (610.5 g/mol) among the polyphenols, and therefore, it is regarded as significant for clinical applications. Interestingly, the non-flavonoid compound resveratrol was identified in red wine, though its presence was very low. Among other compounds, resveratrol has numerous biological activities. Resveratrol has shown antiviral activity against several viruses, such as polyomavirus, influenza virus, respiratory syncytial virus, MERS-CoV, and the emerging SARS-CoV [61]. The mechanisms of action of three important compounds (resveratrol, rapamycin, and metformin) on the cellular pathways involved in viral replication and the mechanisms of virus-related diseases, as well as the current status of their clinical use, were discussed [62]. The antiviral effect of resveratrol against MERS-CoV was evaluated through antiapoptotic and viral titer assays. The results revealed that resveratrol targeted caspase 3 in Vero E6 cells, as well as the nucleocapsid protein of MERS CoV. A concentration ranging from 31.5 to 125 µM and 0.5 mg/mL (2.05 mM) was shown to have a positive effect against MERS-CoV and SARS-CoV, respectively, in Vero E6 cells [63,64]. Moreover, it reduced the viral RNA levels and infectious titers. In the case of SARS-CoV, resveratrol derivatives suppressed the replication of SARS-CoV and decreased the cytopathic effects [63]. Resveratrol was demonstrated to be a stimulator of fetal hemoglobin and a potent antioxidant by trapping reactive oxygen species. Resveratrol could be proposed as potential therapeutic in the treatment of SARS-CoV-2 [65]. Resveratrol has been reported for its excellent pharmacokinetic characteristics, which have marked it for use as a potential drug candidate. When ingested, resveratrol can be easily absorbed in the gastrointestinal tract and rapidly metabolized in the liver into sulfate and glucuronide conjugates. Later, the metabolites are easily excreted through urine [63]. Thus, with its good absorption, low molecular weight (228.24 g/mol), low bioavailability, and extensive metabolism and excretion, resveratrol is considered to be a suitable drug candidate for oral administration during clinical ailments. To summarize the presented information about polyphenols and flavonoids in red wines, their high affinity and improvement of pharmacological actions, it is important to verify the exact mechanisms of their action. As discussed, polyphenols are among the molecules with pharmacological activity produced by plants. As drugs, these, due to their molecular structure, also have the ability to interact with molecules in our body, presenting various pharmacological properties [66]. Tannic acid and flavonoids play a role in the prevention of some diseases, and specifically in atherosclerosis. As was shown above, tannic acid is present in varying concentrations in plant foods and in relatively high concentrations in green teas and red wines. Human ether-a-go-go-related gene (hERG) channels are expressed in multiple tissues and play important roles in modulating the potential repolarization of cardiac action. Lung Ching green tea and red wine inhibited hERG currents with IC50 values of 0.04% and 0.19%, respectively. The effects of tannic acid, teas, and red wine on hERG currents were found to be irreversible. These results suggest that tannic acid is a novel hERG channel blocker, and consequently, it provides new mechanistic evidence for the understanding of the effects of tannic acid and the potential pharmacological basis of tea-and red-wine-induced biological activity [67]. As previously discussed, resveratrol first came to our attention in 1992, following reports of the cardioprotective effects of red wine. Thereafter, resveratrol was shown to exert antioxidant, anti-inflammatory, anti-proliferative, and angio-regulatory effects against atherosclerosis, ischemia, and cardiomyopathy. Although resveratrol has been claimed to be a master anti-aging agent that is effective against several age-associated diseases, a further detailed mechanistic investigation is still required in order to thoroughly unravel the therapeutic value of resveratrol against cardiovascular diseases at different stages of disease development [68]. Rutin and hesperidin were investigated in vitro for their anticoagulant activity through coagulation tests in terms of activated partial thromboplastin time (aPTT), prothrombin time (PT), and thrombin time (TT). Only an ethanolic solution of rutin at the concentration of 830 μM prolonged the aPTT, while the TT and PT were unaffected [69]. According to the results obtained and based on numerous recent reports [7,66–69], it was declared that wines possess a high bioactivity that allows them to be settled in the industries of food additives and medicinal products. In addition to phenolics, a number of other compounds are responsible for the quality of wine, as they determine the overall organoleptic impression of the wine in addition to the content of phenolic compounds, which are also present in large amounts in many other food products. Very recent publications mostly cited the results of phenolic compounds as secondary metabolites that are known to play crucial roles in important chemical reactions that impact the mouthfeel, color, and aging potential of red wine [70]. The phenolic profile—particularly the tannin concentration and structure—was the most important predictor of astringency and its subcomponents [71]. There were attempts to fingerprint the quality of red wines by using the ratio of the polymeric pigments and malvidin-3-glycoside, which was demonstrated to be a promising and practical chemical parameter for the rapid and simple assessment of the age of dry red wines. [72]. Volatile substances also play an important role in the quality of red wines, and it was found that 2-phenylethyl acetate, ethyl nonanoate, 2-hexanol, isoamyl octanoate, and ethyl 2-hydroxymethylbutanoate were the primary compounds responsible for wine classification [73]. The effects of various wine polyphenolic compounds were evaluated; among the wine phenolics tested, quercetin and resveratrol, in a dose-dependent manner, suppressed cytokine-induced C-reactive protein expression. Wine phenolics inhibited CRP expression [14]. Oversimplifications of the above results were indicated in reports [2,7,14] that the consumption of antioxidant-/polyphenol-rich foods might, therefore, impart anti-thrombotic- and cardiovascular-protective effects via their inhibition of platelet hyperactivation or aggregation. As a result of the ability of polyphenols to target additional pathways of platelet activation, they may have the potential to substitute or complement currently used anti-platelet drugs in sedentary, obese, pre-diabetic, or diabetic populations who can be resistant or sensitive to pharmacological anti-platelet therapy [2]. The health-promoting effects of red wine have been supported by epidemiological evidence, indicating that its components could improve endothelial dysfunction and hypertension, dyslipidemia, and metabolic disorders. The positive role of red wine in human health has been attributed to its phytochemical compounds, including polyphenols, as suggested by several clinical trials, including our previous studies in vivo [8]. The alcohol content of the investigated wines was 14.5%. The fluorescence measurements and calculations showed a significant difference between the original wine samples, tannic acid, and ethanol that were investigated in their interactions with fibrinogen and HSA. The binding properties of the tannic acid and wine samples were about 10–20 times higher than with those of ethanol, showing that the dry matter of the samples had higher bioactivity than the samples containing alcohol (Table 2). The biological role of red wine polymers remains largely unknown, although in vivo and in vitro antioxidant activity has been shown, which contributes against oxidative stress. The biological properties of red wine are therefore explained by the presence of phenolic compounds that are able to interact with physiological targets. The potential cardioprotective activity of red wine polyphenols has mainly been linked to the inhibition of platelet aggregation.

4. Materials and Methods

4.1. Reagents and Chemicals

Caffeic and gallic acids, catechin, epicatechin, quercetin, Trolox, Folin-Ciocalteu reagent, human serum albumin, fibrinogen, sodium nitrite, iron (III) chloride hexahydrate, 2,2-diphenyl-1-picrylhydrazyl (DPPH) and 2, 4, 6-tripyridyl-s-triazine (TPTZ) aluminum chloride, potassium peroxodisulfate and 2,2-azino-bis(3-ethylbenzothiazoline-6-sulfonic acid) diammonium salt (ABTS), copper (II) chloride dihydrate, sodium hydroxide, hydrochloric acid (37% w/w), 2,9-dimethyl-1,10-phenanthroline (neocuproine), and glacial acetic acid were obtained from Sigma (St. Louis, MO, USA). Standard phenolics were dissolved in methanol (1 mg/mL) and stored at -80 °C. All reagents and chemicals were of analytical grade.

4.2. Samples

Commercial wine bottles were purchased at wine shops and were investigated in this study. Every sample was bought in five bottles from wine shops in different locations; each had the same vintage, was from the same wine company and the same batch, and had an identical shelf life. The samples with a range of alcohol in same volume and from the same bottle were frozen at -80 °C to assess their antioxidant status and bioactivity. Cabernet Sauvignon wine samples of 2017 and 2019 vintages were purchased: Carmel Selected Cabernet Sauvignon 2017 (CSCarmel1) and 2019 (CSCarmel2) with 14.5% alcohol and Golan Heights Winery Yarden Cabernet Sauvignon 2017 (CSYarden1) and 2019 (CSYarden2) with 14.5% alcohol. The wines were transferred into 50 mL conical centrifuge tubes and stored at 8 °C. Samples were taken out of the refrigerator prior to, yesanalysis and were diluted according to the methods explained here.

4.3. Analyses of Bioactive Compounds

The total phenolic content (TPC) was measured by using the Folin–Ciocalteu method [74]. In brief, 250 µL of wine was mixed with 1000 µL of sodium carbonate (7.5%) and 1250 µL of Folin–Ciocalteu's (10% in water) reagent. The mixture was incubated for 15 min at 50 °C in the dark (water bath) and measured at 765 nm using a spectrophotometer (Hewlett-Packard, model 8452A, Rockvile, MD, USA). Gallic acid was used as the standard, and the results were expressed as milligrams of gallic acid equivalent per liter (mg GAE/L).

The total flavonoid content (TFC) in the wine was measured by following the $AlCl_3$ complexation method. Briefly, 31 µL of the sample was mixed in a microplate with 125 µL of water, 9.3 µL of sodium nitrate (5%), 9.3 µL of aluminium chloride (10%), and 125 µL of sodium hydroxide (0.5 M). The mixture was incubated for 30 min at room temperature in the dark. The reaction was measured at 510 nm. Catechin was used as the standard [75], and the results were expressed as milligrams of catechin equivalent per liter (mg CE/L).

The total flavanol content was determined [76] by measuring at 640 nm using 0.1 mL of the wine sample and 3 mL of 0.1% p-dimethylaminocinnamaldehyde (DMACA) solution (0.1% in 1 mol/L HCl in methanol). The results were expressed as catechin equivalent (CE).

The total tannins (TNs) were estimated by using spectrophotometric measurements of 0.5 mL of wine, where 3 mL of a 4% methanol vanillin solution and 1.5 mL of concentrated hydrochloric acid were added. The mixture was allowed to stand for 15 min. The absorption of the samples and a blank against water was measured at 500 nm [77].

The anthocyanin content (AC) in the wines was measured according to the following method [20,78]. An aliquot of 250 µL of the wine sample was poured into a tube with 2 mL of potassium chloride solution (0.025 M) and then adjusted to pH 1 with concentrated HCl. The mixture was incubated at room temperature for 20 min. In another tube, 250 μL of wine was mixed with 2 mL of sodium acetate solution (0.4 M, pH 4.5) and incubated at room temperature for 20 min. The absorbance of an aliquot of 300 μL of each sample was measured at 520 and 700 nm. The anthocyanin content in the samples was calculated in the following way: mg C3G/L = A × MW × D.F. × 103/ε × 1.

Here, A = (Abs520−Abs700), pH 1−(Abs520−Abs700) pH 4.5, MW (molecular weight) = 449.2 g/mol− for cyaniding 3-glucoside, D.F. = the dilution factor used, 103 = factor conversion of g to mg, ε = 26,900: molar extinction coefficient in L/mol cm, and 1 = the path length in cm. The results were expressed as milligrams of cyanidin 3-glucoside equivalent per liter (mg C3G/L).

The 2, 2′-azino-bis (3-ethyl-benzothiazoline-6-sulfonic acid) diammonium salt (ABTS) radical cation was formed by the ABTS solution (7 mM) with potassium persulfate (2.45 mM) in distilled water at room temperature at 16 h before use. A working solution (ABTS reagent) was diluted to obtain absorbance values of 0.7 at 734 nm and equilibrated at 30 °C. After the addition of ABTS solution, the absorbance reading was taken 1 min after the initial mixing and for up to 6 min; the percentage inhibition of absorbance was calculated with reference to a Trolox calibration curve and evaluated as mM Trolox equivalent/L of wine [20,79].

The antioxidant capacity in the wines was measured by using FRAP in 24 μL of the sample, which was mixed with 180 μL of FRAP reagent (TPTZ 10 mM in HCl 40 mM, iron chloride hexahydrate 20 mM, acetate buffer 0.3 M, pH 3 in a ratio of 1:1:10, prepared daily). The reaction was carried out at 37 °C, and the absorbance was measured at 595 nm every min for 30 min [80].

The antioxidant capacity was measured byusing 1, 1-diphenyl-2-picrylhydrazyl (DPPH) in 25 μL of the sample, which was mixed with 180 μL of DPPH radical at 6 mM and measured at 517 nm every 30 s for 10 min. Trolox was used as a standard for all antioxidant methods, and the results were expressed as millimoles of Trolox equivalent per liter (mmol TE/L) [20,81].

For the Cupric-Reducing Antioxidant Capacity (CUPRAC) assay [82,83], the red wines were diluted in a ratio of 1:10 (*v*/*v*) with dH$_2$O. Prior to the determination, 1.0 mL of each of three solutions containing 0.010 M Cu (II), ammonium acetate buffer at pH 7.0, and 0.0075 M neocuproine (2,9-dimethyl-1,10-phenanthroline) in EtOH was mixed with 0.5 mL of the appropriately diluted sample together with 0.6 mL of dH$_2$O in a tube. The reaction mixture was left for 1 h in the dark, and then the absorption was measured at 450 nm.

Some bioactive compounds, such as rutin, resveratrol, quercetin, caffeic acid, catechin, and epicatechin, were determined with an HPLC system [84–86]. A volume of 50 mL of each of the wine samples was extracted three times with 25 mL of diethyl ether and then three times with 25 mL of diethyl acetate, and the organic fractions were combined. After 30 min of drying with anhydrous Na$_2$SO$_4$, the extract was filtered through a Whatman-40 filter and evaporated to dryness in a rotary evaporator. The residue was dissolved in 2 mL of methanol/water (1:1, *v*/*v*) and analyzed by using high-performance liquid chromatography (HPLC). A Waters (Milford, MA, USA) chromatograph equipped with a 600-MS controller, a 717 plus autosampler, and a 996 photodiode-array detector was used. A gradient of solvent A (water/acetic acid, 98:2, *v*/*v*) and solvent B (water/acetonitrile/acetic acid, 78:20:2, *v*/*v*/*v*) was applied to a reverse-phase Nova-pack C18 column (30 cm × 3.9 mm internal diameter (I. D.) as follows: 0–55 min, 80% B linear, 1.1 mL/min; 55–57 min, 90% B linear, 1.2 mL/min; 57–70 min, 90% B isocratic, 1.2 mL/min; 70–80 min, 95% B linear, 1.2 mL/min; 80–90 min, 100% B linear, 1.2 mL/min; 90–120 min. For the HPLC analysis, an aliquot (50 μL) was injected into the column and eluted at the temperature of 20 °C. The samples were prepared and analyzed.

4.4. Fluorometric Studies

The profiles and properties of the polyphenols in wine were determined by using the two- (2D-FL) and three-dimensional (3D-FL) fluorescence (model FP-6500, Jasco spectrofluorometer, serial N261332, Tokyo, Japan). The 2D-FL measurements were taken at emission wavelengths from 310 to 500 nm and at an excitation of 295 nm. The 3D-FL was measured at emission wavelengths between 200 and 795 nm, and the initial excitation wavelength was 200 nm. For comparison of the obtained results, caffeic acid, quercetin, and tannic acid were used [9,10]. The binding properties of wine to human serum albumin (HSAlb) and fibrinogen were evaluated by using 2D- and 3D-FL. For the fluorescence measurements, 3.0 mL of 1.0×10^{-5} mol/L HSA was prepared in 0.05 mol/L Tris–HCl buffer (pH 7.4) containing 0.1 mol/L NaCl. Fibrinogen stock solution was made by dissolving in phosphate buffer (10 mM, pH 7.4) to obtain a concentration of 20 µM. Standard phenolic solutions, such as tannic acid, quercetin, and caffeic acid stock solution, were prepared daily by dissolving at a concentration of 10 mM in methanol and then diluting with 10 mM phosphate buffer at pH 7.4. All samples were kept at 4 °C before the analysis. The initial fluorescence intensities of albumin and fibrinogen were measured before their interactions with the investigated samples, as pure substances, and after their interactions with the samples (quenching of fluorescence emissions of proteins—in our case, albumin, fibrinogen, and polyphenols of wines). As mentioned above, the changes in the fluorescence intensities were used in the estimation of the binding activities [87].

4.5. Molecular Docking of Ligands with the Serum Proteins

Molecular docking studies of the docking of the compounds identified in red wine into serum protein targets were carried out by using AutoDock Vina [88]. For the studies, receptor proteins, such as human C-reactive protein (CRP) (PDB ID: 1B09), human serum albumin (HSA) (PDB ID: 1H9Z), human glutathione peroxidase 3 (GPX3) (PDB ID: 2R37), and human fibrinogen (PDB ID: 3GHG)—with resolutions of 2.5, 2.5, 1.85, and 2.9 Å, respectively—were obtained in pdb format from the PDB database. The crystal structures were pre-processed through the removal of all water and the addition of Kollman charges. The Gasteiger charges and hydrogen bond optimization was performed with AutoDock MGL tools version 1.5.6. Chain A of the grid-box of the receptor proteins CRP (144 Å × 157 Å × 27 Å), GPX3 (25 Å × 43 Å × 37 Å), and HSA (87 Å × 51 Å × 86 Å) was generated by AutoGrid4. For fibrinogen (213 Å × 55 Å × 87 Å), the gamma chain was selected for the study. The outputs of the best ligands and their interactions in the the molecular docking were analyzed through the BIOVIA-DS 17 R2 client (Dassault Systèmes Biovia Corp®, San Diego, CA, USA).

4.6. Data Analysis

All data obtained were calculated on the basis of a statistical analysis of Duncan's multiple range test. Values were means ± SD per liter of 25 measurements, representing the commercial status of the wines and their replicates. Five replications of five wine samples from each vintage were used. To determine the statistical significance at the 95% interval of reliability, one-way analysis of variance (ANOVA) was used.

5. Conclusions

Previous reports have shown that the consumption of wine has several health benefits. However, there are different types of wine; in the present study, red wine was investigated for its composition of active ingredients. The interaction of the each component was investigated for its binding modes with different serum protein, thus implicating them as drug candidates in clinical settings. Overall, the study indicates that red wine has a composition of flavonoids, non-flavonoids, and phenolic acids that can interact with the amino acids in the key regions of the proteins to enhance their biological activity. Among them, rutin, resveratrol, and tannic acid showed good binding affinity. Therefore, the protein–ligand complexes provide the basis for further studies and are envisaged to possess beneficial properties that can enhance their role in clinical applications.

Author Contributions: Conceptualization, methodology, visualization, writing—original draft, R.M.B.S., S.K.P. and S.G.; data curation, formal analysis, S.A.L. and Y.-M.K.; software, validation, resources, J.D. and E.K.; supervision, S.G.; reviewing and editing, R.M.B.S., S.K.P. and S.G.; All authors have read and agreed to the published version of the manuscript.

Funding: This research received no external funding.

Institutional Review Board Statement: Not applicable.

Informed Consent Statement: Not applicable.

Data Availability Statement: Not applicable.

Acknowledgments: S.G. is grateful with admiration to Abraham J. Domb (Institute for Drug Research, School of Pharmacy, Faculty of Medicine, Hebrew University of Jerusalem, Israel) for his support with FTIR and fluorescence studies. He always encouraged all S.G.'s activities. The authors R.M.B.S., S.A., and S.K.P. sincerely acknowledge the computational and bioinformatics facility provided by the Bioinformatics Infrastructure Facility (funded by DBT, GOI; File No. BT/BI/25/012/2012, BIF). R.B.S. is thankful to RUSA 2.0 (F.24-51/2014-U, Policy (TN Multi-Gen), Dept. of Edn, GoI] for supporting her as RUSA-PDF.

Conflicts of Interest: The authors declare that they have no known competing financial interests or personal relationships that could have appeared to influence the work reported in this paper.

Sample Availability: Samples of the compounds are not available from the authors.

References

1. Giglio, R.V.; Patti, A.M.; Cicero, A.F.; Lippi, G.; Rizzo, M.; Toth, P.P.; Banach, M. Polyphenols: Potential Use in the Prevention and Treatment of Cardiovascular Diseases. *Curr. Pharm. Des.* **2018**, *24*, 239–258. [CrossRef] [PubMed]
2. Santhakumar, A.B.; Bulmer, A.C.; Singh, I. A review of the mechanisms and effectiveness of dietary polyphenols in reducing oxidative stress and thrombotic risk. *J. Hum. Nutr. Diet.* **2013**, *27*, 1–21. [CrossRef] [PubMed]
3. Gorinstein, S.; Zachwieja, Z.; Katrich, E.; Pawelzik, E.; Haruenkit, R.; Trakhtenberg, S.; Martin-Belloso, O. Comparison of the contents of the main antioxidant compounds and the antioxidant activity of white grapefruit and his new hybrid. *LWT* **2004**, *37*, 337–343. [CrossRef]
4. Gorinstein, S.; Zemser, M.; Weisz, M.; Halevy, S.; Martin-Belloso, O.; Trakhtenberg, S. The influence of alcohol-containing and alcohol-free beverages on lipid levels and lipid peroxides in serum of rats. *J. Nutr. Biochem.* **1998**, *9*, 682–686. [CrossRef]
5. Garaguso, I.; Nardini, M. Polyphenols content, phenolics profile and antioxidant activity of organic red wines produced without sulfur dioxide/sulfites addition in comparison to conventional red wines. *Food Chem.* **2015**, *179*, 336–342. [CrossRef] [PubMed]
6. Natella, F.; Nardini, M.; Virgili, F.; Scaccini, C. Role of dietary polyphenols in the platelet aggregation network—A review of the in vitro studies. *Curr. Top. Nutraceutical Res.* **2006**, *4*, 1–21.
7. Tekos, F.; Makri, S.; Skaperda, Z.-V.; Patouna, A.; Terizi, K.; Kyriazis, I.; Kotseridis, Y.; Mikropoulou, E.; Papaefstathiou, G.; Halabalaki, M.; et al. Assessment of Antioxidant and Antimutagenic Properties of Red and White Wine Extracts In Vitro. *Metabolites* **2021**, *11*, 436. [CrossRef] [PubMed]
8. Castaldo, L.; Narváez, A.; Izzo, L.; Graziani, G.; Gaspari, A.; Di Minno, G.; Ritieni, A. Red Wine Consumption and Cardiovascular Health. *Molecules* **2019**, *24*, 3626. [CrossRef]
9. Shafreen, R.; Lakshmi, S.; Pandian, S.; Park, Y.; Kim, Y.; Paśko, P.; Deutsch, J.; Katrich, E.; Gorinstein, S. Unraveling the Antioxidant, Binding and Health-Protecting Properties of Phenolic Compounds of Beers with Main Human Serum Proteins: In Vitro and In Silico Approaches. *Molecules* **2020**, *25*, 4962. [CrossRef]
10. Kim, Y.-M.; Abas, F.; Park, Y.-S.; Park, Y.-K.; Ham, K.-S.; Kang, S.-G.; Lubinska-Szczygeł, M.; Ezra, A.; Gorinstein, S. Bioactivities of Phenolic Compounds from Kiwifruit and Persimmon. *Molecules* **2021**, *26*, 4405. [CrossRef] [PubMed]
11. Tedesco, I.; Spagnuolo, C.; Russo, G.; Russo, M.; Cervellera, C.; Moccia, S. The Pro-Oxidant Activity of Red Wine Polyphenols Induces an Adaptive Antioxidant Response in Human Erythrocytes. *Antioxidants* **2021**, *10*, 800. [CrossRef]
12. Zhang, P.; Ma, W.; Meng, Y.; Zhang, Y.; Jin, G.; Fang, Z. Wine phenolic profile altered by yeast: Mechanisms and influences. *Compr. Rev. Food Sci. Food Saf.* **2021**, *20*, 3579–3619. [CrossRef]
13. Giacosa, S.; Parpinello, G.P.; Segade, S.R.; Ricci, A.; Paissoni, M.A.; Curioni, A.; Marangon, M.; Mattivi, F.; Arapitsas, P.; Moio, L.; et al. Diversity of Italian red wines: A study by enological parameters, color, and phenolic indices. *Food Res. Int.* **2021**, *143*, 110277. [CrossRef] [PubMed]
14. Kaur, G.; Rao, L.V.M.; Agrawal, A.; Pendurthi, U.R. Effect of wine phenolics on cytokine-induced C-reactive protein expression. *J. Thromb. Haemost.* **2007**, *5*, 1309–1317. [CrossRef]

15. Bijak, M.; Nowak, P.; Borowiecka, M.; Ponczek, M.B.; Zbikowska, H.; Wachowicz, B. Protective effects of (−)-epicatechin against nitrative modifications of fibrinogen. *Thromb. Res.* **2012**, *130*, e123–e128. [CrossRef]

16. Zhang, Q.-A.; Fu, X.; Martín, J.F.G. Effect of ultrasound on the interaction between (−)-epicatechin gallate and bovine serum albumin in a model wine. *Ultrason. Sonochem.* **2017**, *37*, 405–413. [CrossRef] [PubMed]

17. Liberale, L.; Bonaventura, A.; Montecucco, F.; Dallegri, F.; Carbone, F. Impact of Red Wine Consumption on Cardiovascular Health. *Curr. Med. Chem.* **2019**, *26*, 3542–3566. [CrossRef]

18. Snopek, L.; Mlcek, J.; Sochorova, L.; Baron, M.; Hlavacova, I.; Jurikova, T.; Kizek, R.; Sedlackova, E.; Sochor, J. Contribution of Red Wine Consumption to Human Health Protection. *Molecules* **2018**, *23*, 1684. [CrossRef]

19. Chen, W.-K.; He, F.; Wang, Y.-X.; Liu, X.; Duan, C.-Q.; Wang, J. Influences of Berry Size on Fruit Composition and Wine Quality of Vitis vinifera L. cv. 'Cabernet Sauvignon' Grapes. *S. Afr. J. Enol. Vitic.* **2018**, *39*, 67–76. [CrossRef]

20. Muñoz-Bernal, Ó.A.; Coria-Oliveros, A.J.; Vazquez-Flores, A.A.; de la Rosa, L.A.; Núñez-Gastélum, A.; Rodrigo-García, J.; Ayala-Zavala, J.F.; Alvarez-Parrilla, E. Evolution of phenolic content, antioxidant capacity and phenolic profile during cold pre-fermentative maceration and subsequent fermentation of Cabernet Sauvignon red wine. *S. Afr. J. Enol. Vitic.* **2020**, *41*, 72–82. [CrossRef]

21. Padilha, C.V.D.S.; Miskinis, G.A.; Souza, M.E.; Pereira, G.E.; Oliveira, D.; Bordignon-Luiz, M.; Lima, M.D.S. Rapid determination of flavonoids and phenolic acids in grape juices and wines by RP-HPLC/DAD: Method validation and characterization of commercial products of the new Brazilian varieties of grape. *Food Chem.* **2017**, *228*, 106–115. [CrossRef]

22. Moreno-Montoro, M.; Olalla, M.; Gimenez-Martinez, R.; Navarro-Alarcon, M.; Rufián-Henares, J.A. Phenolic compounds and antioxidant activity of Spanish commercial grape juices. *J. Food Compos. Anal.* **2015**, *38*, 19–26. [CrossRef]

23. Raczkowska, J.; Mielcarz, G.; Howard, A.; Raczkowski, M. UPLC and Spectrophotometric Analysis of Polyphenols in Wines Available in the Polish Market. *Int. J. Food Prop.* **2011**, *14*, 514–522. [CrossRef]

24. Landrault, N.; Poucheret, P.; Ravel, P.; Gasc, F.; Cros, G.; Teissedre, P.-L. Antioxidant Capacities and Phenolics Levels of French Wines from Different Varieties and Vintages. *J. Agric. Food Chem.* **2001**, *49*, 3341–3348. [CrossRef]

25. del Barrio-Galán, R.; Medel-Maraboli, M.; Peña-Neira, A. Effect of different aging techniques on the polysaccharide and phenoliccomposition and sensory characteristics of Syrah red wines fermented using different yeast strains. *Food Chem.* **2015**, *179*, 116–126. [CrossRef] [PubMed]

26. Park, Y.S.; Polovka, M.; Martinez-Ayala, A.L.; González-Aguilar, G.A.; Ham, K.S.; Kang, S.G.; Park, Y.K.; Heo, B.G.; Namiesnik, J.; Gorinstein, S. Fluorescence studies by quenching and protein unfolding on the interaction of bioactive compounds in water extracts of kiwifruit cultivars with human serum albumin. *J. Lumin.* **2015**, *160*, 71–77. [CrossRef]

27. Ren, M.; Du, G.; Tian, C.; Song, X.; Zhu, D.; Wang, X.; Zhang, J. Influence of Different Phenolic Fractions on Red Wine Astringency Based on Polyphenol/Protein Binding. *S. Afr. J. Enol. Vitic.* **2017**, *38*. [CrossRef]

28. Mondal, P.; Bose, A. Spectroscopic overview of quercetin and its Cu(II) complex interaction with serum albumins. *BioImpacts* **2019**, *9*, 115–121. [CrossRef] [PubMed]

29. Zhang, L.; Liu, Y.; Hu, X.; Xu, M.; Wang, Y. Studies on interactions of pentagalloyl glucose, ellagic acid and gallic acid with bovine serum albumin: A spectroscopic analysis. *Food Chem.* **2020**, *324*, 126872. [CrossRef] [PubMed]

30. Zhu, Y.-Y.; Zhao, P.; Wang, X.-Y.; Zhang, J.; Wang, X.H.; Tian, C.-R.; Ren, M.-M.; Chen, T.-G.; Yuan, H.-H. Evaluation of the potential astringency of the skins and seeds of different grape varieties based on polyphenol/protein binding. *Food Sci. Technol.* **2019**, *39*, 930–938. [CrossRef]

31. Gorinstein, S.; Caspi, A.; Libman, I.; Trakhtenberg, S. Cardioprotective effect of alcohol consumption: Contemporary concepts. *Nutr. Res.* **2002**, *22*, 659–666. [CrossRef]

32. Gorinstein, S.; Caspi, A.; Libman, I.; Trakhtenberg, S. Mechanism of cardioprotective effect and the choice of alcoholic beverage. *Am. J. Cardiol.* **2000**, *85*, 280–281. [CrossRef]

33. Roumenina, L.T.; Ruseva, M.M.; Zlatarova, A.; Ghai, R.; Kolev, M.; Olova, N.; Gadjeva, M.; Agrawal, A.; Bottazzi, B.; Mantovani, A.; et al. Interaction of C1q with IgG1, C-reactive Protein and Pentraxin 3: Mutational Studies Using Recombinant Globular Head Modules of Human C1q A, B, and C Chains. *Biochemistry* **2006**, *45*, 4093–4104. [CrossRef] [PubMed]

34. Altieri, D.; Plescia, J.; Plow, E. The structural motif glycine 190-valine 202 of the fibrinogen gamma chain interacts with CD11b/CD18 integrin (alpha M beta 2, Mac-1) and promotes leukocyte adhesion. *J. Biol. Chem.* **1993**, *268*, 1847–1853. [CrossRef]

35. Ugarova, T.P.; Lishko, V.K.; Podolnikova, N.P.; Okumura, N.; Merkulov, S.M.; Yakubenko, V.P.; Yee, V.C.; Lord, A.S.T.; Haas, T.A. Sequence γ377−395(P2), but Not γ190−202(P1), Is the Binding Site for the αMI-Domain of Integrin αMβ2 in the γC-Domain of Fibrinogen. *Biochemistry* **2003**, *42*, 9365–9373. [CrossRef] [PubMed]

36. Petersen, M.A.; Ryu, J.K.; Akassoglou, K. Fibrinogen in neurological diseases: Mechanisms, imaging and therapeutics. *Nat. Rev. Neurosci.* **2018**, *19*, 283–301. [CrossRef] [PubMed]

37. Adams, R.A.; Bauer, J.; Flick, M.J.; Sikorski, S.L.; Nuriel, T.; Lassmann, H.; Degen, J.L.; Akassoglou, K. The fibrin-derived γ377-395 peptide inhibits microglia activation and suppresses relapsing paralysis in central nervous system autoimmune disease. *J. Exp. Med.* **2007**, *204*, 571–582. [CrossRef]

38. Song, J.; Yü, Y.; Xing, R.; Guo, X.; Liu, D.; Wei, J.; Song, H. Unglycosylated recombinant human glutathione peroxidase 3 mutant from Escherichia coli is active as a monomer. *Sci. Rep.* **2014**, *4*, 6698. [CrossRef] [PubMed]

39. Ren, B.; Huang, W.; Åkesson, B.; Ladenstein, R. The crystal structure of seleno-glutathione peroxidase from human plasma at 2.9 Å resolution. *J. Mol. Biol.* **1997**, *268*, 869–885. [CrossRef] [PubMed]

40. Pinto, A.F.; Nascimento, J.M.D.; Sobral, R.V.D.S.; de Amorim, E.L.C.; Silva, R.; Leite, A.C.L. Tannic acid as a precipitating agent of human plasma proteins. *Eur. J. Pharm. Sci.* **2019**, *138*, 105018. [CrossRef] [PubMed]

41. Xiao, J.; Zhao, Y.; Wang, H.; Yuan, Y.; Yang, F.; Zhang, C.; Yamamoto, K. Noncovalent Interaction of Dietary Polyphenols with Common Human Plasma Proteins. *J. Agric. Food Chem.* **2011**, *59*, 10747–10754. [CrossRef]

42. Coşan, D.T.; Saydam, F.; Özbayer, C.; Doğaner, F.; Soyocak, A.; Güneş, H.V.; Değirmenci, I.; Kurt, H.; Üstüner, M.C.; Bal, C. Impact of tannic acid on blood pressure, oxidative stress and urinary parameters in L-NNA-induced hypertensive rats. *Cytotechnology* **2013**, *67*, 97–105. [CrossRef]

43. Zhang, J.; Cui, L.; Han, X.; Zhang, Y.; Zhang, X.; Chu, X.; Zhang, F.; Zhang, Y.; Chu, L. Protective effects of tannic acid on acute doxorubicin-induced cardiotoxicity: Involvement of suppression in oxidative stress, inflammation, and apoptosis. *Biomed. Pharmacother.* **2017**, *93*, 1253–1260. [CrossRef] [PubMed]

44. Tikoo, K.; Sane, M.S.; Gupta, C. Tannic acid ameliorates doxorubicin-induced cardiotoxicity and potentiates its anti-cancer activity: Potential role of tannins in cancer chemotherapy. *Toxicol. Appl. Pharmacol.* **2011**, *251*, 191–200. [CrossRef] [PubMed]

45. Salman, M.; Tabassum, H.; Parvez, S. Tannic Acid Provides Neuroprotective Effects against Traumatic Brain Injury through the PGC-1α/Nrf2/HO-1 Pathway. *Mol. Neurobiol.* **2020**, *57*, 2870–2885. [CrossRef] [PubMed]

46. Ono, K.; Hasegawa, K.; Naiki, H.; Yamada, M. Anti-amyloidogenic activity of tannic acid and its activity to destabilize Alzheimer's β-amyloid fibrils in vitro. *Biochim. Biophys. Acta (BBA) Mol. Basis Dis.* **2004**, *1690*, 193–202. [CrossRef] [PubMed]

47. Wu, Y.; Zhong, L.; Yu, Z.; Qi, J. Anti-neuroinflammatory effects of tannic acid against lipopolysaccharide-induced BV2 microglial cells via inhibition of NF-κB activation. *Drug Dev. Res.* **2018**, *80*, 262–268. [CrossRef]

48. Kim, H.-Y.; Kim, J.; Jeong, H.-J.; Kim, H.-M. Potential anti-inflammatory effect of Madi-Ryuk and its active ingredient tannic acid on allergic rhinitis. *Mol. Immunol.* **2019**, *114*, 362–368. [CrossRef]

49. Dong, G.; Liu, H.; Yu, X.; Zhang, X.; Lu, H.; Zhou, T.; Cao, J. Antimicrobial and anti-biofilm activity of tannic acid against Staphylococcus aureus. *Nat. Prod. Res.* **2017**, *32*, 2225–2228. [CrossRef] [PubMed]

50. Kaczmarek, B. Tannic Acid with Antiviral and Antibacterial Activity as A Promising Component of Biomaterials—A Minireview. *Materials* **2020**, *13*, 3224. [CrossRef]

51. Fei, J.; Sun, Y.; Duan, Y.; Xia, J.; Yu, S.; Ouyang, P.; Wang, T.; Zhang, G. Low concentration of rutin treatment might alleviate the cardiotoxicity effect of pirarubicin on cardiomyocytes via activation of PI3K/AKT/mTOR signaling pathway. *Biosci. Rep.* **2019**, *39*, BSR20190546. [CrossRef] [PubMed]

52. Teixeira, F.M.; Coelho, M.N.; José-Chagas, F.D.N.; Malvar, D.D.C.; Kanashiro, A.; Cunha, F.Q.; Vianna-Filho, M.D.M.; Pinto, A.D.C.; Vanderlinde, F.A.; Costa, S.S. Oral treatments with a flavonoid-enriched fraction from Cecropia hololeuca and with rutin reduce articular pain and inflammation in murine zymosan-induced arthritis. *J. Ethnopharmacol.* **2020**, *260*, 112841. [CrossRef] [PubMed]

53. Macedo, A.S.; Quelhas, S.; Silva, A.M.; Souto, E.B. Nanoemulsions for delivery of flavonoids: Formulation and in vitro release of rutin as model drug. *Pharm. Dev. Technol.* **2014**, *19*, 677–680. [CrossRef] [PubMed]

54. Ganeshpurkar, A.; Saluja, A. Immunomodulatory effect of rutin, catechin, and hesperidin on macrophage function. *Indian J. Biochem. Biophys.* **2020**, *57*, 58–63.

55. Na, J.Y.; Song, K.; Kim, S.; Kwon, J. Rutin protects rat articular chondrocytes against oxidative stress induced by hydrogen peroxide through SIRT1 activation. *Biochem. Biophys. Res. Commun.* **2016**, *473*, 1301–1308. [CrossRef]

56. Luo, M.; Sui, Y.; Tian, R.; Lu, N. Formation of a bovine serum albumin diligand complex with rutin for the suppression of heme toxicity. *Biophys. Chem.* **2020**, *258*, 106327. [CrossRef]

57. Naseer, B.; Iqbal, S.; Wahid, N.; Qazi, H.J.; Nadeem, M.; Nawaz, M. Evaluation of antioxidant and antimicrobial potential of rutin in combination with butylated hydroxytoluene in cheddar cheese. *J. Food Process. Preserv.* **2020**, *45*, e15046. [CrossRef]

58. Lee, L.-C.; Hou, Y.-C.; Hsieh, Y.-Y.; Chen, Y.-H.; Shen, Y.-C.; Lee, I.-J.; Shih, M.-C.M.; Hou, W.-C.; Liu, H.-K. Dietary supplementation of rutin and rutin-rich buckwheat elevates endogenous glucagon-like peptide 1 levels to facilitate glycemic control in type 2 diabetic mice. *J. Funct. Foods* **2021**, *85*, 104653. [CrossRef]

59. Rakshit, S.; Shukla, P.; Verma, A.; Nirala, S.K.; Bhadauria, M. Protective role of rutin against combined exposure to lipopolysaccharide and D-galactosamine-induced dysfunctions in liver, kidney, and brain: Hematological, biochemical, and histological evidences. *J. Food Biochem.* **2021**, *45*, e13605. [CrossRef]

60. Sharma, S.; Ali, A.; Ali, J.; Sahni, J.K.; Baboota, S. Rutin: Therapeutic potential and recent advances in drug delivery. *Expert Opin. Investig. Drugs* **2013**, *22*, 1063–1079. [CrossRef]

61. Lin, S.C.; Ho, C.T.; Chuo, W.H.; Li, S.M.; Wang, T.T.; Lin, C.C. Effective inhibition of MERS-CoV infection by resveratrol. *BMC Infect. Dis.* **2017**, *17*, 144. [CrossRef] [PubMed]

62. Benedetti, F.; Sorrenti, V.; Buriani, A.; Fortinguerra, S.; Scapagnini, G.; Zella, D. Resveratrol, Rapamycin and Metformin as Modulators of Antiviral Pathways. *Viruses* **2020**, *12*, 1458. [CrossRef] [PubMed]

63. Filardo, S.; Di Pietro, M.; Mastromarino, P.; Sessa, R. Therapeutic potential of resveratrol against emerging respiratory viral infections. *Pharmacol. Ther.* **2020**, *214*, 107613. [CrossRef] [PubMed]

64. Ellen, B.M.; Kumar, N.D.; Bouma, E.M.; Troost, B.; van de Pol, D.P.I.; Van der Ende-Metselaar, H.H.; Ap-perloo, L.; van Gosliga, D.; van den Berge, M.; Nawijn, M.C. Resveratrol and Pterostilbene inhibit SARS-CoV-2 replication in air-liquid interface cultured human primary bronchial epithelial cells. *Viruses* **2021**, *13*, 1335. [CrossRef] [PubMed]

65. Ramdani, L.H.; Bachari, K. Potential therapeutic effects of resveratrol against SARS-CoV-2. *Acta Virol.* **2020**, *64*, 276–280. [CrossRef] [PubMed]
66. Stromsnes, K.; Lagzdina, R.; Olaso-Gonzalez, G.; Gimeno-Mallench, L.; Gambini, J. Pharmacological Properties of Polyphenols: Bioavailability, Mechanisms of Action, and Biological Effects in In Vitro Studies, Animal Models, and Humans. *Biomedicines* **2021**, *9*, 1074. [CrossRef] [PubMed]
67. Chu, X.; Guo, Y.; Xu, B.; Li, W.; Lin, Y.; Sun, X.; Ding, C.; Zhang, X. Effects of Tannic Acid, Green Tea and Red Wine on hERG Channels Expressed in HEK293 Cells. *PLoS ONE* **2015**, *10*, e0143797. [CrossRef] [PubMed]
68. Cheng, C.K.; Luo, J.; Lau, C.W.; Chen, Z.; Tian, X.Y.; Huang, Y. Pharmacological basis and new insights of resveratrol action in the cardiovascular system. *Br. J. Pharmacol.* **2019**, *177*, 1258–1277. [CrossRef] [PubMed]
69. Kuntić, V.; Filipovic, I.; Vujić, Z. Effects of Rutin and Hesperidin and their Al(III) and Cu(II) Complexes on in Vitro Plasma Coagulation Assays. *Molecules* **2011**, *16*, 1378–1388. [CrossRef]
70. dos Santos, I.; Bosman, G.; Aleixandre-Tudo, J.L.; du Toit, W. Direct quantification of red wine phenolics using fluorescence spectroscopy with chemometrics. *Talanta* **2022**, *236*, 122857. [CrossRef] [PubMed]
71. Araujo, L.D.; Parr, W.V.; Grose, C.; Hedderley, D.; Masters, O.; Kilmartin, P.A.; Valentin, D. In-mouth attributes driving perceived quality of Pinot noir wines: Sensory and chemical characterization. *Food Res. Int.* **2021**, *149*, 110665. [CrossRef] [PubMed]
72. Han, G.; Dai, L.; Sun, Y.; Li, C.; Ruan, S.; Li, J.; Xu, Y. Determination of the age of dry red wine by multivariate techniques using color parameters and pigments. *Food Control.* **2021**, *129*, 108253. [CrossRef]
73. Bordiga, M.; Perestrelo, R.; Câmara, J.S.; Yang, Q.; Corke, H.; Travaglia, F.; Locatelli, M.; Arlorio, M.; Coïsson, J.D. Global volatile signature and polyphenols patterns in Vespolina wines according to vintage. *Int. J. Food Sci. Technol.* **2020**, *56*, 1551–1561. [CrossRef]
74. Singleton, V.L.; Rossi, J.A. Colorimetry of total phenolics with phosphomolybdic-phosphotungstic acid reagents. *Am. J. Enol. Vitic.* **1965**, *16*, 144–158.
75. Zhishen, J.; Mengcheng, T.; Jianming, W. The determination of flavonoid contents in mulberry and their scavenging effects on superoxide radicals. *Food Chem.* **1999**, *64*, 555–559. [CrossRef]
76. Feucht, W.; Polster, J. Nuclei of Plants as a Sink for Flavanols. *Z. Nat. C* **2001**, *56*, 479–482. [CrossRef]
77. Broadhurst, R.B.; Jones, W.T. Analysis of condensed tannins using acidified vanillin. *J. Sci. Food Agric.* **1978**, *29*, 788–794. [CrossRef]
78. Lee, J.; Durst, R.W.; Wrolstad, E.R.; Eisele, T.; Giusti, M.M.; Hofsommer, H.; Koswig, S.; Krueger, A.D.; Kupina, S.; Martin, S.K.; et al. Determination of Total Monomeric Anthocyanin Pigment Content of Fruit Juices, Beverages, Natural Colorants, and Wines by the pH Differential Method: Collaborative Study. *J. AOAC Int.* **2005**, *88*, 1269–1278. [CrossRef] [PubMed]
79. Re, R.; Pellegrini, N.; Proteggente, A.; Pannala, A.; Yang, M.; Rice-Evans, C. Antioxidant activity applying an improved ABTS radical cation decolorization assay. *Free. Radic. Biol. Med.* **1999**, *26*, 1231–1237. [CrossRef]
80. Benzie, I.; Strain, J. The Ferric Reducing Ability of Plasma (FRAP) as a Measure of "Antioxidant Power": The FRAP Assay. *Anal. Biochem.* **1996**, *239*, 70–76. [CrossRef] [PubMed]
81. Brand-Williams, W.; Cuvelier, M.; Berset, C. Use of a free radical method to evaluate antioxidant activity. *LWT* **1995**, *28*, 25–30. [CrossRef]
82. Apak, R.; Güçlü, K.; Özyürek, M.; Karademir, S.E. Novel Total Antioxidant Capacity Index for Dietary Polyphenols and Vitamins C and E, Using Their Cupric Ion Reducing Capability in the Presence of Neocuproine: CUPRAC Method. *J. Agric. Food Chem.* **2004**, *52*, 7970–7981. [CrossRef] [PubMed]
83. Mitrevska, K.; Grigorakis, S.; Loupassaki, S.; Calokerinos, A.C. Antioxidant Activity and Polyphenolic Content of North Macedonian Wines. *Appl. Sci.* **2020**, *10*, 2010. [CrossRef]
84. Lamuela-Raventos, R.M.; Waterhouse, A.L. A direct HPLC separation of wine phenolics. *Am. J. Enol. Vitic.* **1994**, *45*, 1–5.
85. Arsa, S.; Wipatanawin, A.; Suwapanich, R.; Makkerdchoo, O.; Chatsuwan, N.; Kaewthong, P.; Pinsirodom, P.; Taprap, R.; Haruenkit, R.; Poovarodom, S.; et al. Properties of Different Varieties of Durian. *Appl. Sci.* **2021**, *11*, 5653. [CrossRef]
86. Shahidi, F.; Ambigaipalan, P. Phenolics and polyphenolics in foods, beverages and spices: Antioxidant activity and health effects—A review. *J. Funct. Foods* **2015**, *18*, 820–897. [CrossRef]
87. Pozo-Bayón, M.; Hernández, M.T.; Martín-Álvarez, A.P.J.; Polo, M.C. Study of Low Molecular Weight Phenolic Compounds during the Aging of Sparkling Wines Manufactured with Red and White Grape Varieties. *J. Agric. Food Chem.* **2003**, *51*, 2089–2095. [CrossRef] [PubMed]
88. Trott, O.; Olson, A.J. AutoDock Vina: Improving the speed and accuracy of docking with a new scoring function, efficient optimization, and multithreading. *J. Comput. Chem.* **2009**, *31*, 455–461. [CrossRef] [PubMed]

 *molecules*

Article

Kinetics of Phenolic Compounds Modification during Maize Flour Fermentation

Oluwafemi Ayodeji Adebo [1],*, Ajibola Bamikole Oyedeji [1],*, Janet Adeyinka Adebiyi [1], Chiemela Enyinnaya Chinma [2,3], Samson Adeoye Oyeyinka [1], Oladipupo Odunayo Olatunde [4], Ezekiel Green [1], Patrick Berka Njobeh [1] and Kulsum Kondiah [1],*

1   Department of Biotechnology and Food Technology, Doornfontein Campus, Faculty of Science, University of Johannesburg, Doornfontein, P.O. Box 17011, Johannesburg 2028, South Africa; janetaadex@gmail.com (J.A.A.); sartf2001@yahoo.com (S.A.O.); egreen@uj.ac.za (E.G.); pnjobeh@uj.ac.za (P.B.N.)
2   Department of Food Science and Technology, Federal University of Technology, P.M.B 65, Minna 920001, Nigeria; chinmachiemela@futminna.edu.ng
3   Africa Center of Excellence for Mycotoxin and Food Safety, Federal University of Technology, P.M.B 65, Minna 920001, Nigeria
4   Department of Food and Human Nutritional Sciences, Faculty of Agricultural and Food Sciences, University of Manitoba, Winnipeg, MB R3T 2N2, Canada; oladipupo.olatunde177@gmail.com
*   Correspondence: oadebo@uj.ac.za (O.A.A.); jibanky2@gmail.com (A.B.O.); kulsumk@uj.ac.za (K.K.); Tel.: +27-115596261 (O.A.A.); +27-744113712 (A.B.O.); +27-115596915 (K.K.)

Citation: Adebo, O.A.; Oyedeji, A.B.; Adebiyi, J.A.; Chinma, C.E.; Oyeyinka, S.A.; Olatunde, O.O.; Green, E.; Njobeh, P.B.; Kondiah, K. Kinetics of Phenolic Compounds Modification during Maize Flour Fermentation. *Molecules* **2021**, *26*, 6702. https://doi.org/10.3390/molecules26216702

Academic Editors: Mirella Nardini and Daniel Granato

Received: 10 September 2021
Accepted: 1 November 2021
Published: 5 November 2021

Publisher's Note: MDPI stays neutral with regard to jurisdictional claims in published maps and institutional affiliations.

Abstract: This study aimed to investigate the kinetics of phenolic compound modification during the fermentation of maize flour at different times. Maize was spontaneously fermented into sourdough at varying times (24, 48, 72, 96, and 120 h) and, at each point, the pH, titratable acidity (TTA), total soluble solids (TSS), phenolic compounds (flavonoids such as apigenin, kaempferol, luteolin, quercetin, and taxifolin) and phenolic acids (caffeic, gallic, ferulic, *p*-coumaric, sinapic, and vanillic acids) were investigated. Three kinetic models (zero-, first-, and second-order equations) were used to determine the kinetics of phenolic modification during the fermentation. Results obtained showed that fermentation significantly reduced pH, with a corresponding increase in TTA and TSS. All the investigated flavonoids were significantly reduced after fermentation, while phenolic acids gradually increased during fermentation. Among the kinetic models adopted, first-order (R^2 = 0.45–0.96) and zero-order (R^2 = 0.20–0.82) equations best described the time-dependent modifications of free and bound flavonoids, respectively. On the other hand, first-order (R^2 = 0.46–0.69) and second-order (R^2 = 0.005–0.28) equations were best suited to explain the degradation of bound and free phenolic acids, respectively. This study shows that the modification of phenolic compounds during fermentation is compound-specific and that their rates of change may be largely dependent on their forms of existence in the fermented products.

Keywords: first order; flavonoids; kinetic modelling; phenolic acids; zero order

1. Introduction

Phenolic compounds are vital constituents of food and secondary metabolites of plants derived from several biosynthetic precursors through the action of shikimate, phenylpropanoid, and pentose phosphate metabolism pathways [1]. These phenolic compounds (also called phenolics) are often encountered in food products, especially those derived from plants and cereals, and are known to exert health benefits such as anticarcinogenic potential, the prevention and counteraction of oxidative stress, chemo-preventive effects, and the reduction of free radical-related cellular damage [2–5].

Maize (*Zea mays*) is one of the main cereals produced worldwide, providing 30% of food calories to over 4 billion people in the world [6,7]. It is also considered a major staple in 125 developing countries [6,7]. Maize is transformed into other food forms using a wide range of processing techniques, one of them being fermentation. Subsequent fermented maize-based food products play a significant role in the nutrition and diets of inhabitants of developing and underdeveloped countries. The fermentation process not only improves food composition, palatability, sensory properties, digestibility, and nutritional constituents, but also reduces antinutritional properties [8–10]. Although fermentation is known to positively impact the phenolic constituents of cereals [4], it is nonetheless important to study the changes in the levels of these beneficial food components during the fermentation process. This is vital, considering the purported roles of these phenolic compounds as health beneficial components in foods.

According to van Boekel [11], kinetic modelling has numerous applications in food, including its use as a tool to understand biochemical modifications in food. Natural (spontaneous) fermentation is a complex process involving interactions between microorganisms, the food substrate, and inherent constituents. It is thus desirable to adopt mathematical equations for kinetic simulations of the phenolic compound modification, as a function of fermentation time, to fully understand the process. To the best of our knowledge, there is a dearth of information on this in the literature. This study aimed to investigate the modifications in the phenolic compounds of maize flour over different fermentation times.

2. Results and Discussion

2.1. pH, Titratable Acidity (TTA) and Total Soluble Solids (TSS)

The pH and TTA are important biochemical parameters particular to fermented foods. A decrease in pH with a corresponding increase in TTA signifies a progression in microbial activity, i.e., the metabolism of fermenting microorganisms coupled with an accumulation of organic acids produced by fermenting microorganisms. Table 1 shows the pH, TTA, and TSS values of unfermented and fermented maize flour samples (24 to 120 h). The values obtained herein are similar to the values of fermented cereals with a pH between 3.6 and 4.8 [12–14]. The pH values of the maize flour declined with an increase in fermentation time, from an initial pH of 6.30 to a final pH of 3.89, after 120 h of fermentation. While a significant ($p \leq 0.05$) linear decrease was observed till 72 h, a steady pH value of between 3.88–3.90 was observed from 72–120 h. The initial drop in pH could be associated with the actions of fermenting microorganisms causing production and accumulation in organic acids [15], while the insignificant modification in pH between 72–120 h could possibly suggest a saturation of the activity of the fermentation organisms. An inverse relationship between the TTA and pH values was observed, with a significant increase in TTA as fermentation progressed. Both reduced pH and corresponding higher TTA values were also shown to be desirable against pathogenic microorganisms that would not survive the fermented product.

Table 1. pH, TTA, and TSS values of maize sourdough over different fermentation periods.

Fermentation Time (h)	pH	TTA (g/kg)	TSS (°Brix)
0	6.30 [c] ± 0.14	0.27 [a] ± 0.06	0.20 [a] ± 0
24	5.21 [b] ± 0.11	0.57 [b] ± 0.03	0.20 [a] ± 0.10
48	4.01 [a] ± 0.03	1.05 [c] ± 0.18	0.53 [ab] ± 0.25
72	3.88 [a] ± 0.11	1.65 [d] ± 0.22	0.50 [ab] ± 0.10
96	3.90 [a] ± 0.04	1.45 [d] ± 0.05	0.87 [c] ± 0.15
120	3.89 [a] ± 0.08	1.53 [d] ± 0.06	0.63 [bc] ± 0.06

TTA—titratable acidity; TSS—total soluble solids. Values represent mean ± standard deviation of triplicate measurements. Values with differing letters within a row are significantly different at $p \leq 0.05$.

The TSS value, as measured using a refractometer, is an approximate measure of the sugar content. Fermentation of the maize flours also resulted in changes to the soluble solids (Table 1). Generally, TSS values significantly increased over the fermentation periods. However, after 48 and 72 h of fermentation, there were no significant changes in the TSS values, but a 74% increase was recorded after an additional day of fermentation (96 h samples). A further decline of TSS was later recorded after 120 h of fermentation, with a similar pattern being reported by Yousif et al. [16] during the fermentation of sorghum. These modifications in the TSS values suggest sugar metabolism during the fermentation process and further biotransformation of these components at the latter stage of fermentation. The initially observed increase in TSS suggests the release of sugars and carbohydrate-related compounds, and a continued increase with the fermentation time indicates that the rate of sugar utilization by the fermenting microbiota is lower than the rate of sugars released [17]. Similar trends of increase in TSS have also been reported in the fermentation of rice, maize, and sorghum [18–20]. Pearson correlation was also used to show the relationship between the pH, TTA, and TSS data. A positive correlation coefficient indicates that a positive increase in the parameter will lead to a positive increase in the other variable. On the contrary, a negative correlation coefficient means that for a positive increase in the parameter, there would be a negative decrease in the other variable. A negative (-0.818) and significant ($p \leq 0.05$) correlation was recorded between the pH and TSS values, which suggests that a decrease in pH would significantly increase the TSS values. Such an increase could also be attributed to enzymatic activities of the fermenting organisms that hydrolyse complex polysaccharides into simpler ones [21]. A similar negative (-0.936) and significant ($p \leq 0.05$) correlation between the pH and TTA also alludes to the trend observed for pH and TTA in Table 1. The TTA, TSS, and pH values suggest that, as fermentation progresses, the metabolism of the fermenting microorganisms increases, albeit reaching a saturated point at longer fermentation times.

2.2. Free and Bound Flavonoids Contents and Kinetics of Modification

Phenolic compounds exist in both free and bound forms of plant cells, and the free phenolic compounds are generally solvent extractable. Bound phenolics, on the other hand, remain after extraction and adhere to the food matrix after extraction of the free fraction [22]. These bound fractions in food have numerous health potentials, especially for gut health, as well as other benefits documented in the literature [1,4,5,22].

Flavonoids are plant secondary metabolites and significant non-nutritive dietary components, as well as a sub-class of phenolic compound groups [4,5,22]. The consumption of these dietary flavonoids from plant-based foods has also been related to the prevention of several chronic diseases, and they are thus seen as vital primary sources for consumption [22]. The trend of modification in the quantified free and bound flavonoid compounds is presented in Figure 1A,B. All five flavonoids (free and bound) investigated in this study were reduced after 120 h of fermentation. This general decrease in the flavonoids (free and bound) suggests that they were rapidly metabolized till they were non-detectable from 48 h onwards for bound apigenin, kaempferol, luteolin, and quercetin (Figure 1B). Apigenin is a naturally occurring flavonoid in yellow maize [23] and, similar to other pigment-related compounds, confers colour to the maize kernel. This compound was significantly higher than other flavonoid compounds in maize, with an initial concentration of 338.42 µg/g to a final concentration of 1.3 µg/g, after 120 h of fermentation (Figure 1A). A similar trend was observed in its bound form from an initial concentration of 0.27 µg/g to it not being detected after 48 h and onwards. Such total degradation of pigment-related compounds has also been reported in fermented sorghum, with the authors ascribing this to the conversion of apigeninidin, and methoxyapigeninidin into novel adducts of deoxyanthocyanidins [24]. Further investigation into the product(s) from such degradation of apigenin in this study should be elucidated in future research and would provide an understanding of the mechanisms involved, as well as any subsequent products formed.

Such decreases in the bound and free flavonoids have been reported during the fermentation of maize and other cereals and can be ascribed to the metabolism of these compounds by fermenting microorganisms, degradation/polymerization of flavonoid compounds into dihydroxyl flavone analogues or derivatives, and the activities of endogenous grain enzymes such as glycosidases, glycosyltransferases, cellulase, tannase, esterase, and hydrolases [4,5,25–27]. The noted significant decrease of flavonoid content after longer fermentation periods coincides with the marked decrease in pH values of the fermented samples and is well-reflected in the strong correlation coefficient of these flavonoids with pH. This also suggests that these flavonoids are liable to acidic pH, thus resulting in further hydrolysis at lower pH [28].

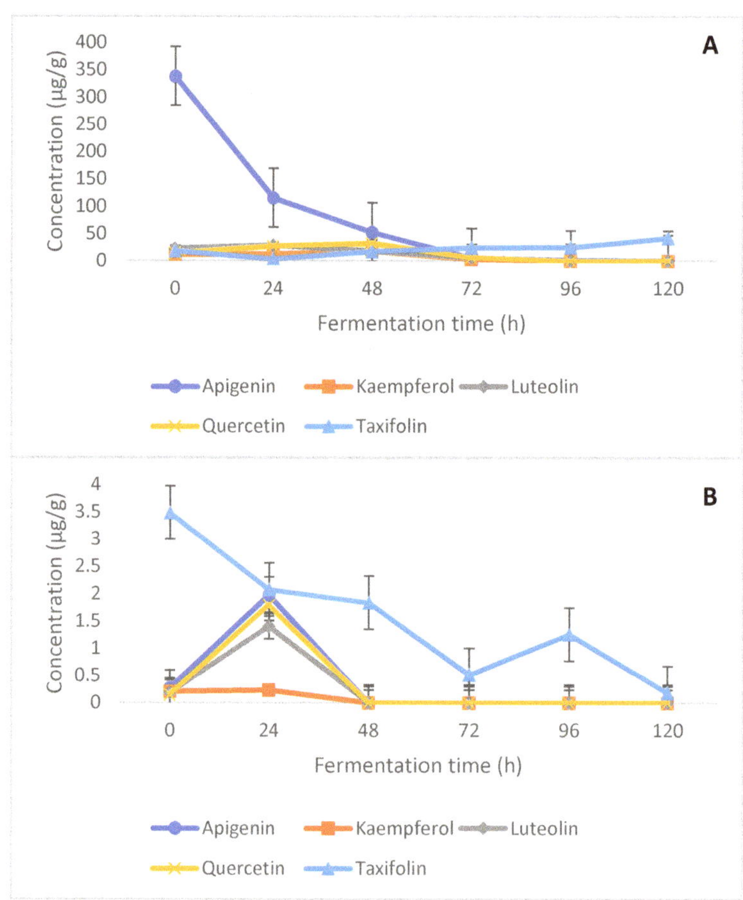

Figure 1. Modification of (**A**) free and (**B**) bound flavonoids in fermented maize flour over time.

The parameters (rate of degradation, *k*, and the coefficient of determination, R^2) of each of the models used to describe the kinetic degradation of free and bound flavonoids are presented in Table 2. The first-order kinetics model ($R^2 = 0.45$–0.96) best describes the time-dependent degradation of free flavonoids, while the zero-order model ($R^2 = 0.20$–0.82) is most suited for describing the kinetics of reduction in bound flavonoids for the fermented maize products, as they both present the highest R^2 values in each case (Figure 2; Figure 3). The low R^2 values for the kinetics of some free and bound flavonoids can be attributed to the drastic reduction after 48 h. While there were no other significant changes in the phenolic contents after 48 h, fermentation continued to occur, as reflected in the TTA and TSS values. First-order models have been successfully used to describe the degradation of phenolic compounds in different food systems subjected to various processing operations [29–31] and zero-order [32,33]. Low R^2 values generally obtained for the kinetic models, especially for bound flavonoids, is reflective of the oscillating degradation patterns of these phenolic compounds at the different fermentation times investigated. This can be an effect of various biological reactions, such as decarboxylation, esterification, and hydrolysis occurring simultaneously [4,5,15] and leading to changes at the different fermentation times [29], which could be degradation or synthesis of bioactive components, including phenolic compounds.

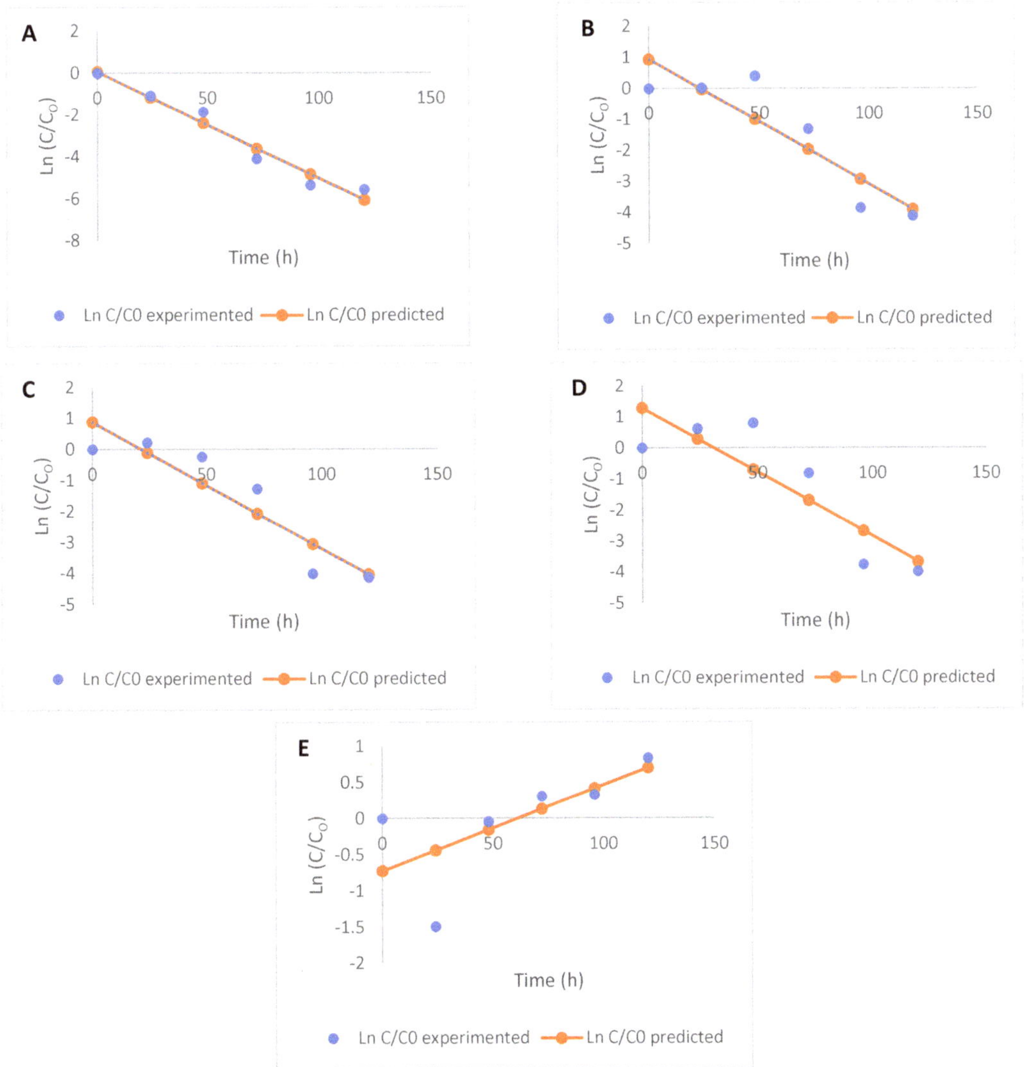

Figure 2. Kinetics (first-order) of free flavonoid degradation during the fermentation of maize flour (**A**) apigenin, (**B**) kaempferol, (**C**) luteolin, (**D**) quercetin, and (**E**) taxifolin.

Table 2. Kinetics of phenolic degradation using zero-, first-, and second-order reactions.

	Free Phenolic Compounds						Bound Phenolic Compounds					
	Zero Order		First Order		Second Order		Zero Order		First Order		Second Order	
	k (mgh^{-1})	R^2	k (h^{-1})	R^2	k (mgh^{-1} g^{-1} h^{-1})	R^2	k (mgh^{-1})	R^2	k (h^{-1})	R^2	k (mgh^{-1} g^{-1})	R^2
Flavonoids												
Apigenin	−0.015	0.71	−2.470	0.96	−0.007	0.82	−0.009	0.24	−0.007	0.15	−0.007	0.15
Kaempferol	−0.040	0.62	−0.134	0.79	−6.655	0.74	−0.002	0.66	−0.0004	0.15	−0.0002	0.15
Luteolin	−0.041	0.85	−0.248	0.84	−13.268	0.73	−0.006	0.24	−0.007	0.15	−0.005	0.15
Quercetin	−0.214	0.48	−0.041	0.74	−0.214	0.48	−0.007	0.20	−0.009	0.15	−0.006	0.15
Taxifolin	0.012	0.66	0.227	0.45	0.538	0.40	−0.024	0.82	−0.021	0.76	−0.432	0.60
Phenolic Acids												
Caffeic Acid	0.046	0.46	3.125	0.44	3.125	0.46	−0.005	0.004	−0.005	0.02	−0.006	0.005
Ferulic Acid	0.146	0.44	0.012	0.53	0.321	0.43	0.032	0.20	−0.016	0.18	0.033	0.20
Gallic Acid	0.056	0.68	4.107	0.68	4.112	0.68	0.083	0.28	0.043	0.59	0.085	0.28
p-Coumaric Acid	0.012	0.60	0.139	0.47	0.312	0.44	−0.0004	0.19	−0.002	0.19	−0.0007	0.19
Sinapic Acid	0.009	0.30	0.174	0.20	0.384	0.22	0.006	0.20	0.0114	0.19	0.007	0.20
Vanillic Acid	−0.038	0.69	11.110	0.70	11.227	0.70	0.074	0.02	0.009	0.07	5.160	0.15

The rate constant k, for the degradation of free flavonoids, described by the first-order kinetics model, ranged from −2.470 to 0.227 h^{-1}, while that of bound flavonoids described by zero-order kinetics ranged from −0.024 to −0.002 h^{-1}. The nature of change (degradation or increase) in flavonoids is depicted by the direction of the kinetic curves, hence the different 'k' values obtained for each model. Rate constants are indications of the speed of progress of biological reactions with time, as authors [34,35] have previously used 'k' values to predict the progress and compare different biological reactions. Generally, rate constants of zero-order kinetics were the lowest for bound flavonoids compared to the other models used and degradation of free flavonoids also proceeded at lower rates, using first-order kinetics (Table 2).

2.3. Free and Bound Phenolic Acids and Kinetics of Modification

Similar to the flavonoids, all the investigated phenolic acids were present in the initial maize samples prior to fermentation. The free phenolic acid content in the unfermented maize flour ranged from 0.16 µg/g (caffeic acid) to 20.25 µg/g (vanillic acid). The least bound phenolic acid in the unfermented flour was gallic acid (0.10 µg/g), while the highest was vanillic acid (44.01 µg/g). While a consistent decrease trend was observed for the flavonoids, the phenolic acids were generally observed to increase, albeit not in a linear manner (Figure 4A). As posited by several studies, the influence of fermenting microorganisms on the levels of individual phenolics can differ, as it is dependent on the microbial strain and possible genes for phenolic metabolism, as well as fermentation conditions [4,36,37]. Nevertheless, remarkable increases of over 10 000-fold were particularly noted for caffeic, gallic, and vanillic acids in their free forms. Similar increases in phenolic acids after fermentation have been reported in other cereal-based products such as sorghum sourdough [38], fermented barley and oat [39], and spelt wheat *tempe* [40].

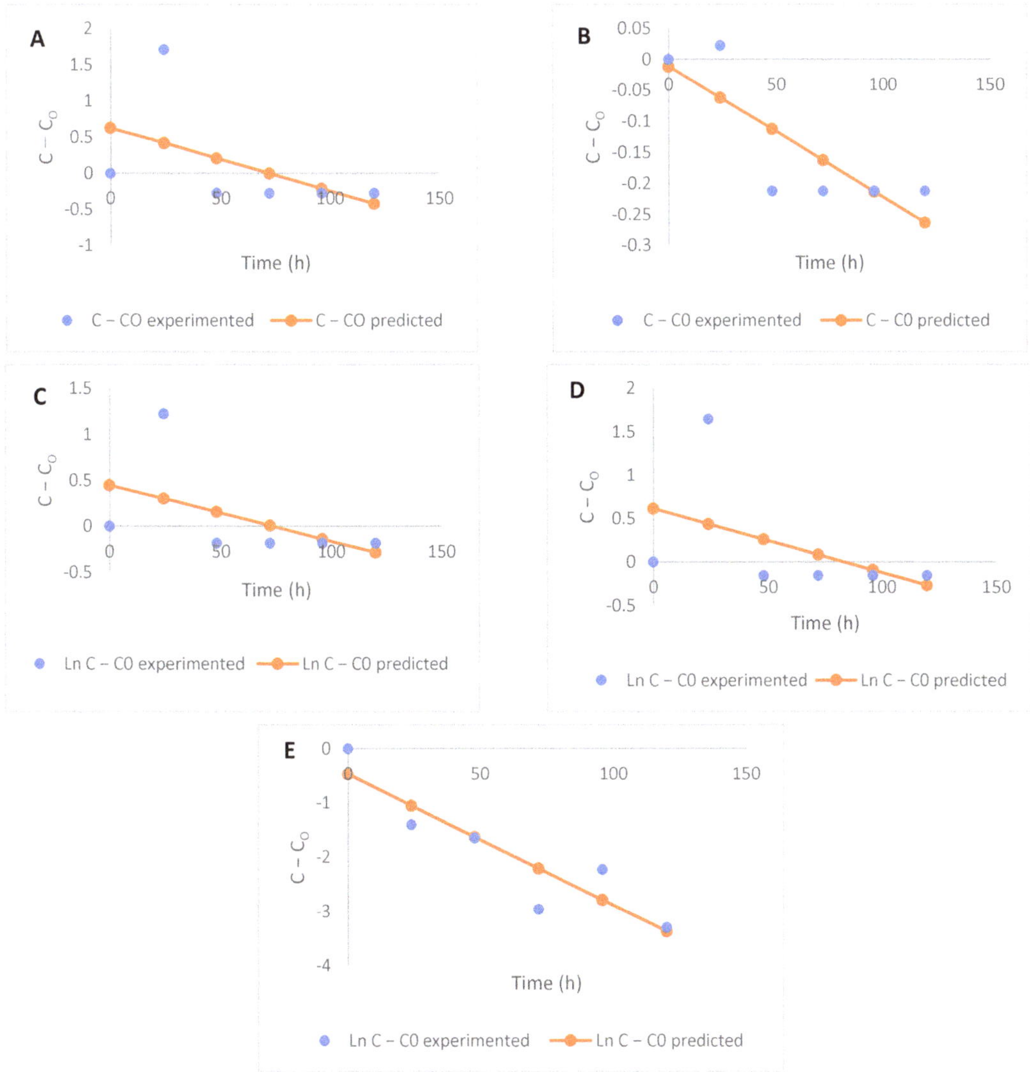

Figure 3. Kinetics (zero-order) of bound flavonoid degradation during the fermentation of maize flour (**A**) apigenin, (**B**) kaempferol, (**C**) luteolin, (**D**) quercetin, and (**E**) taxifolin.

Although not as pronounced as in the free form, similar increases were noted for the bound phenolic acids (Figure 4B). The observed general increase in these phenolic acids can be ascribed to the activities of some microbially secreted enzymes that hydrolysed the ester, polymeric, and glycosidic bonds of the phenolic acids, as well as the structural breakdown of cell walls, which invariably resulted in improved bioavailability and extractability [41–43]. These mechanisms also contributed to the decrease of the flavonoids (Section 2.2) and efficiently released these phenolic acids from the cell wall of the maize grain. The remarkable decrease in apigenin noted in this study may also have contributed to the increase in phenolic acids. Vernhet et al. [44] also reported this observation in their study on the fermentation of red musts, noting that the interconversion between phenolic compounds and the breakdown of pigment-related compounds could contribute to an increase in phenolic acids.

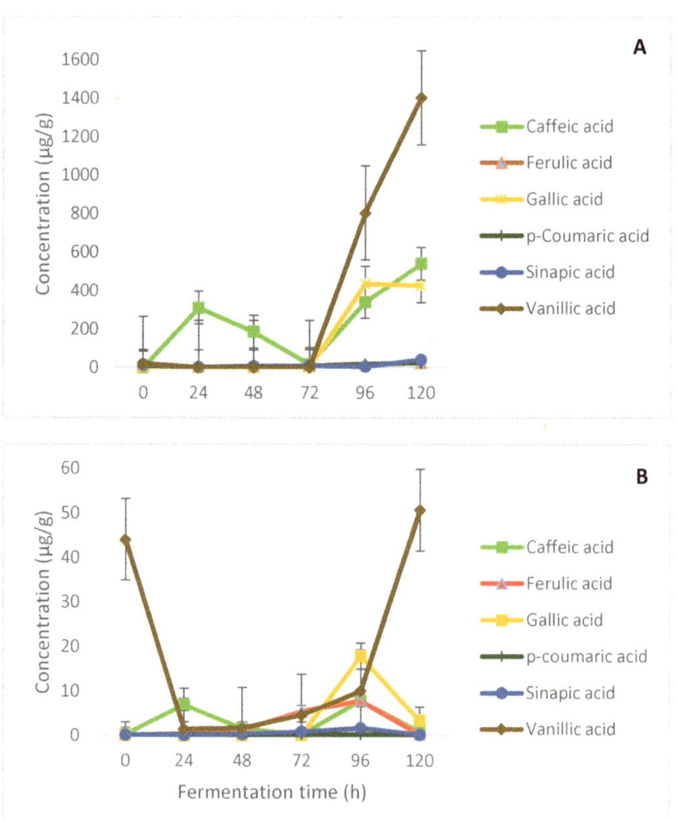

Figure 4. Modification of (**A**) free and (**B**) bound phenolic acids in fermented maize flour over time.

Contrary to the kinetics of bound flavonoids, the first-order kinetic model more appropriately described the time-dependent changes in bound phenolic acids. Both first- and second-order kinetics better described the effect of fermentation on free phenolics with comparable R^2 and k values, as compared to zero-order kinetics (Table 2). However, second-order kinetics were preferable in describing the changes in free phenolic acids, with the highest reaction rate constants of the three models used. The R^2 values obtained for bound phenolic acids using first-order kinetics (0.02–0.59) were higher than those obtained when zero-order (0.004–0.28) and second-order (0.005–0.28) were used. The release of bound phenolics from fermented maize products was depicted to have proceeded at faster rates for all bound phenolic acids by zero- and second-order kinetic models, evident by the higher k values compared to first-order kinetics which produced the lowest k values. However, the higher coefficients of determination (R^2) produced by the first-order kinetics showed that the model better described the changes in bound phenolic acids, despite lower rates than were described by the zero- and second-order kinetics—hence the choice of first-order kinetics.

For free phenolic acids, first-order rate constant 'k' ranged between 0.012–11.110 h^{-1} (R^2 = 0.46–0.69), while for zero- and second-order kinetics, k ranged between −0.038–0.146 mgh^{-1} (R^2 = 0.44–0.70) and 0.312–11.227 mgh^{-1} g^{-1} h^{-1} (R^2 = 0.46–0.70), respectively. Slight similarities in k values for the first- and second-order reaction for free phenolic compounds suggests that the reactions proceeded at similar rates for both models. Also, rate constants showed that changes in free phenolic acids proceeded at faster rates, as compared to bound phenolic compounds (Table 2)—a trend also shown by the kinetic model curves (Figures 5 and 6). These faster rates could be because of their bioavailability as free phenolics in the fermenting matrices. The degradation of free vanillic acid had the highest 'k' and R^2 values, as described by the first- and second-order kinetics. De Beer et al. [29] had previously used zero-, first-, and second-order kinetics to describe the rate of changes in the phenolic compounds of rooibos tea fermented at different temperatures, and found that first-order kinetics was the most suitable model for describing the degradation of these compounds.

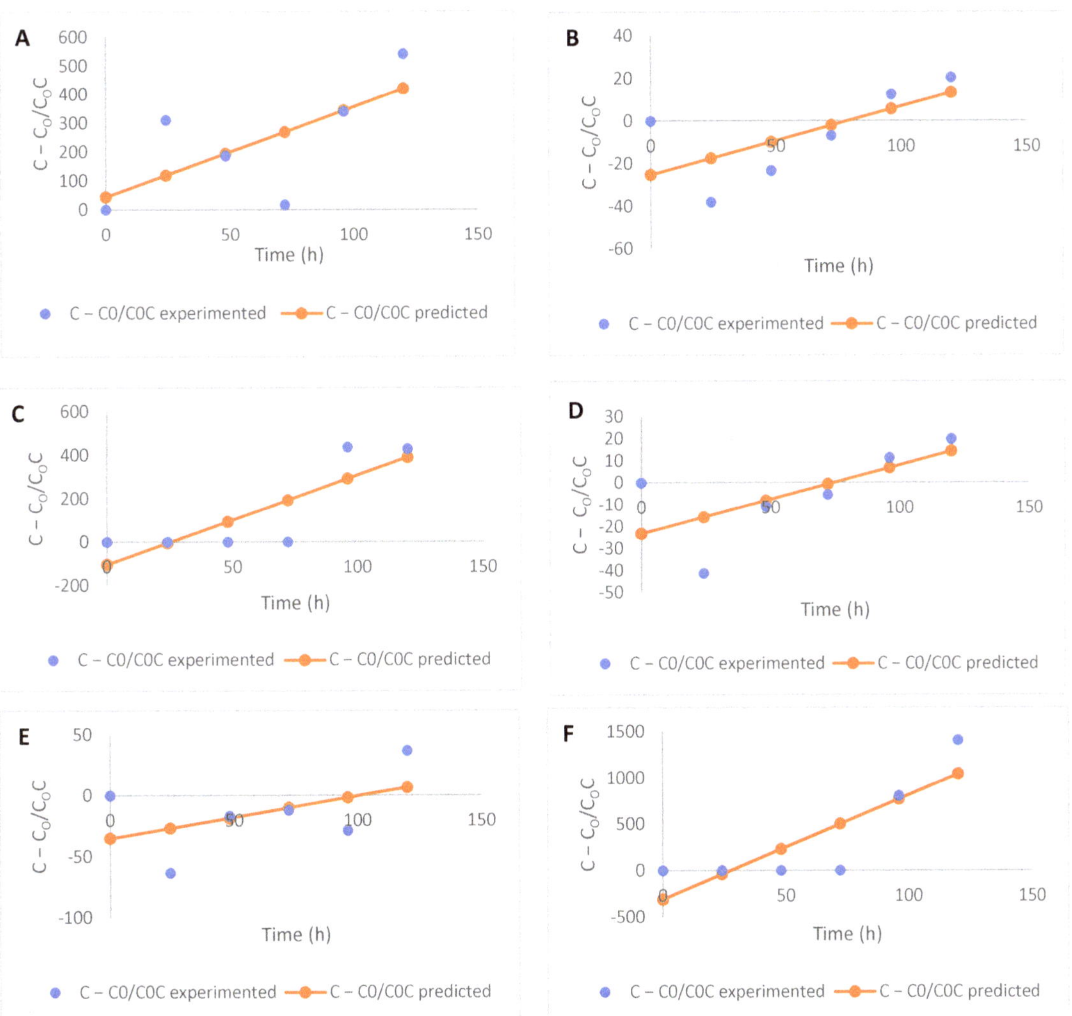

Figure 5. Kinetics (second-order) of free phenolic acid degradation during the fermentation of maize flour (**A**) caffeic acid, (**B**) ferulic acid, (**C**) gallic acid, (**D**) *p*-coumaric acid, (**E**) sinapic acid, and (**F**) vanillic acid.

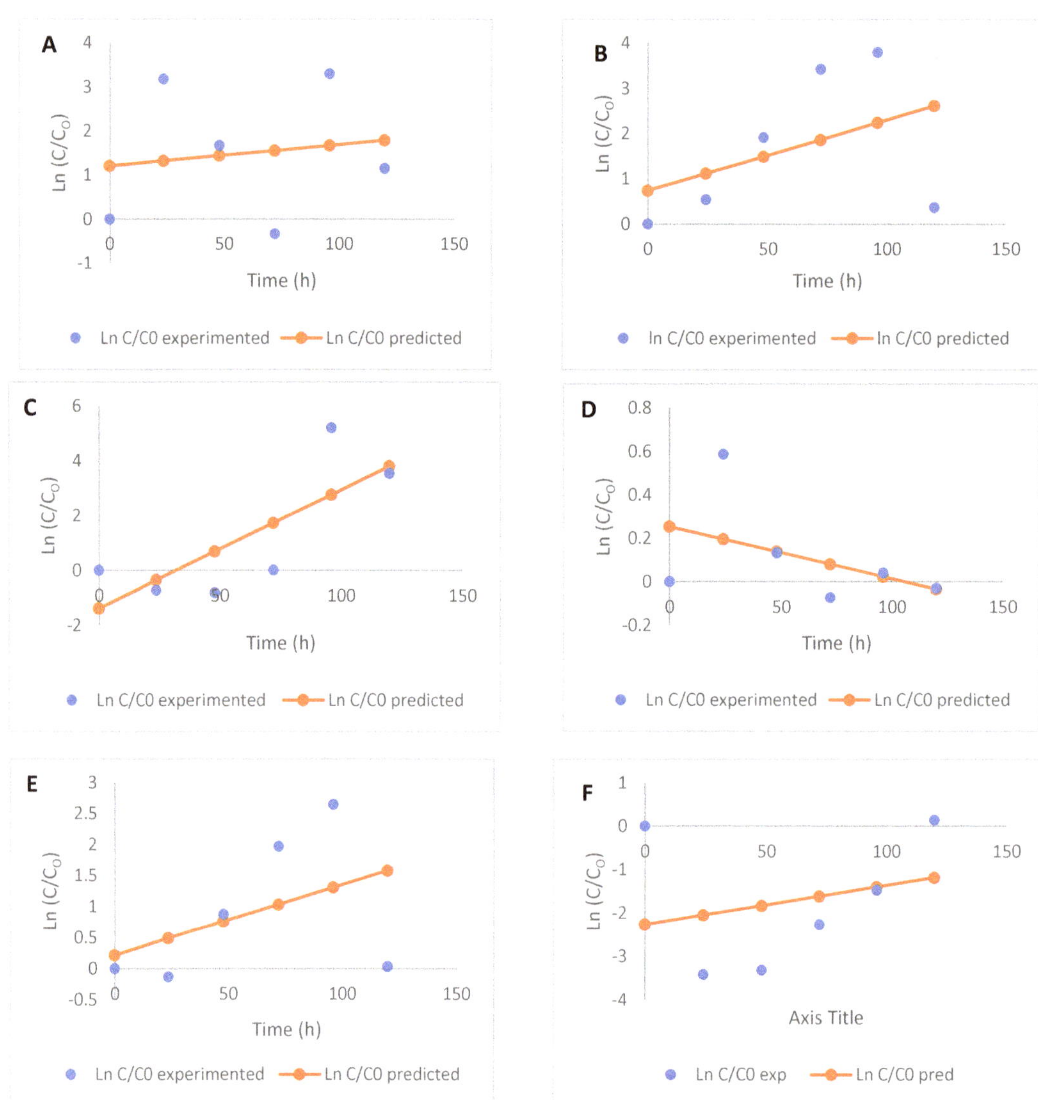

Figure 6. Kinetics (first-order) of bound phenolic acid degradation during the fermentation of maize flour (**A**) caffeic acid, (**B**) ferulic acid, (**C**) gallic acid, (**D**) *p*-coumaric acid, (**E**) sinapic acid, and (**F**) vanillic acid.

2.4. Principal Component Analysis

The PCA was used to explore the distribution and highlight natural groupings, as well as the trends and relationship of the parameters investigated in the unfermented and fermented maize flour samples. The first two principal components (PCs), PC1 and PC2, explained 49.8% and 17.5%, respectively, of the variation (total of 67.3%). The unfermented maize flour (0 h) on the bottom right quadrant showed a clear separation from the fermented samples (grouped in other parts on the PC plot (Figure 7A). A further classification based on the fermentation periods (Figure 7B) showed the different groupings as a result of the fermentation time. Clusters of the 24 and 48 h fermented samples moved on the right of the PC1 quadrant, while clusters for the later stages (72–120 h) of fermentation moved to the left part of the PC1 quadrant. The observed separation and clusters of the parameters investigated in the maize samples are suggestive of differential changes during fermentation and correlate with the results of the pH, TTA, and TSS (Table 1), as well as the phenolic compounds (Figure 1; Figure 4).

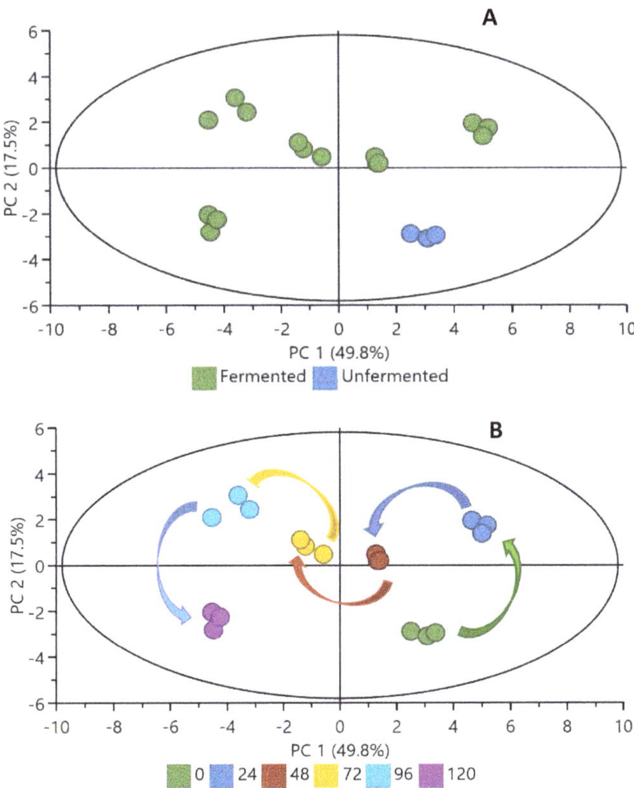

Figure 7. Exploratory data analysis with PCA. (**A**) Relationship of unfermented and fermented samples; (**B**) relationship of samples fermented at different times.

3. Materials and Methods

Yellow maize (*Zea mays*) grains (variety IMP50BR) were obtained from the Agricultural Research Council (Grain Crops, Potchefstroom, South Africa). The grains were milled using a Perten Laboratory Mill 3310 (Perten Instruments AB, Helsinki, Finland) and passed through a 0.5 mm aperture size sieve (Analysette 3 Spartan, Fritsch, Germany) to obtain the flour.

3.1. Fermentation

Maize flour was processed into sourdough by mixing it with sterile distilled water (1:1, *v/w*). The mixture was then incubated (Incotherm, Labotec, Johannesburg, South Africa) at 37 °C and fermentation was separately done for 24, 48, 72, 96, and 120 h. For each time, the fermentation process was performed in triplicates.

3.2. Total Soluble Solids (TSS), pH, and Titratable Acidity (TTA)

A refractometer (HI 96801, HANNA Instruments, Inc., Cluj-Napoca, Romania) was used to measure the total soluble solids of samples, while an initially calibrated pH meter (pH 510, Eutech Pte Ltd., Taus, Singapore) was used to obtain their pH values. For titratable acidity (TTA), the method described by Aguilar et al. [45] was used. It involved mixing 10 g of each sample in 90 mL of distilled water, with the mixture being allowed to rest to obtain the supernatant. Thereafter, the supernatant was titrated against 0.1 M NaOH (pH 8.3).

3.3. Extraction and Quantification of Free and Bound Phenolic Compounds

Prior to quantification, phenolic compounds were extracted using the modified methods of Xiang et al. [46] and Ravisankar et al. [47]. A 0.25 g of the raw ground and fermented maize samples, as well as 2.5 mL of acidified methanol (1% HCl in 80% aqueous methanol), were added. The mixture was sonicated in an ultrasonic bath (AU 220, Argo Lab, Carpi, Italy) for 1 h at 4 °C, followed by centrifugation (Eppendorf 5702R, Merck, Germiston, South Africa) at $2100\times g$ at 4 °C for 10 min. The supernatant (free phenolics) was evaporated using a vacuum concentrator (Eppendorf Concentrator Plus, Analytical Solutions, Johannesburg, South Africa) and the dried extract was reconstituted with 1 mL 50% liquid chromatographic-grade methanol (Merck, Johannesburg, South Africa). The residue earlier obtained was then hydrolysed (for 30 min at 60 °C) with 2.5 mL of 2 M NaOH and 2.5 mL of ethyl acetate. A similar process of sonication and centrifugation was performed. The supernatant (bound phenolics) was evaporated using a vacuum concentrator, and the dried extract was reconstituted with 1 mL 50% liquid chromatographic-grade methanol.

Quantification of the free and bound phenolics was performed on an ultra-high pressure liquid chromatography (UHPLC) system (Shimadzu, Kyoto, Japan) equipped with a degassing unit (DGU-403), binary pumps (LC-40B XR), solvent delivery module (LC-40B XR), auto-sampler (SIL-40C XR), column oven (CTO-40C), and a diode array detector (DAD) (SPD-M40). A two (2) μL extract was injected into the system, and separation (Supplementary Materials Figure S1) was carried out on a Raptor C18 column (2.7 μm × 100 mm × 2.1 mm ID, Restek, Bellefonte, USA) at an oven temperature of 40 °C. Mobile phase A consisted of 1% formic acid in Milli-Q water and B was 1% formic acid in a mixture of 50% methanol and acetonitrile, with the total run time being 15 min. Quantification of each phenolic compound, including the apigenin, caffeic acid, gallic acid, ferulic acid, kaempferol, luteolin, *p*-coumaric acid, quercetin, sinapic acid, taxifolin, and vanillic acid (Sigma Aldrich, Johannesburg, South Africa), was executed through extrapolation from the calibration curves of analytical standards at different concentrations (2.5, 5, 10, 20, 40, 80, and 160 μg/mL) (Supplementary Materials Figure S2). Table 3 shows the retention times and wavelengths of each phenolic compound.

Table 3. Retention times and wavelengths of the quantified phenolic compounds.

Compound	Retention Time (min)	Wavelength (nm)
Flavonoids		
Apigenin	11.324	336
Kaempferol	11.347	366
Luteolin	10.670	348
Quercetin	10.597	371
Taxifolin	9.300	290
Phenolic acids		
Caffeic acid	8.597	323
Ferulic acid	9.447	322
Gallic acid	8.528	271
p-Coumaric acid	9.348	309
Sinapic acid	9.016	319
Vanillic acid	8.713	261

3.4. Kinetic Modelling

To understand the kinetics of changes in the phenolic compounds in this study, zero-, first-, and second-order mathematical models were adopted. The use of these models has been widely adopted to describe reaction processes in biological systems, especially foods [14]. Linearized forms of these models' Equations (1)–(3) were used to determine the kinetic parameters using Datafit 9.1 software (Oakdale Engineering, Oakdale, PA, USA).

$$\text{Zero order}: C - C_0 = kt \tag{1}$$

$$\text{First order}: ln\frac{C}{C_0} = kt \tag{2}$$

$$\text{Second order}: \frac{C - C_0}{C_0 C} = kt \tag{3}$$

where C is the concentration of phenolic acids at a specific experimental time (t), C_0 is the initial concentration at 0 h, k is the rate constant, and t is the time.

3.5. Statistical Analysis

All analyses were done in triplicates and the differences between the means were established using analysis of variance, followed by Tukey's test (SPSS 22, IBM, Armonk, NY, USA). Variances at a 5% confidence level were considered to be statistically different. Pearson's correlation test and the principal component analysis (PCA) (SIMCA 16, Umetrics, Umea, Sweden) were also conducted on the investigated parameters.

4. Conclusions

Fermentation of maize over a period of 120 h showed modifications in both the biochemical parameters (pH, TTA, and TSS) and the phenolic compounds (flavonoids and phenolic acids). While the flavonoids followed a general decrease, there was an increase in the phenolic acids, suggestive of different metabolism and modification routes of the phenolic compounds. Although higher levels of both flavonoids and phenolic acids are desirable due to the health benefits they confer, the general decrease in the flavonoids might have led to other equally bioactive monomers of significance. Among the three different kinetic models used, first-order was notably adequate for bound flavonoids and phenolic acids, with zero-order and second-order effectively describing the time-dependent degradation of other phenolic compound groups. Trends of such modifications could also be important in deciding fermentation times during the processing of maize-based products. However, further studies are required on the elucidation of the degraded phenolic compounds using liquid chromatography coupled with mass spectrometry systems, perhaps in combination with metabolomic techniques, to understand the mechanisms and metabolism of degradation and metabolic products. Equally important is the isolation of the fermenting microorganisms and in vitro investigation into the specific phenolic metabolism of the respective microorganisms.

Supplementary Materials: The following are available online, Figure S1: An example of chromatographic separation of analytes in the sample (with peaks of each phenolic compound); Figure S2: Standard calibration curves and R^2 values for (A) apigenin, (B) kaempferol, (C) luteolin, (D) quercetin, (E) taxifolin, (F) caffeic acid, (G) ferulic acid, (G) gallic acid, (H) p-coumaric acid, (I) sinapic acid and (J) vanillic acid.

Author Contributions: Conceptualization, O.A.A.; methodology, O.A.A., A.B.O. and J.A.A.; software, O.A.A., S.A.O. and A.B.O.; validation, O.A.A., A.B.O., C.E.C., P.B.N. and K.K.; formal analysis, O.A.A., A.B.O., J.A.A., C.E.C. and K.K.; investigation, O.A.A.; resources, O.A.A.; data curation, O.A.A., A.B.O., J.A.A., S.A.O. and E.G.; writing—original draft preparation, O.A.A., A.B.O., J.A.A., O.O.O. and K.K.; writing—review and editing, O.A.A., A.B.O., J.A.A., C.E.C., S.A.O., O.O.O., E.G., P.B.N. and K.K.; visualization, O.A.A., E.G., P.B.N. and K.K.; project administration, O.A.A.; funding acquisition, O.A.A. All authors have read and agreed to the published version of the manuscript.

Funding: Funding from the University of Johannesburg (UJ) Global Excellence and Stature (GES) 4.0 Catalytic Initiative Grant, the UJ Research Committee (URC) Grant, and the National Research Foundation of South Africa Thuthuka Grant (Grant number: 121826) are duly acknowledged.

Institutional Review Board Statement: Not applicable.

Informed Consent Statement: Not applicable.

Data Availability Statement: Not applicable.

Acknowledgments: This work was also supported by the University of Johannesburg (UJ) Faculty of Science and UJ Research Committee (URC) Postdoctoral Research Fellowship and National Research Foundation (NRF) of South Africa Scarce Skills Fellowship (Grant number: 120751). Appreciation also goes to the members of the Food Innovation Research Group (FIRG) at the University of Johannesburg, South Africa.

Conflicts of Interest: The authors declare no conflict of interest.

Sample Availability: Samples studied herein are available on reasonable request from the authors.

References

1. Li, W.; Beta, T. Food sources of phenolics compounds. In *Natural Products*; Ramawat, K., Mérillon, J.M., Eds.; Springer: Berlin, Germany, 2013; pp. 2527–2558.
2. Olatunde, O.O.; Benjakul, S.; Vongkamjan, K. Antioxidant and antibacterial properties of guava leaf extracts as affected by solvents used for prior dechlorophyllization. *J. Food Biochem.* **2018**, *42*, e12600. [CrossRef]
3. Bonta, R.K. Dietary phenolic acids and flavonoids as potential anti-cancer agents: Current state of the art and future perspectives. *Anti-Cancer Agents Med. Chem* **2019**, *20*, 29–48. [CrossRef]
4. Adebo, O.A.; Medina-Meza, I. Impact of fermentation on the phenolic compounds and antioxidant activity of whole cereal grains: A mini review. *Molecules* **2020**, *25*, 927. [CrossRef] [PubMed]
5. Leonard, W.; Zhang, P.; Ying, D.; Adhikari, B.; Fang, Z. Fermentation transforms the phenolic profiles and bioactivities of plant-based foods. *Biotechnol. Adv.* **2021**, *49*, 107763. [CrossRef] [PubMed]
6. FAOSTAT. Crop Yields. Available online: http://www.fao.org/faostat/en/#data/QC (accessed on 8 June 2021).
7. Nyirenda, H.; Mwangomba, W.; Nyirenda, E.M. Delving into possible missing links for attainment of food security in Central Malawi: Farmers' perceptions and long term dynamics in maize (*Zea mays* L.) production. *Heliyon* **2021**, *7*, e07130. [CrossRef]
8. Adebo, O.A.; Njobeh, P.B.; Adeboye, A.S.; Adebiyi, J.A.; Sobowale, S.S.; Ogundele, O.M.; Kayitesi, E. Advances in fermentation technology for novel food products. In *Innovations in Technologies for Fermented Food and Beverage Industries*; Panda, S., Shetty, P., Eds.; Springer: Berlin, Germany, 2018; pp. 71–87.
9. Adebiyi, J.A.; Obadina, A.O.; Adebo, O.A.; Kayitesi, E. Fermented and malted millet products in Africa: Expedition from traditional/ethnic foods to industrial value added products. *Crit Rev. Food Sci. Nutr.* **2018**, *58*, 463–474. [CrossRef] [PubMed]
10. Adebo, O.A. African sorghum-based fermented foods: Past, current and future prospects. *Nutrients* **2020**, *12*, 1111. [CrossRef] [PubMed]
11. van Boekel, M.A.J.S. On the pros and cons of Bayesian kinetic modeling in food science. *Trends Food Sci. Technol.* **2020**, *99*, 181–193. [CrossRef]
12. Sekwati-Monang, B.; Gänzle, M.G. Microbiological and chemical characterization of *ting*, a sorghum-based sourdough product from Botswana. *Int. J. Food Microbiol.* **2011**, *150*, 115–121. [CrossRef]
13. Edema, M.O. A modified sourdough procedure for non-wheat bread from maize meal. *Food Bioproc. Technol.* **2011**, *4*, 1264–1272. [CrossRef]
14. Decimo, M.; Quattrini, M.; Ricci, G.; Fortina, M.G.; Brasca, M.; Silvetti, T.; Manini, F.; Erba, D.; Criscuoli, F.; Casiraghi, M.C. Evaluation of microbial consortia and chemical changes in spontaneous maize bran fermentation. *AMB Express* **2017**, *7*, 205. [CrossRef] [PubMed]
15. Adebo, O.A.; Njobeh, P.B.; Adebiyi, J.A.; Kayitesi, E. Co-influence of fermentation time and temperature on physicochemical properties, bioactive components and microstructure of *ting* (a Southern African food) from whole grain sorghum. *Food Biosci.* **2018**, *25*, 118–127. [CrossRef]
16. Yousif, N.E.; El-Tinay, A.H. Effect of fermentation on sorghum protein fractions and in vitro protein digestibility. *Plant. Foods Human Nutr.* **2001**, *56*, 175–182. [CrossRef]
17. Chai, K.F.; Adzahan, N.M.; Karim, R.; Rukayadi, Y.; Ghazali, H.M. Effects of fermentation time and turning intervals on the physicochemical properties of rambutan (*Nephelium lappaceum* L.) fruit sweatings. *Sains Malays.* **2018**, *47*, 2311–2318.
18. Padhye, V.W.; Salunkhe, D.K. Biochemical studies on black gram (*Phaseolus mungo* L.) III. Fermentation of black gram and rice blend and its influence on the in vitro protein digestibilities of proteins. *J. Food Biochem.* **1979**, *2*, 327–347. [CrossRef]
19. El Tinay, A.H.; El Mahdi, Z.M.; El Soubki, A. Supplementation of fermented sorghum *kisra* bread with legume protein. *J. Food Technol.* **1985**, *20*, 679–687.

20. Yousif, N.E.; El Tinay, A.H. Effect of fermentation on protein fractions and in vitro protein digestibility of maize. *Food Chem.* **2000**, *70*, 181–184. [CrossRef]
21. Hansen, C.E.; del Olmo, M.; Burri, C. Enzyme activities in cocoa beans during fermentation. *J. Sci Food Agric.* **1998**, *77*, 273–281. [CrossRef]
22. Pérez-Jiménez, J.; Torres, J.L. Analysis of nonextractable phenolic compounds in foods: The current state of the art. *J. Sci Food Agric.* **2011**, *59*, 12713–12724. [CrossRef] [PubMed]
23. Righini, S.; Rodriguez, E.J.; Berosich, C.; Grotewold, E.; Casati, P.; Ferreyra, M.L.F. Apigenin produced by maize flavone synthase I and II protects plants against UV-B-induced damage. *Plant. Cell Environ.* **2019**, *42*, 495–508. [CrossRef]
24. Bai, Y.; Findlay, B.; Maldonado, A.F.S.; Schieber, A.; Vederas, J.C.; Ganzle, M.G. Novel pyrano and vinylphenol adducts of deoxyanthocyanidins in sorghum sourdough. *J. Agric. Food Chem.* **2014**, *62*, 11536–11546. [CrossRef]
25. Huynh, N.T.; Van Camp, J.; Smagghe, G.; Raes, K. Improved release and metabolism of flavonoids by steered fermentation processes: A review. *Int. J. Mol. Sci.* **2014**, *15*, 19369–19388. [CrossRef] [PubMed]
26. Ali, F.; Rahul; Naz, F.; Jyoti, S.; Siddique, Y.H. Health functionality of apigenin: A review. *Int. J. Food Prop.* **2017**, *20*, 1197–1238. [CrossRef]
27. Oladeji, B.S.; Akanbi, C.T.; Gbadamosi, S.O. Effects of fermentation on antioxidant properties of flours of a normal endosperm and quality protein maize varieties. *J. Food Measur. Character.* **2017**, *11*, 1148–1158. [CrossRef]
28. Yang, L.; Allred, K.; Dykes, L.; Allred, C.; Awika, J.M. Enhanced action of apigenin and naringenin combination on estrogen receptor activation in non-malignant colonocytes: Implications on sorghum-derived phytoestrogens. *Food Funct.* **2015**, *6*, 749–755. [CrossRef] [PubMed]
29. De Beer, D.; Tobin, J.; Walczak, B.; Van der Rijst, M.; Joubert, E. Phenolic composition of rooibos changes during simulated fermentation: Effect of endogenous enzymes and fermentation temperature on reaction kinetics. *Food Res. Int.* **2019**, *121*, 185–196. [CrossRef] [PubMed]
30. Karaaslan, M.; Yilmaz, F.M.; Cesur, Ö.; Vardin, H.; Ikinci, A.; Dalgiç, A.C. Drying kinetics and thermal degradation of phenolic compounds and anthocyanins in pomegranate arils dried under vacuum conditions. *Int. J. Food Sci. Technol.* **2014**, *49*, 595–605. [CrossRef]
31. Turturică, M.; Stănciuc, N.; Bahrim, G.; Râpeanu, G. Effect of thermal treatment on phenolic compounds from plum (*Prunus domestica*) extracts–A kinetic study. *J. Food Eng.* **2016**, *171*, 200–207. [CrossRef]
32. Rękas, A.; Ścibisz, I.; Siger, A.; Wroniak, M. The effect of microwave pretreatment of seeds on the stability and degradation kinetics of phenolic compounds in rapeseed oil during long-term storage. *Food Chem.* **2017**, *222*, 43–52. [CrossRef]
33. Ali, A.; Chong, C.H.; Mah, S.H.; Abdullah, L.C.; Choong, T.S.Y.; Chua, B.L. Impact of storage conditions on the stability of predominant phenolic constituents and antioxidant activity of dried piper betle extracts. *Molecules* **2018**, *23*, 484. [CrossRef]
34. Oyedeji, A.B.; Sobukola, O.P.; Henshaw, F.O.; Adegunwa, M.O.; Sanni, L.O.; Tomlins, K.I. Kinetics of mass transfer during deep fat frying of yellow fleshed cassava root slices. *Heat Mass Transf.* **2016**, *52*, 1061–1070. [CrossRef]
35. Oyedeji, A.B.; Sobukola, O.A.; Green, E.; Adebo, O.A. Physical properties and water absorption kinetics of three varieties of Mucuna beans. *Sci. Rep.* **2021**, *11*, 5450. [CrossRef]
36. Moore, J.; Cheng, Z.; Hao, J.; Guo, G.; Liu, J.-G.; Lin, C.; Yu, L. Effects of solid-state yeast treatment on the antioxidant properties and protein and fiber compositions of common hard wheat bran. *J. Agric. Food Chem.* **2007**, *55*, 10173–10182. [CrossRef]
37. Starzyńska-Janiszewska, A.; Stodolaka, B.; Gómez- Caravaca, A.M.; Mickowska, B.; Martin-Garcia, B.; Byczyńskia, L. Mould starter selection for extended solid-state fermentation of quinoa. *LWT Food Sci. Technol.* **2019**, *99*, 213–237. [CrossRef]
38. Svensson, L.; Sekwati-Monang, B.; Lutz, D.L.; Schieber, A.; Gänzle, M.G. Phenolic acids and flavonoids in nonfermented and fermented red sorghum (*Sorghum bicolor* (L.) Moench). *J. Agric. Food Chem.* **2010**, *58*, 9214–9220. [CrossRef] [PubMed]
39. Hole, A.S.; Rud, I.; Grimmer, S.; Sigl, S.; Narvhus, J.; Sahlstrøm, S. Improved bioavailability of dietary phenolic acids in whole grain barley and oat groat following fermentation with probiotic *Lactobacillus acidophilus*, *Lactobacillus johnsonii*, and *Lactobacillus reuteri*. *J. Agric. Food Chem.* **2012**, *60*, 6369–6375. [CrossRef] [PubMed]
40. Starzyńska-Janiszewska, A.; Stodolak, B.; Socha, R.; Mickowska, B.; Wywrocka-Gurgul, A. Spelt wheat *tempe* as a value-added whole-grain food product. *LWT Food Sci. Technol.* **2019**, *113*, 108250. [CrossRef]
41. Zhang, L.; Gao, W.; Chen, X.; Wang, H. The effect of bioprocessing on the phenolic acid composition and antioxidant activity of wheat bran. *Cereal Chem.* **2014**, *91*, 255–261. [CrossRef]
42. Kadiri, O. A review on the status of the phenolic compounds and antioxidant capacity of the flour: Effects of cereal processing. *Int. J. Food Prop.* **2017**, *20*, S798–S809. [CrossRef]
43. Skrajda-Brdak, M.; Konopka, I.; Tańska, M.; Czaplicki, S. Changes in the content of free phenolic acids and antioxidative capacity of wholemeal bread in relation to cereal species and fermentation type. *Eur. Food Res. Technol.* **2019**, *245*, 2247–2256. [CrossRef]
44. Vernhet, A.; Carrillo, S.; Rattier, A.; Verbaere, A.; Cheynier, V.; Nguela, J.M. Fate of anthocyanins and proanthocyanidins during the alcoholic fermentation of thermovinified red musts by different *Saccharomyces cerevisiae* strains. *J. Agric. Food Chem.* **2020**, *68*, 3615–3625. [CrossRef]
45. Aguilar, N.; Albanell, E.; Miñarro, B.; Capellas, M. Chestnut flour sourdough for gluten-free bread making. *Eur. Food Res. Technol.* **2016**, *242*, 1795–1802. [CrossRef]

46. Xiang, J.; Apea-Bah, F.B.; Ndolo, V.U.; Katundu, M.C.; Beta, T. Profile of phenolic compounds and antioxidant activity of finger millet varieties. *Food Chem.* **2019**, *275*, 361–368. [CrossRef] [PubMed]
47. Ravisankar, S.; Abegaz, K.; Awika, J.M. Structural profile of soluble and bound phenolic compounds in teff (*Eragrostis tef*) reveals abundance of distinctly different flavones in white and brown varieties. *Food Chem.* **2018**, 265–274. [CrossRef] [PubMed]

Article

Antioxidant, Antimicrobial and Metmyoglobin Reducing Activity of Artichoke (*Cynara scolymus*) Powder Extract-Added Minced Meat during Frozen Storage

Tuğba Demir * and Sema Ağaoğlu

Department of Food Hygiene and Technology, Faculty of Veterinary Medicine, Sivas Cumhuriyet University, Sivas 58140, Turkey; sagaoglu@cumhuriyet.edu.tr
* Correspondence: tugba@cumhuriyet.edu.tr; Tel.: +90-346-219-1010-3618; Fax: +90-346-219-1812

Abstract: The present study aimed to investigate the bioactive compounds in artichoke (*Cynara scolymus*) powder, having antioxidant and antimicrobial activity, and to determine the effectiveness of artichoke (*C. scolymus*) powder extract within the minced meat. *C. scolymus* was extracted using two different methods. The method incorporating high phenolic and flavonoid content levels was used in other analyses and the phenolic and flavonoid contents in *C. scolymus* extract was determined using LC-QTOF-MS. Antioxidant, antimicrobial, and metmyoglobin (metMb) reducing activities and pH values of the extract-added minced meat samples were measured for 10 days during storage. DPPH, FRAP, and ABTS were used in the antioxidant analyses. The antimicrobial activity of *C. scolymus* extract was evaluated on five different food pathogens by using the disc diffusion method. The most resistant bacterium was found to be *Listeria monocytogenes* (18.05 mm $\pm$ 0.24). The amount of metMb was measured in the minced meat sample that was added to the extract during storage ($p < 0.05$). MetMb formation and pH value on the sixth day of storage were found to be at lower levels than in the control group. In conclusion, *C. scolymus* exhibited a good antimicrobial and antioxidant effect and can be used in storing and packaging the food products, especially the meat and meat products.

Keywords: meat quality; antioxidant activity; antimicrobial activity; functional food; *C. scolymus*; food quality

Citation: Demir, T.; Ağaoğlu, S. Antioxidant, Antimicrobial and Metmyoglobin Reducing Activity of Artichoke (*Cynara scolymus*) Powder Extract-Added Minced Meat during Frozen Storage. *Molecules* **2021**, *26*, 5494. https://doi.org/10.3390/molecules26185494

Academic Editor: Mirella Nardini

Received: 12 August 2021
Accepted: 6 September 2021
Published: 9 September 2021

Publisher's Note: MDPI stays neutral with regard to jurisdictional claims in published maps and institutional affiliations.

1. Introduction

Meat is one of the foods with the highest protein content and provides essential amino acids needed in human nutrition. Biochemically, meat contains proteins, essential amino acids, water, and low amounts of minerals, vitamins and carbohydrates [1]. The oxidative degradation of lipids during processing and storage affects the quality characteristics of meat and meat products. Primary and secondary oxidative decomposition products reduce the nutritional quality of meat, and they create an important health risk [2].

As a result of the advancing technology and increasing fast-food consumption, natural or synthetic antioxidants are used in order to expand the shelf life and enhance the quality properties of meat and meat products.

Antioxidants are defined as the compounds that are capable of binding the free radicals to give hydrogen (H·) radicals in order to prevent the oxidation reaction [3].

Antioxidative agents delay the oxidative degradation of lipids, increase the quality, and maintain the nutritional value. These substances prevent cell damage and tumor formation by neutralizing free radicals [4]. In recent years, synthetic antioxidants in meat and meat products were aimed at increasing the preservation time and improving the quality [3]. However, there also are safety concerns about the use of synthetic antioxidants. Consumers consider the natural antioxidants more acceptable than the synthetic ones. Natural antioxidant agents delay the lipid oxidation reaction and the quality and shelf life can preserve without damaging the nutritional value of the meat [5].

Meat and meat products offer a suitable environment for the growth and propagation of bacterial spoilage and foodborne pathogens. Meat and meat products can be contaminated by pathogenic bacteria in any of the steps between the "farm and the fork", such as processing, packaging, and pre-cooking storages [6]. For this reason, hygiene and preservation methods are reinforced using natural bioactive compounds in the meat industry [7].

The antimicrobial action mechanisms of plant components differ from those of synthetic antimicrobials, and they inhibit the bacterial growth through a series of metabolic reactions [7,8]. Phenolic compounds commonly found in their structures show an antimicrobial effect by lowering the intracellular pH, by chelating some metals that are necessary for the survival of microorganisms, or by changing the permeability of the cell membrane and disrupting the substrate transport [9,10]. Plant-based extracts protect meat and meat products against bacteria, which cause spoilage and expand preservation time and nutritional value thanks to their phenolic and flavonoid compound contents. Natural phenolic compounds also play an effective role in taste and flavor, since they have at least one aromatic ring [11]. Previous research showed that the addition of antioxidants to fresh red meat inhibited lipid oxidation and delayed metMb formation. They reported that the antioxidants preserved the fresh meat color by preventing the lipid oxidation of the hemoproteins and/or acting on the enzymic reducing systems [12].

Artichoke (*Cynara scolymus*) is a plant that is widely grown and consumed throughout the world. *C. scolymus* has a strong potential in terms of scavenging the reactive oxygen species and free radicals, which contain many natural compounds. In many studies, it was emphasized that the caffeoylquinic acid derivatives and luteolin and apigenin glycosides found in the structure of *C. scolymus* showed strong antioxidant effects [13]. There also are many studies underlining the antibacterial and antioxidant properties of *C. scolymus* extract. The bioactive property of artichoke is related to its high-level luteolin and chlorogenic acid contents [14]. The main phenolics of *C. scolymus* are cinnamic acid derivatives including caffeic acid, cynarin, 1,5-*O*-di-o-cafeoylquinic acid and 3,4-*O*-di-o-cafeoylquinic acid and 1,5-*O*-di-o-cafeoylquinic acid, whereas the major flavonoids were reported to be 4′,5,7-trihydroxyflavone (apigenin), 3′,4′,5,7-tetrahydroxyflavone (luteolin), and glycosidic derivatives [15]. There are limited studies on marinating meat and meat products with *C. scolymus*. The present study aims to extract and characterize the total phenolic and total flavonoid compounds from *C. scolymus* powder applied to minced meat in order to improve quality characteristics during the storage process. For this purpose, antioxidant, antimicrobial, and metMb reducing activities of artichoke powder extract added into minced meat were investigated.

2. Materials and Methods

2.1. Materials

Artichoke flowers (*C. scolymus*) were obtained from an organic market, ground and stored at +4 °C until analyzed. Minced beef was obtained from a local butcher in Sivas, Turkey. Mueller–Hinton Broth (MHB) and Mueller–Hinton Agar (MHA) were obtained from Merck (Merck KGaA, Darmstadt, Germany). The microorganisms (*Enterococcus faecalis* (ATCC 029212), *Escherichia coli* (ATCC 025922), *Staphylococcus aureus* (ATCC 029213), *Listeria monocytogenes* (ATCC 07644) and *Salmonella typhimurium* (ATCC 014028)) were provided from the Microbiology Laboratory of Sivas Cumhuriyet University Research Hospital. The ampicillin disc was obtained from Oxoid (Oxoid Ltd., Thermo Fisher Scientific, Inc., Basingstoke, Hampshire, United Kingdom). The phenolic compounds were purchased from Sigma-Aldrich (St. Louis, MO, USA). All other chemicals used are of analytical standard and obtained from Merck (MerckKGaA, Darmstadt, Germany) or Sigma-Aldrich (USA).

2.2. Preparation of the C. scolymus Extract

We used mature artichoke flowers (*C. scolymus*) in an organic market from the Ege Region of Turkey. The drying process was carried out in a lyophilizer for 24 h in the following conditions: pressure 0.945 mbar, initial temperature $-30\,°C$, final temperature $+30\,°C$. After drying, the samples were ground in a laboratory mill to a fine powder and stored vacuum-packed in a freezer at $-80\,°C$ until analysis. Powdered samples (ratio; 1:3 g/mL water) were extracted for 24 h in the Soxhlet extractor (30 g plant: 90 mL distilled water). This group was filtered and used directly (Method 1). In another group, the extract was concentrated by a rotary evaporator under a low pressure (174 mbar) and controlled temperature (40–50 $°C$) for 4 h (Method 2). The working condition with the highest phenolic and flavonoid content (Method 2) was chosen for chromatographic analysis. All analysis were performed in triplicate. All methods were performed in accordance with the relevant guidelines and regulations (institutional, national, and international guidelines and legislation for use of plant material).

2.3. Determination of Total Phenolic and Flavonoid in the Extract

The total phenolic content of the extract was determined according to the method described using the Folin–Ciocalteu phenol reagent [16]. The quantitative determination of total phenolic content of extract was obtained by mixing 100 μL of 2 N Folin–Ciocalteu's phenol reagent with 100 μL of extract/100 μL of the gallic acid solution 1 mL of 7% sodium carbonate, and 2.3 mL of water. A standard curve was created with gallic acid $(0-0.5$ mg/mL). Then, (25 $°C$, 2 h), the absorbance at 750 nm was measured in the spectrophotometer. The determination of the total amount of the flavonoid of the extracts were applied according to the method provided by Zhishen et al. [17]. The extract was reacted with in order of 5% $NaNO_2$, 10% $AlCl_3$ and NaOH (1 M). Quercetin was used as a standard to determine the total flavonoid content of the extracts. A standard curve was created with quercetin $(0-100$ mg/L) and the absorbance was measured at 510 nm. Total phenolic compounds were expressed as gallic acid equivalent (GAE), and total flavonoid compounds as quercetin equivalent (QE). All analysis were performed in triplicate.

2.4. Characterization of Phytochemical Composition with LC/QTOF-MS

The phenolic and flavonoid contents in extract of *C. scolymus* was characterized with LC-QTOF-MS. Phytochemical composition was defined using an Agilent Technologies 6530 OHD Accurate-Mass Q-TOF-MS&MS system (Agilent Technologies, Santa Clara, CA, USA) accoutered with the Agilent 1290UPLC system and an electrospray ionization source. For the characterization of phytochemical compounds, many parameters were analyzed and applied in combination (ion, molecular weight, fragmentation pattern, retention time, LOD and LOQ). Method calibration and validation parameters for analysis of the phenolic standards are presented in Table 2 and compared with reference data. Characterizations of components were made using formic acid (0.1%) and acetonitrile (0.1%) as mobile phase. The flow rate was 0.5 mL/min while the sample injection volume was 0.5 μL. Analysis were made on the programmed MS/MS system. The collision energy conditions are given below. Conditions; $m/z\ 0-200$, 20 eV, $(0-10)$; $m/z \geq 200$, 3 eV, $(10-20)$; $m/z \geq 400$, 40 eV, $(20-30)$; $m/z\ 600-700$, 50 eV, $(30-40)$; $m/z \geq 700$, 60 eV, $(40-50)$. Finally, MS data were operated with software. MS spectra of the compounds were identified by comparison with standards [18].

2.5. Preparation of Meat Samples

Beef rounds (*Semimembranosus* muscle) from different steers ($n = 4$) were purchased from a local butcher (Sivas, Turkey) at 3 days postmortem. The lean beef round was ground using a meat grinder through a plate with an 8 mm steel plate twice. After mixing, the minced meat round was divided into 2 batches (approximately 1 kg/batch) for study. The minced meat without fat was divided into two groups which are for control and treatment. Twenty-four samples (two groups) of 30 g of raw minced meat were placed in plastic bags. Twenty samples of 30 g of minced meat were added with the extract. Other samples (4 samples) were used as negative controls (not treated with *C. scolymus*). Minced meat in the treatment groups were immersed in 8% (v/w) of extract [19]. Treatment and control groups were stored ($-18\ ^{\circ}$C) before being ready to be analyzed for a storage period (day 0, day 3, day 6, day 10). Each test was performed in triplicate.

2.6. Antimicrobial Activity

Antimicrobial activity of *C. scolymus* extract was defined by the agar diffusion method [20]. *E. coli, S. aureus, L. monocytogenes, S. typhimurium and E. faecalis* pathogens used. The fresh inoculum of bacteria (10^5 cfu/mL) was prepared in sterile-saline water (2 mL). The turbidity of inoculum suspensions were set to a 0.5 McFarland standard. Sterile antimicrobial discs (6 mm) were impregnated with 30 μL of *C. scolymus* extract, followed by waiting until the disk absorbed the extract. Distilled water was used as the negative control, and ampicillin as the positive control. Finally, all treatment plates were left for 10 min at 25 $^{\circ}$C to allow the diffusion of the plant extract, and all plates were last-incubated at 37 $^{\circ}$C for 24 h. The diameters of the inhibition zones were measured after the incubation period. Measurements were conducted in triplicate.

2.7. Preparation of Minced Meat Agar-Solution

Peptone water (90 mL) was added (buffered) to 10 g of minced meat. The prepared mix was vortexed and blended for 5 min. It was blended until no particles were left and smoothed. After that, the solution was centrifuged ($5000\times g$, 10 min) to obtain the clarified extract by removed the solid particle. The final solid media with agar (1.5%) was prepared until 1 L. Then, prepared meat agar solution were autoclaved (121 $^{\circ}$C, 15 min) [18,21].

2.8. Antioxidant Activity

2.8.1. DPPH: 2,2 Diphenyl-1-picrylhydrazyl Radical Scavenging Activity

Free radical scavenging activity was evaluated by the DPPH assay using the standard method [22]. Minced meat with added extract (0.05 g) was continually mixed with 3 mL of DPPH working solution (1.95 mL, 100 μM) that has been prepared using methanol in a test tube at 25 $^{\circ}$C for 10 min. Then, all treatments were centrifuged ($1500\times g$, 10 min). Trolox solution was used as standard. By using methanol as blank, absorbance of the supernatants was measured at 517 nm. All results were expressed as μmol "Trolox Equivalent" (TE) after constructing a TE standard curve. Measurements were performed in triplicate.

2.8.2. TEAC Trolox Equivalent Antioxidant Capacity

Antioxidant activity of the samples was also measured using an improved ABTS procedure [18]. The ABTS radical cation (ABTS+) solution (7 mM ABTS, 2.45 mM) was prepared through the reaction potassium persulphate, then the pre-incubation in the dark for 20 h at 25 $^{\circ}$C. The ABTS+ solution was then diluted with 80% ethanol to obtain an absorbance of 0.700 ± 0.005 at 734 nm. Then, 2.9 mL of ABTS working solution at absorbance 700 nm was added to 0.05 g of the treatments and the solution was mixed strongly. This working mixture was stored at 30 $^{\circ}$C for 25 min, then centrifugated at $1500\times g$ for 10 min. Trolox solution was used as standard. The supernatant's absorbance was measured at 734 nm. All results were expressed as μmol "Trolox Equivalent" (TE) after constructing a TE standard curve. Measurements were performed in triplicate.

2.8.3. FRAP: Ferric Reducing Antioxidant Power

Minced meat with 0.05 g was mixed with 2.5 mL of 200 mM sodium phosphate buffer (pH 6.6) and 2.5 mL of 1% potassium ferric cyanide $\{K_4[Fe\,(CN)_6]\}$ and incubated at 45 $^{\circ}$C for 30 min. Then, 2.5 mL of 10% trichloroacetic acid (*w/v*) was added on mix. The mixture was centrifuged at $1500\times g$ for 10 min and deionized water was added with an equal volume of resulting supernatant and $1/5$ volume of 0.1% $FeCl_3$ and stored at 25 $^{\circ}$C for 10 min. Absorbance of mix-solution was read at 700 nm [21]. Measurements were performed in triplicate.

2.9. Measurement of metMb Reducing Activity

Firstly, 0.5 g minced meat was homogenized with added 3 mL phosphate buffer (0.04 M; pH 6.8), and was stored at a cooled temperature (4 °C). The homogenate was centrifuged at 2000× g for 10 min and left to pre-incubate at 4 °C for 1 h. MetMb accumulation in minced meat was evaluated by spectrophotometrically as defined by Huang et al. [19]. The homogenate was centrifuged at 2500× g for 5 min at 4 °C and the supernatant filtered through Whatman No. 1 filter paper to remove fat. The standard mixture contained 0.1 mL EDTA (5 Mm), 0.1 mL phosphate buffer (50 mM; pH 7.0), 0.1 mL [K_4 Fe(CN)$_6$]; 3.0 mM, 0.1 mL water, and 0.2 mL 0.75 metMb Fe(III) in 2.0 mM phosphate buffer (pH 7.0). MetMb reducing activity was calculated from a calibration curve using standard mixture. Finally, metMb reducing activity was estimated from the absorbance at 580 nm. Measurements were performed in triplicate.

2.10. Instrumental Color Measurements

Instrumental color analysis was executed using a Hunterlab colorimeter (Mini Scan XEPlus, Virginia, WV, USA). Before each application, the colorimeter was calibrated on the CIE (Commission internationale de l'éclairage) color space scheme using a tile (black and white). The L* value designates lightness (L* = 0–100; darkness–lightness); a* value designates redness ((+) 60; red, (−) 60; green)) and b* value designates yellowness ((+) 60; yellow, (−) 60; blue). Color measurements were taken at 4 °C with illuminant D65 and a 0° angle observer. All measurements were taken on the outer surface of minced meat from randomly chosen selected locations [23]. Measurements were performed in triplicate.

2.11. pH Determination

For the pH value measurement, a 1.00 g sample was homogenized by adding 5 mL sterile-distilled water and the homogenate was centrifuged at 2000× g for 10 min. The supernatant was filtered through Whatman No. 1 filter paper, and the pH of the supernatant recorded using a pH meter (Hanna Edge, Hanna Instruments, Woonsocket, RI, USA).

2.12. Statistical Analysis

All the tested sample data (mean values) were statistically analyzed with the SPSS analysis of variance (SPSS version 19.0 software, SPSS; Chicago, IL, USA). A Duncan's multiple range test was used for the study of studentized range distribution in order to determine critical values for comparisons between means. The significance for all comparisons were determined at the $p < 0.05$ level.

3. Results and Discussion

3.1. Total Phenolic and Flavonoid Analysis in Extracts

Phenolic compounds are considered to be natural sources of antioxidants required by metabolism, as well as their antioxidant activities; they are demonstrate by binding free radicals or chelating with metals [2]. These effects increase with the increase in the number of OH groups in the phenol ring they contain in their structure. Moreover, phenolic compounds have the ability to delay, slow, or prevent the oxidation at low concentrations and to remain in a stable form when converted to free radicals [11]. In the present study, the phenolic compounds were calculated using standard gallic acid, and total flavonoid content was measured using standard quercetin. Quercetin and derivatives are available in various foods and plant-based products. Quercetin and gallic acid were reported to play effective roles against oxidative stress. Quercetin is among the compounds that play a role in the inhibition of lipoxygenase that is responsible for inflammation [9].

Total phenolic and total flavonoid contents of *C. scolymus* powder extracts, which were extracted using one of the extraction methods, are presented in Table 1. Calculations were made on the standard curve formed by the measured absorbance values (y = 1.8902x + 0.0128, R^2 = 0.99; y = 0.0039x − 0.003, R^2 = 0.99). The results obtained using evaporator were higher in terms of both phenolic and flavonoid contents. In the literature, the amount of bioactive components passing into the solvent in concentrated extracts was reported to be high [10].

Table 1. Total phenolic and total flavonoid compounds.

Method of Extract.	Total Phenolic Compounds (mg GAE/g extract)	Total Flavonoid Compounds (mg QE/g extract)
Method 1	17.37 ± 0.07 [aB]	6.45 ± 0.03 [bB]
Method 2	98.26 ± 0.05 [aA]	19.74 ± 0.04 [bA]

[a,b]: Means followed by different letters within the same line represent significant differences ($p < 0.05$). [A,B]: Means followed by different letters within the same column represent significant differences. Data are the average of triplicates.

A past study about total phenolic content in ethanol extracts of the artichoke demonstrated that the extracts were rich in total phenolic (32.2 mg GAE/g) [24,25]. Curadi et al. [26] found that total phenolic content in artichoke byproducts were found to be between 7.31 and 13.05 mg/g in the methanol extracts as chlorogenic acid equivalent. In this other study, the highest phenolic content of artichoke extract was determined as 4.39 mg GAE/100 g extract [23]. In another study, total phenolics were found to be between 1.60−9.80% of chlorogenic acid, equivalent in different species of artichoke [27]. When all the findings are interpreted together with the results of this study, our study are parallel to the literature studies.

Gallic acid is a phenolic acid and it belongs to the hydroxyl-benzoic acid group. It has a high level of antioxidant and antimicrobial activity. Gallic acid is a natural antioxidant substance that can be isolated from plants and is used in a wide range of medications and cosmetic products, as well as usage in the food industry [9]. Gallic acid can be used as a food additive, especially in food preservation, in food technology. The bioactive role of this phenolic acid is to prevent rancidity in fats and oils. Many studies examine natural food additives [4]. Comparing the antioxidant capacities of natural and synthetic antioxidants, gallic acid showed properties close to natural antioxidants (BHT, BHA) [6]. For this reason, gallic acid and quercetin were used as standards in determining total phenolic and total flavonoid compounds, and the results were calculated as gallic acid and quercetin equivalents. As can be seen in Table 1, the extraction method affects the amounts of total phenolic and total flavonoid substances. Extraction method, method validation, and extraction under optimum conditions are among the purposes of obtaining the highest amount of phenolic compounds. In the present study, the extract obtained using a rotary evaporator was used in all the experiments below.

3.2. Identification and Quantification of Polyphenols by LC-QTOF-MS

LC-QTOF-MS analysis was performed to specifically identify and characterize phenolic and flavonoid compounds found in *C. scolymus* extracts. In addition, the bioactive components of the extract were determined quantitatively and using standards (Figure 1). Figure 1 shows LC-QTOF-MS photodiode array detector analysis spectra (Figure 1a) and MS spectra analysis of *C. scolymus* powder extracts (Figure 1b). Table 2 shows method calibration and validation parameters for analysis of the phenolic standards, and Figure 2 illustrates the correlation matrix of the phenolic compound compositions. The linearity range as well as the slope and the intercept of calibration graph with the respective square of correlation coefficient (R^2) was defined for each of the 19 phenolic compounds (Table 2). The R^2 value of each phenolic compound analyzed was found to be higher than 0.9964, and this result indicates the high correlation of data in the concentration range analyzed here. The use of higher concentrations for the chosen phenolic compounds significantly decreased their linearity. The correlation matrix of the phenolic compound composition obtained is presented in Figure 2. Pearson's correlation analysis was performed to determine the direction and level of the relationship between phenolic compounds. A correlation coefficient between the variables close to −1 indicates a strong negative relationship, whereas those closer to +1 indicate a strong positive relationship. Values closer to zero indicate that there is no significant relationship.

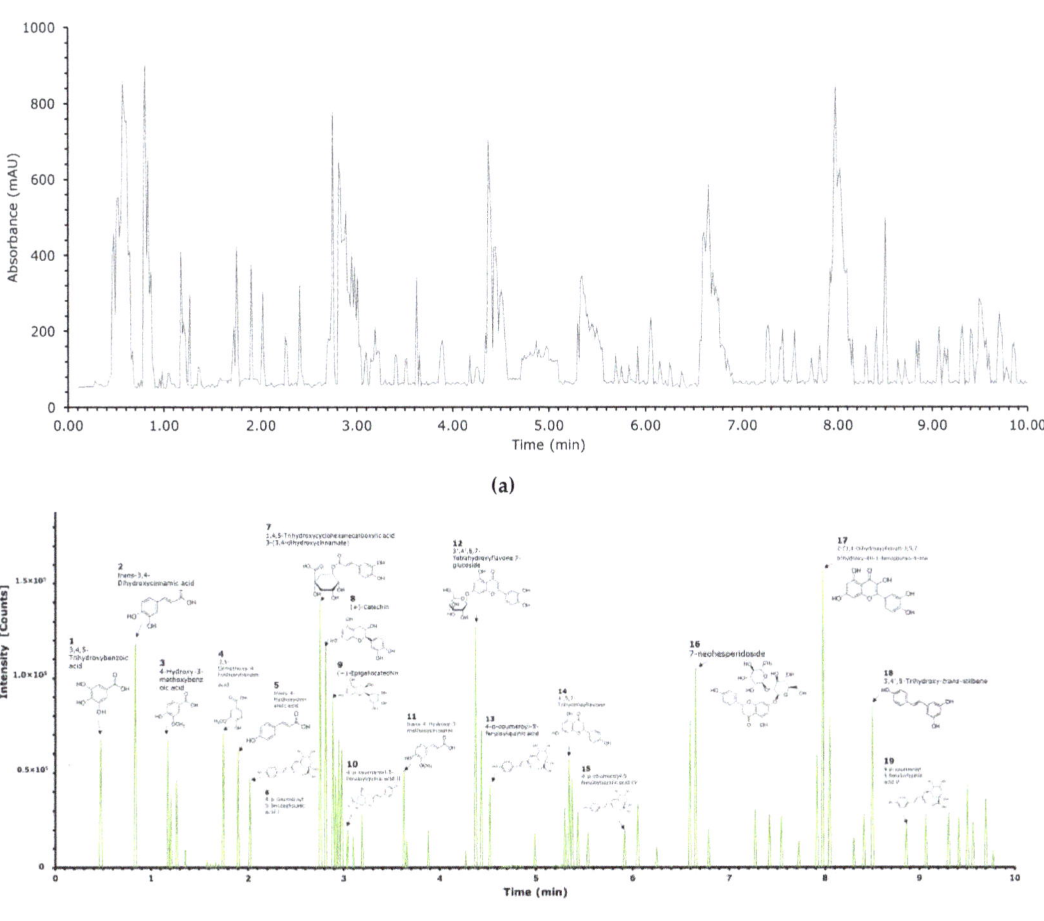

Figure 1. LC-QTOF-MS (**a**) photodiode array detector analysis spectra (**b**) MS spectra analysis of *C. scolymus* powder extracts.

Table 2. Method calibration and validation parameters for analysis of the phenolic standards.

Peak No	Phenolic Compounds	R_t (min)	Observed m/z	Recovery (%)	Rec.RSD (%)	R^2	LOD (µg/mL)	LOQ (µg/mL)
1	3,4,5-Trihydroxybenzoic acid	0.475	400.1022	90.28 ± 2.24	1.76	0.9967	0.07	0.18
2	*trans*-3,4-Dihydroxycinnamic acid	0.850	455.1171	88.27 ± 4.37	3.42	0.9996	0.11	0.35
3	4-Hydroxy-3-methoxybenzoic acid	1.18	388.2636	90.51 ± 2.10	1.63	0.9985	0.06	0.13
4	3,5-Dimethoxy-4-hydroxybenzoic acid	1.75	392.2331	90.38 ± 2.16	1.69	0.9967	0.06	0.16
5	*trans*-4-Hydroxycinnamic acid	1.9	371.1338	90.53 ± 2.03	1.58	0.9988	0.06	0.13
6	p-Coumaroyl-*O*-feruloylquinic acid I	2.02	266.2013	90.99 ± 0.36	0.27	0.9978	0.005	0.01
7	1,4,5-Trihydroxycyclohexanecarboxylic acid 3-(3,4-dihydroxycinnamate)	2.75	481.1711	85.18 ± 6.27	4.92	0.9994	0.15	0.51
8	(+)-Catechin	2.83	461.1035	87.32 ± 4.90	3.85	0.9991	0.12	0.42
9	(−)-Epigallocatechin	2.93	423.1521	89.39 ± 3.33	2.61	0.9967	0.08	0.27
10	p-Coumaroyl-*O*-feruloylquinic acid II	3.04	362.2082	90.70 ± 1.49	1.17	0.9980	0.05	0.11
11	*trans*-4-Hydroxy-3-methoxycinnamic acid	3.62	298.1291	90.83 ± 1.30	1.00	0.9985	0.03	0.07
12	*trans*-4-Hydroxy-3-methoxycinnamic acid	4.33	466.1588	85.96 ± 5.88	4.61	0.9964	0.13	0.48
13	p-Coumaroyl-*O*-feruloylquinic acid III	4.52	403.1022	90.04 ± 2.67	2.09	0.9985	0.07	0.20
14	4′,5,7-Trihydroxyflavone	5.36	321.3371	90.77 ± 1.41	1.09	0.9980	0.04	0.09
15	p-Coumaroyl-*O*-feruloylquinic acid IV	5.92	286.3122	90.90 ± 0.88	0.68	0.9966	0.01	0.06
16	7-neohesperidoside	6.65	448.1822	88.53 ± 3.94	3.11	0.9973	0.10	0.35
17	2-(3,4-Dihydroxyphenyl)-3,5,7-trihydroxy-4H-1-benzopyran-4-one	7.97	488.1088	84.95 ± 6.52	5.11	0.9973	0.15	0.52
18	3,4′,5-Trihydroxy-*trans*-stilbene	8.5	411.2633	89.74 ± 3.20	2.49	0.9988	0.08	0.22
19	p-Coumaroyl-*O*-feruloylquinic acid V	8.84	273.1023	90.95 ± 0.63	0.49	0.9985	0.01	0.04

LOD: limit of detection; LOQ: limit of quantification; Rec.: recovery; RSD: relative standard deviation.

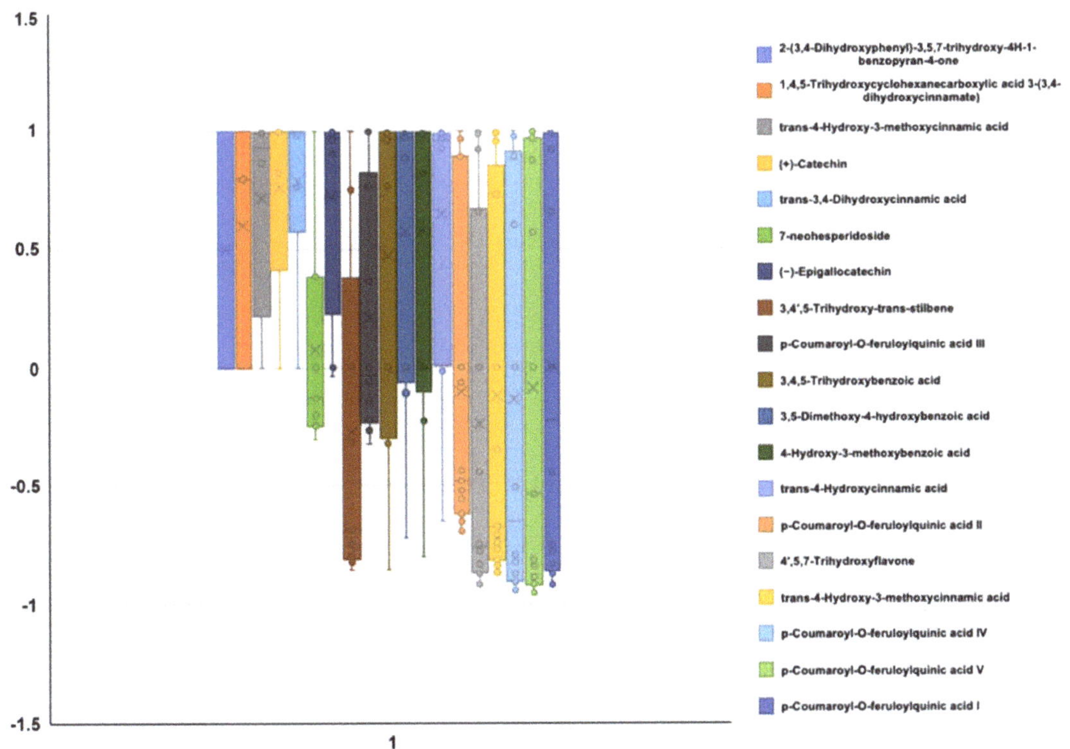

Figure 2. Correlation matrix of the obtained phenolic compound compositions.

In recent years, LC-QTOF-MS chromatography can be used in many fields, especially for identifying the bioactive compounds [27]. When the analysis results were evaluated, a total of nineteen components were identified and their chemical formulations were presented (Figure 1). Quercetin [2-(3,4-Dihydroxyphenyl)-3,5,7-trihydroxy-4H-1-benzopyran-4-one] was found at the highest density (11.63%). In addition, the following compounds were identified, respectively; chlorogenic acid [(1,4,5-Trihydroxycyclohexanecarboxylic acid 3-(3,4-dihydroxycinnamate)] (11.18%), luteolin (trans-4-Hydroxy-3-methoxycinnamic acid) (10.49%), (+)-catechin (8.75%) and caffeic acid (7.79%) (trans-3, 4-dihydroxycinnamic acid). With the addition of quercetin to meat products, the antioxidant potential has increased and lipid peroxidation and fatty acid composition were improved [28]. On the other hand, quercetin had the potential of improving the quality of meat products. This process improved storage stability and inhibited the formation of lipid oxidation products, which was associated with quercetin [29]. Another bioactive compound having a positive effect on meat quality is chlorogenic acid. This compound improves meat quality and oxidative stress [30] and can be used as an antioxidant and antimicrobial agent in meat products [9]. Besides the antioxidant properties of luteolin (trans-4-Hydroxy-3-methoxycinnamic acid), catechin and caffeic acid, which were isolated from plants, these compounds also have anti-inflammatory and anticancer activities [15]. The other compounds identified and quantified in the present study are as follows; 7-neohesperidoside (7.05%), (−)-Epigallocatechin (5.95%), 3,4',5-Trihydroxy-trans-stilbene (5.70%), 3,4,5-Trihydroxybenzoic acid (4.0%), 3,5-Dimethoxy-4-hydroxybenzoic acid (3.86%), 4-Hydroxy-3-methoxybenzoic acid (3.72%), trans-4-Hydroxycinnamic acid (3.62%), 4',5,7-Trihydroxyflavone (2.50%), trans-4-Hydroxy-3-methoxycinnamic acid (2.30%), p-Coumaroyl-O-feruloylquinic acid I-V (11.45%).

3.3. Antimicrobial Activity

The antimicrobial activity of *C. scolymus* extract was examined on five different food pathogens. In selecting the pathogens, the risk on meat and meat products was determined among the microorganisms. All the analyses were performed using the agar disk diffusion method. Two concentrations of the extract were used to measure the antimicrobial activity of the minced meat (5% and 10%). Evaluating the results, the most resistant bacterium was found to be *L. monocytogenes* (18.05 ± 0.45) (Table 3). The extract showed the highest inhibition on *S. typhimurium* (30.95 ± 0.47), followed by *E. faecalis* (25.37 ± 0.53), *S. aureus* (21.10 ± 0.92) and *E. coli* (20.07 ± 0.66). As the concentration of extract increased (*C. scolymus* (10%)), the zone of inhibition increased. Standard ampicillin disc was used as positive control and distilled was used as negative control. Results are presented in Table 3.

Table 3. Antimicrobial activity of the added *C. scolymus* powder on meat extract solution.

Treatments	Inhibition Zone (mm)				
	E. coli	*S. aureus*	*L. monocytogenes*	*S. typhimurium*	*E. faecalis*
Added + *C. scolymus* (5%)	19.74 ± 0.08 [cB]	19.75 ± 1.09 [cC]	17.32 ± 0.83 [dB]	29.33 ± 1.13 [aB]	25.20 ± 0.95 [bB]
Added + *C. scolymus* (10%)	20.07 ± 0.66 [cB]	21.10 ± 0.92 [cB]	18.05 ± 0.45 [dB]	30.95 ± 0.88 [aB]	25.37 ± 0.53 [bB]
Control(+)	29.00 ± 1.41 [bA]	30.00 ± 0.46 [bA]	27.00 ± 0.95 [cA]	32.00 ± 0.62 [aA]	31.50 ± 1.15 [aA]
Control(−)	0	0	0	0	0

Values are expressed as means $\pm$ SD. Standard disk; 6mm, control(+); ampicillin, control(−); distilled water, [a,b,c,d]: Means followed by different letters within the same line represent significant differences ($p < 0.05$). [A,B,C]: Means followed by different letters within the same column represent significant differences. Data are the average of triplicates.

There are many studies reporting that the five pathogen bacteria selected in our study cause food poisoning and foodborne infections and intoxications. Various inhibition rates were achieved in pathogens [5,6]. Stop the development of microorganisms and prevent a secondary infection are the expected properties of plant-based antimicrobial agents [31]. The antimicrobial activity of extracts can be attributed to the identified phenolic compositions. A high level of correlation was found between antimicrobial properties and phenolic and flavonoid compounds (r = 0.9678; $p \leq 0.001$). Meat and meat products are exposed to many potential risks in the post-slaughter production process. These risks include foodborne pathogenic microorganisms and environmental equipment contamination [29,32]. The meat products added with polyphenols such as luteolin-7-*O*-rutinoside, caffeic acid, quercetin and epigallocatechin-3-gallate can reduce these complications [32]. They are responsible for microbial inactivation depending on the position of the hydroxyl groups of phenolic compounds. Therefore, the more phenolic compounds the extract contains, the higher the effect will be. Thus, the amount of phenolic compound in extract plays an effective role in scavenging the food pathogens [13]. Nineteen different phenolic compounds were identified in the present study. The high-level antimicrobial activity is associated with this finding.

3.4. Antioxidant Activity

Antioxidant activity of *C. scolymus* extracts was determined by DPPH, ABTS and FRAP methods. The analysis were completed in 10 days and run in triplicate. Absorbance readings were made on days 0, 3, 6 and 10 and average values were calculated. The results are presented in Figure 3. As seen in Figure 3, the samples containing *C. scolymus* showed a statistically significant difference since the third day comparison to the control group ($p < 0.05$). The decrease in the value of absorbance indicated the increase in antioxidant activities (Figures 3 and 4). ABTS results were found to be higher than DPPH results (Figure 4). This finding might be explained by the faster electron transfer of ABTS• radical reactions when compared to DPPH. Figure 5 shows that increased absorbance of the reaction mix-solution increased the reducing power and antioxidant capacity ($p < 0.05$). Another useful and fast screening method for measuring antioxidant capacity is the FRAP method. The phenolic compounds found in *C. scolymus* extract are capable of blocking the radical chain reaction during the oxidation process and converting the free radicals into stable molecules by donating free electrons or hydrogen [13]. Various spectrometric and chromatographic studies have confirmed that quercetin and catechin are effective scavengers of singlet oxygen and peroxyl radicals [28]. The studies about *C. scolymus* extracts have reported their strong radical scavenging activities [23,33]. In the present study, the high antioxidant capacity of extracts is thought to be because of high quercetin and catechin content. A past study reported high antioxidant activity toward DPPH (8.3–49.7%), which was correlated with their polyphenol contents (1.7−9.86%) in the different artichoke samples [15]. Moreover, the other study about the artichoke byproducts showed a relatively high free radical scavenging activity and capacity to inhibit lipid peroxidation in artichoke byproducts [34]. The differences in antioxidant activities between this study and previous studies can arise from the use of different antioxidant measurement methods, different species (family) and different extraction conditions.

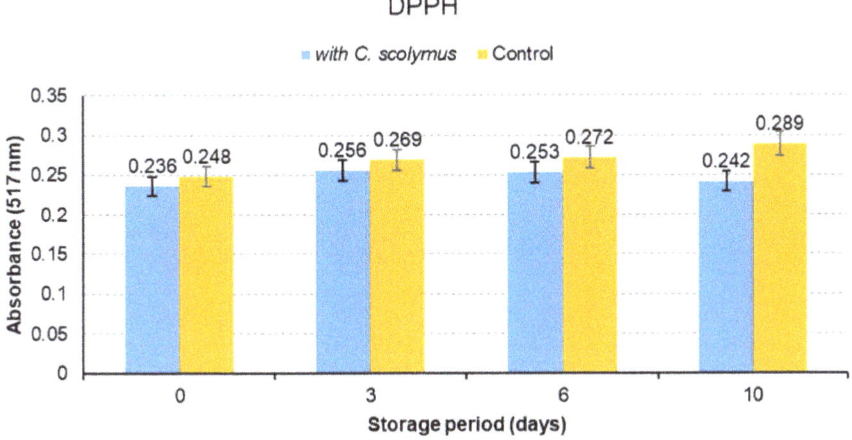

Figure 3. The antioxidant activity of control and marinated meat samples (DPPH assay).

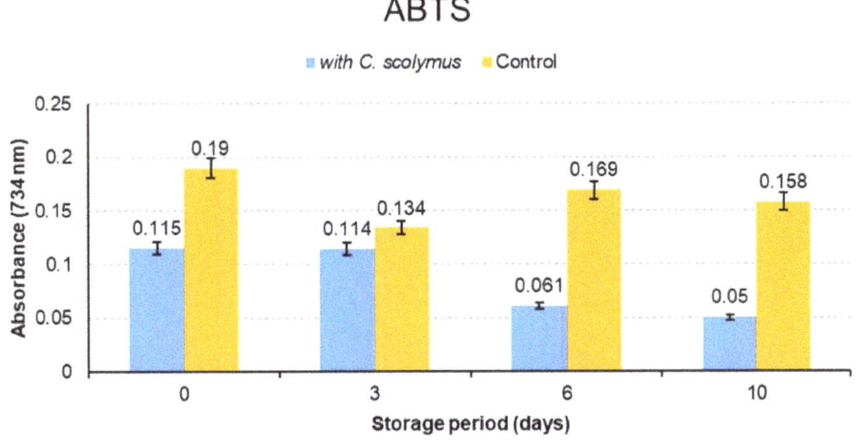

Figure 4. The antioxidant activity of control and marinated meat samples (ABTS assay).

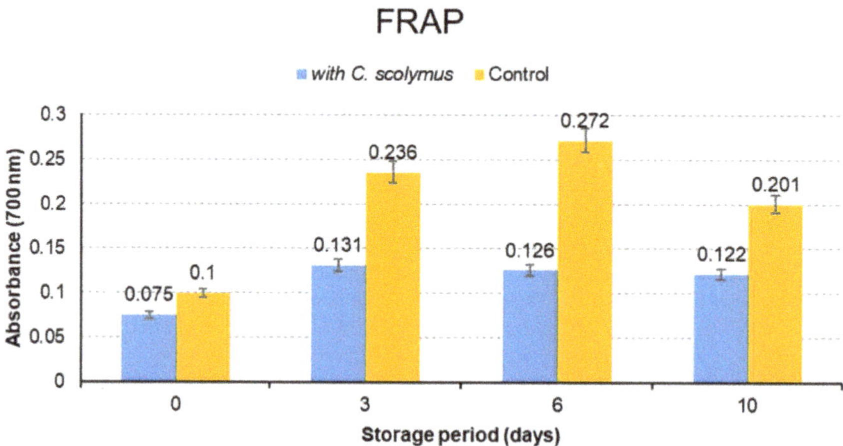

Figure 5. The antioxidant activity of control and marinated meat samples (FRAP assay).

3.5. Measurement of metMb and Color Measurements

The amount of metMb is correlated with the level of protein oxidation in the meat products. The amount of metMb increases with the storage time [18]. Two extract concentrations were used in measuring the metMb reducing activity of the minced meat (5% and 10%). Figure 6 illustrates the metMb values. As can be seen in Figure 6, the highest metMb ratios were observed in extract-added samples. Amounts of extract-added samples during storage are different compared to the control group. The visual sign of freshness and quality is the red color of meat. The reduction of red color with the formation of brown metMb occurs in parallel with the oxidative degradation of meat during storage [12]. MetMb is a pigment that is undesirable to occur on the meat surface. When an amount (60%) of the myoglobin present in meat is converted to metMb, the brown color of the meat can be detected by the eye [27]. Discoloration of meat occurs during the oxidation process because the aftereffect of lipid oxidation is the formation of pro-oxidants capable of reacting with oxymyoglobin, which leads to the formation of metMb [35]. Long-term storage at refrigeration temperatures or short-term storage at high temperatures causes surface drying; salt concentration increases and metMb formation is encouraged [18]. Liu et al. [36] reported that, compared to the control group, the treatment of beef patties with added natural antioxidants (vitamin E, carnosine, grape seed extract and tea catechins) resulted in lower metMb after eight days of storage. In this study, metMb formation was found to be lower than in the control group until the sixth day of storage (Figure 6). Oxidation inhibition was observed in the extract-added meat samples. The lower result compared to the literature can be explained by the larger surface area of minced meat compared to meatballs. This situation had an effect on the metMb formation of minced meat with natural antioxidant added.

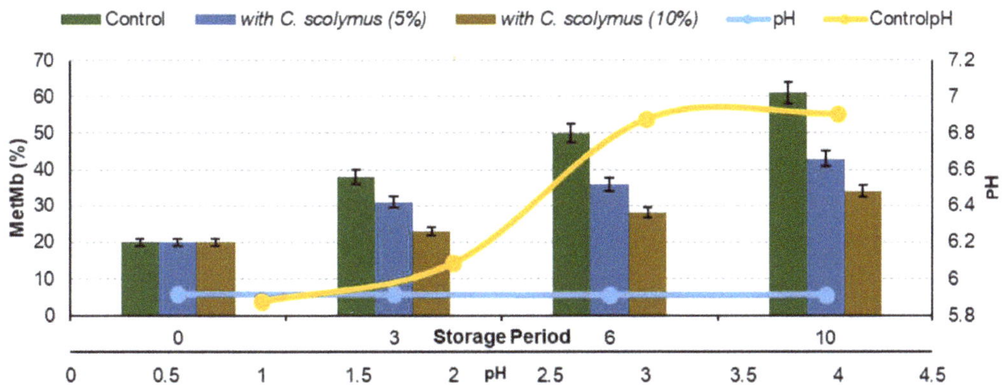

Figure 6. MetMb (%) and pH changes in meat samples marinated with *C. scolymus* powder.

Comparing the metMb and pH values, the efficiency of extract on quality was found to be significant (Figure 6). pH is effective on color, aroma and water retaining capacity. Consumers frequently associate bright-red color with beef freshness and wholesomeness. Higher than normal pH conditions are an example of a color deviation in which beef failed to have a bright-red color, leading to discounted carcasses and economic losses to the meat industry [37]. The change in pH of extract-marinated meat positively affected the quality characteristics. On the other hand, in studies on experimental model systems, the chlorogenic acid was reported to be among the effective polyphenols playing a role in metMb stability [9]. In the present study, a high rate of chlorogenic acid was found in the extract (11.18%).

Table 4 shows that effect from different references of artichoke (*C. scolymus*) powder on the quality performances on meat and meat products. Studies have had positive effects on quality at different levels in all meat products with added artichoke compared to our study. In our study, storage times of unprocessed meat products are less resistant to spoilage. When all the findings are interpreted together with the results of this study, phytochemicals can be used as a natural antioxidant agent of control during the storage period. Color plays an important role in both the quality and consumer acceptance of meat and meat products [38]. The instrumental color values were given in Table 5. The L* value was found to be slightly higher in added *C. scolymus* treatments than control samples (day zero and day three). Previous studies showed that the addition of various plant-based antioxidants did not change the L* values of meat and meat products [23]. The most important standard for the assessment of the oxidation is the "a*" value (redness), and decreased redness in meat is imagined as an index of oxidation. The a* values (redness) of minced meat ranges between 10.13 and 15.12. During the storage from the initial to the tenth day, redness of control and control samples decreased significantly ($p < 0.05$), but extract added treatments were higher than control samples and kept their redness. In a study in which plant-based antioxidants were used, including *C. scolymus* extract, decreased a* values were found, similar to our study findings. The b* values (yellowness) of minced meat ranged between 12.11 and 15.92. Significant reduction in the yellowness (b*) values were recorded in minced meat due to the added extract during storage. Different studies reported that the natural phenolic extracts showed no significant difference in L* and b* values but significant decrease in a* values during storage [23]. When the findings were compared with the literature, the color properties of the *C. scolymus* extract were found to be similar to the literature since they contained different plant-based extracts of the aforementioned compounds.

Table 4. Effect from different references of artichoke (*C. scolymus*) powder on the quality performances on meat products.

Added-Extract	Foodstuffs	Storage Period	Quality Measurements	Results	Reference
Artichoke (*C. scolymus*) powder extract	Minced Meat on the During Frozen Storage	10 days	• Antioxidant Activity (DPPH, FRAP, ABTS) • Antimicrobial Activity • metMb Reducing Activity	A potential to improve the meat quality	In this study
Artichoke (*C. scolymus*) leaf powder	Meat Quality of Japanese Quail	21 days	• TBARS • WHC	A potential to improve the oxidative stability and meat quality	[39]
Artichoke (*C. scolymus*) leaf powder	Frozen Meat Quality of Japanese Quail	21 days	• TBARS • WHC	A potential to improve the oxidative stability and meat quality	[40]
Artichoke (*C. scolymus*) extract	Chicken thigh meat	35 days	• Antioxidant activity (DPPH)	Decreases GPx and CAT activities in Chicken meat	[33]
Artichoke (*C. scolymus*) byproducts extracts	Raw beef patties during refrigerated storage	7 days	• Antioxidant activity (DPPH) • TBARS	As natural antioxidant in meat products	[23]

TBARS: 2-thiobarbituric acid-reactive substance, WHC: Water holding capacity, GPx: glutathione peroxidase, CAT: catalase.

Table 5. Color parameters (L*, a*, b*) of minced meat samples during storage.

Storage Period	Storage Period	L*	a*	b*
Control	Day 0	35.62 ± 0.32 [aB]	14.67 ± 0.48 [bA]	15.03 ± 0.57 [bA]
Added + *C. scolymus*		36.84 ± 0.51 [aA]	15.12 ± 0.37 [bA]	15.92 ± 0.63 [bA]
Control	Day 3	35.74 ± 0.43 [aB]	12.28 ± 0.51 [bA]	14.22 ± 0.39 [bA]
Added + *C. scolymus*		36.21 ± 0.31 [aA]	12.88 ± 0.31 [bA]	14.20 ± 0.46 [bA]
Control	Day 6	37.13 ± 0.46 [aA]	10.67 ± 0.54 [bB]	13.44 ± 0.63 [bB]
Added + *C. scolymus*		35.24 ± 0.36 [aB]	10.75 ± 0.49 [bB]	12.88 ± 0.38 [bB]
Control	Day 10	38.56 ± 0.34 [aA]	9.93 ± 0.48 [bB]	12.82 ± 0.51 [bB]
Added + *C. scolymus*		34.88 ± 0.38 [aB]	10.13 ± 0.56 [bB]	12.11 ± 0.34 [bB]

[a,b]: Means followed by different letters within the same line represent significant differences ($p < 0.05$). [A,B]: Means followed by different letters within the same column represent significant differences. Data are the average of triplicates.

4. Conclusions

Results of the present study demonstrate using LC-QTOF-MS that artichoke (*C. scolymus*) powder extract contains high levels of quercetin and chlorogenic acid. The artichoke, which was found to have antioxidant properties, positively affected the chemical characteristics of the minced meat. It can be concluded that the added extract (as natural antioxidants) could successfully preserve antioxidant activity in minced meat stored at −18 °C for up to 10 days. *C. scolymus* powder with highly bioactive compounds improved the quality properties of minced meat samples during storage. It exhibited antimicrobial activity against pathogens that spoil food and are important for meat technology. MetMb levels could be maintained during storage. Hence, incorporation of extract stabilized the color of minced meat and had a significant impact on sensory characteristics during storage period. As a result, the *C. scolymus* powder extract can replace the chemical compounds in meat and meat products formulations to control the oxidative change and undesirable microbial activity. It can be applied as natural antioxidants to extend the shelf-life of meat products to achieve and valuable healthy meat products. In addition, the sensory and other quality parameters examined in more detailed research needs to be addressed in further studies.

Author Contributions: In this article, T.D and S.A. conceived and designed the research. T.D. conducted the experiments, and S.A. contributed to biochemical analyses. T.D. analyzed the data and wrote the first manuscript draft, T.D. and S.A. revised the paper up to its final version. All authors have read and agreed to the published version of the manuscript.

Funding: This research received no external funding.

Institutional Review Board Statement: Not applicable.

Informed Consent Statement: Not applicable.

Data Availability Statement: The data are available by the corresponding author upon.

Conflicts of Interest: The authors declare no conflict of interest.

Sample Availability: Samples of the compounds are available from the authors.

References

1. Pereira, P.M.D.C.C.; Vicente, A.F.D.R.B. Meat nutritional composition and nutritive role in the human diet. *Meat Sci.* **2013**, *93*, 586–592. [CrossRef] [PubMed]
2. Falowo, A.B.; Fayemi, P.O.; Muchenje, V. Natural antioxidants against lipid–protein oxidative deterioration in meat and meat products: A review. *Food Res. Int.* **2014**, *64*, 171–181. [CrossRef] [PubMed]
3. Poprac, P.; Jomova, K.; Simunkova, M.; Kollar, V.; Rhodes, C.J.; Valko, M. Targeting free radicals in oxidative stress-related human diseases. *Trends Pharmacol. Sci.* **2017**, *38*, 592–607. [CrossRef] [PubMed]
4. Bensid, A.; El Abed, N.; Houicher, A.; Regenstein, J.M.; Özogul, F. Antioxidant and antimicrobial preservatives: Properties, mechanism of action and applications in food–a review. *Crit. Rev. Food Sci. Nutr.* **2020**, *60*, 1–17. [CrossRef] [PubMed]
5. Turgut, S.S.; Isıkcı, F.; Soyer, A. Antioxidant activity of pomegranate peel extract on lipid and protein oxidation in beef meatballs during frozen storage. *Meat Sci.* **2017**, *129*, 111–119. [CrossRef] [PubMed]
6. Niyonzima, E.; Ongol, M.P.; Kimonyo, A.; Sindic, M. Risk Factors and Control Measures for Bacterial Contamination in the Bovine Meat Chain: A Review on Salmonella and Pathogenic, *E. coli. J. Food Res.* **2015**, *4*, 98–121. [CrossRef]
7. Gassara, F.; Kouassi, A.P.; Brar, S.K.; Belkacemi, K. Green alternatives to nitrates and nitrites in meat-based products—A review. *Crit. Rev. Food Sci. Nutr.* **2016**, *56*, 2133–2148. [CrossRef]
8. Gutiérrez-Larraínzar, M.; Rúa, J.; Caro, I.; de Castro, C.; de Arriaga, D.; García-Armesto, M.R.; del Valle, P. Evaluation of antimicrobial and antioxidant activities of natural phenolic compounds against foodborne pathogens and spoilage bacteria. *Food Control* **2012**, *26*, 555–563. [CrossRef]
9. Papuc, C.; Goran, G.V.; Predescu, C.N.; Nicorescu, V.; Stefan, G. Plant polyphenols as antioxidant and antibacterial agents for shelf-life extension of meat and meat products: Classification, structures, sources, and action mechanisms. *Compr. Rev. Food Sci. Food Saf.* **2017**, *16*, 1243–1268. [CrossRef] [PubMed]
10. Demir, T.; Akpınar, Ö. Biological Activities of Phytochemicals in Plants. *Turk. J. Agric.-Food Sci. Technol.* **2020**, *8*, 1734–1746. [CrossRef]
11. Kalogianni, A.I.; Lazou, T.; Bossis, I.; Gelasakis, A.I. Natural phenolic compounds for the control of oxidation, bacterial spoilage, and foodborne pathogens in meat. *Foods* **2020**, *9*, 794–822. [CrossRef] [PubMed]
12. Bekhit, A.E.D.; Geesink, G.H.; Ilian, M.A.; Morton, J.D.; Bickerstaffe, R. The effects of natural antioxidants on oxidative processes and metmyoglobin reducing activity in beef patties. *Food Chem.* **2003**, *81*, 175–187. [CrossRef]
13. Pandino, G.; Lombardo, S.; Mauromicale, G.; Williamson, G. Profile of polyphenols and phenolic acids in bracts and receptacles of globe artichoke (*Cynara cardunculus* var. *scolymus*) germplasm. *J. Food Compos. Anal.* **2011**, *24*, 148–153. [CrossRef]
14. Miccadei, S.; Di Venere, D.; Cardinali, A.; Romano, F.; Durazzo, A.; Foddai, M.S.; Fraioli, R.; Mobarhan, S.; Maiani, G. Antioxidative and apoptotic properties of polyphenolic extracts from edible part of artichoke (*Cynara scolymus* L.) on cultured rat hepatocytes and on human hepatoma cells. *Nutr. Cancer* **2008**, *60*, 276–283. [CrossRef]
15. Wang, L.; Qiao, K.; Huang, Y.; Zhang, Y.; Xiao, J.; Duan, W. Optimization of beef broth processing technology and isolation and identification of flavor peptides by consecutive chromatography and LC-QTOF-MS/MS. *Food Sci. Nutr.* **2020**, *8*, 4463–4471. [CrossRef]
16. Wootton-Beard, P.C.; Moran, A.; Ryan, L. Stability of the total antioxidant capacity and total polyphenol content of 23 commercially available vegetable juices before and after in vitro digestion measured by FRAP, DPPH, ABTS and Folin–Ciocalteu methods. *Food Res. Int.* **2011**, *44*, 217–224. [CrossRef]

17. Zhishen, J.; Mengcheng, T.; Jianming, W. The determination of flavonoid contents in mulberry and their scavenging effects on superoxide radicals. *Food Chem.* **1999**, *64*, 555–559. [CrossRef]
18. Ramli, A.N.M.; Manap, N.W.A.; Bhuyar, P.; Azelee, N.I.W. Passion fruit (*Passiflora edulis*) peel powder extract and its application towards antibacterial and antioxidant activity on the preserved meat products. *SN Appl. Sci.* **2020**, *2*, 1–11. [CrossRef]
19. Huang, B.; He, J.; Ban, X.; Zeng, H.; Yao, X.; Wang, Y. Antioxidant activity of bovine and porcine meat treated with extracts from edible lotus (*Nelumbo nucifera*) rhizome knot and leaf. *Meat Sci.* **2011**, *87*, 46–53. [CrossRef]
20. Kim, S.; Lee, S.; Lee, H.; Ha, J.; Lee, J.; Choi, Y.; Oh, H.; Hong, J.; Yoon, Y.; Choi, K.H. Evaluation on antimicrobial activity of psoraleae semen extract controlling the growth of gram-positive bacteria. *Korean J. Food Sci. Anim. Resour.* **2017**, *37*, 502–510. [CrossRef] [PubMed]
21. Van Cuong, T.; Chin, K.B. Effects of annatto (*Bixa orellana* L.) seeds powder on physicochemical properties, antioxidant and antimicrobial activities of pork patties during refrigerated storage. *Korean J. Food Sci. Anim. Resour.* **2016**, *36*, 476–486. [CrossRef] [PubMed]
22. Brand-Williams, W.; Cuvelier, M.E.; Berset, C.L.W.T. Use of a free radical method to evaluate antioxidant activity. *LWT-Food Sci. Technol.* **1995**, *28*, 25–30. [CrossRef]
23. Ergezer, H.; Serdaroğlu, M. Antioxidant potential of artichoke (*Cynara scolymus* L.) byproducts extracts in raw beef patties during refrigerated storage. *J. Food Meas. Charact.* **2018**, *12*, 982–991. [CrossRef]
24. Zuorro, A.; Maffei, G.; Lavecchia, R. Reuse potential of artichoke (*Cynara scolymus* L.) waste for the recovery of phenolic compounds and bioenergy. *J. Clean. Prod.* **2016**, *111*, 279–284. [CrossRef]
25. Francavilla, M.; Marone, M.; Marasco, P.; Contillo, F.; Monteleone, M. Artichoke Biorefinery: From Food to Advanced Technological Applications. *Foods* **2021**, *10*, 112–128. [CrossRef]
26. Curadi, M.; Picciarelli, P.; Lorenzi, R.; Graifenberg, A.; Geccarelli, N. Antioxidant activity and phenolic compounds in the edible parts of early and late Italian artichoke (*Cynara scolymus* L.) varieties. *Ital. J. Food Sci.* **2005**, *17*, 33–44. Available online: https://pascal-francis.inist.fr/vibad/index.php?action=getRecordDetail&idt=16660260 (accessed on 1 September 2021).
27. Wang, M.; Simon, J.E.; Aviles, I.F.; He, K.; Zheng, Q.Y.; Tadmor, Y. Analysis of antioxidative phenolic compounds in artichoke (*Cynara scolymus* L.). *J. Agric. Food Chem.* **2003**, *51*, 601–608. [CrossRef]
28. Andrés, S.; Morán, L.; Aldai, N.; Tejido, M.L.; Prieto, N.; Bodas, R.; Giráldez, F.J. Effects of linseed and quercetin added to the diet of fattening lambs on the fatty acid profile and lipid antioxidant status of meat samples. *Meat Sci.* **2014**, *97*, 156–163. [CrossRef]
29. North, M.K.; Dalle Zotte, A.; Hoffman, L.C. The effects of dietary quercetin supplementation on the meat quality and volatile profile of rabbit meat during chilled storage. *Meat Sci.* **2019**, *158*, 1–11. [CrossRef] [PubMed]
30. Zhao, J.S.; Deng, W.; Liu, H.W. Effects of chlorogenic acid-enriched extract from *Eucommia ulmoides* leaf on performance, meat quality, oxidative stability, and fatty acid profile of meat in heat-stressed broilers. *Poult. Sci.* **2019**, *98*, 3040–3049. [CrossRef]
31. Farag, M.A.; Al-Mahdy, D.A.; Salah El Dine, R.; Fahmy, S.; Yassin, A.; Porzel, A.; Brandt, W. Structure-Activity relationships of antimicrobial gallic acid derivatives from pomegranate and acacia fruit extracts against potato bacterial wilt pathogen. *Chem. Biodivers.* **2015**, *12*, 955–962. [CrossRef] [PubMed]
32. Munekata, P.E.S.; Rocchetti, G.; Pateiro, M.; Lucini, L.; Domínguez, R.; Lorenzo, J.M. Addition of plant extracts to meat and meat products to extend shelf-life and health-promoting attributes: An overview. *Curr. Opin. Food Sci.* **2020**, *31*, 81–87. [CrossRef]
33. Mirderikvandi, M.; Kiani, A.; Khaldari, M.; Alirezaei, M. Effects of artichoke (*Cynara scolymus* L.) extract on antioxidant status in chicken thigh meat. *Iran. J. Vet. Med.* **2016**, *10*, 73–81.
34. Llorach, R.; Espin, J.C.; Tomas-Barberan, F.A.; Ferreres, F. Artichoke (*Cynara scolymus* L.) byproducts as a potential source of health-promoting antioxidant phenolics. *J. Agric. Food Chem.* **2002**, *50*, 3458–3464. [CrossRef] [PubMed]
35. Zahid, M.A.; Seo, J.K.; Park, J.Y.; Jeong, J.Y.; Jin, S.K.; Park, T.S.; Yang, H.S. The effects of natural antioxidants on protein oxidation, lipid oxidation, color, and sensory attributes of beef patties during cold storage at 4 °C. *Korean J. Food Sci. Anim. Resour.* **2018**, *38*, 1029. [CrossRef] [PubMed]
36. Liu, F.; Xu, Q.; Dai, R.; Ni, Y. Effects of natural antioxidants on colour stability, lipid oxidation and metmyoglobin reducing activity in raw beef patties. *Acta Sci. Pol. Technol. Aliment.* **2015**, *14*, 37–44. [CrossRef]
37. Ramanathan, R.; Hunt, M.C.; Mancini, R.A.; Nair, M.N.; Denzer, M.L.; Suman, S.P.; Mafi, G.G. Recent updates in meat color research: Integrating traditional and high-throughput approaches. *Meat Muscle Biol.* **2020**, *4*, 1–24. [CrossRef]
38. McCarthy, T.L.; Kerry, J.P.; Kerry, J.F.; Lynch, P.B.; Buckley, D.J. Evaluation of the antioxidant potential of natural food/plant extracts as compared with synthetic antioxidants and vitamin E in raw and cooked pork patties. *Meat Sci.* **2001**, *58*, 45–52. [CrossRef]
39. Abbasi, F.; Samadi, F. Effect of different levels of artichoke (*Cynara scolymus* L.) leaf powder on the performance and meat quality of Japanese quail. *Poult. Sci. J.* **2014**, *2*, 95–111. [CrossRef]
40. Samadi, F.; Abbasi, F.; Samadi, S. Effect of artichoke (*Cynara scolymus*) leaf powder on performance and physicochemical properties of frozen meat of japanese quail. *Iran. J. Appl. Anim. Sci.* **2015**, *5*, 933–940.

Article

Effects of Long-Term Storage on Radical Scavenging Properties and Phenolic Content of Kombucha from Black Tea

Chiara La Torre [1], Alessia Fazio [1,*], Paolino Caputo [2], Pierluigi Plastina [1], Maria Cristina Caroleo [1], Roberto Cannataro [1] and Erika Cione [1]

[1] Department of Pharmacy, Health and Nutritional Sciences, Department of Excellence 2018–2022, University of Calabria, Edificio Polifunzionale, 87036 Rende, Italy; latorre.chiara@libero.it (C.L.T.); pierluigi.plastina@unical.it (P.P.); mariacristinacaroleo@virgilio.it (M.C.C.); r.cannataro@gmail.com (R.C.); erika.cione@unical.it (E.C.)

[2] Department of Chemistry and Chemical Technologies, University of Calabria, 87036 Rende, Italy; paolino.caputo@unical.it

* Correspondence: a.fazio@unical.it; Tel.: +39-0984-493013

Abstract: Kombucha is a fermented beverage. Its consumption has significantly increased during the last decades due to its perceived beneficial effects. For this reason, it has become a highly commercialized drink that is produced industrially. However, kombucha is still also a homemade beverage, and the parameters which, besides its organoleptic characteristics, define the duration of its potential beneficial properties over time, are poorly known. Therefore, this study aimed to determine the effect of 9-month storage at 4 °C with 30-day sampling on the pH, total phenolic, and flavonoid contents, free radical scavenging properties of kombucha fermented from black tea. Our results highlighted that, after four months, the phenolic content decreased significantly from the initial value of 234.1 ± 1.4 µg GAE mL^{-1} to 202.9 ± 2.1 µg GAE mL^{-1}, as well its antioxidant capacity tested by two in vitro models, DPPH, and ABTS assays. Concomitantly, the pH value increased from 2.82 to 3.16. The novel findings of this pilot study revealed that kombucha from sugared black tea can be stored at refrigerator temperature for four months. After this period the antioxidant properties of kombucha are no longer retained.

Keywords: kombucha; black tea; long-term storage; antioxidant scavenging activity; total phenolic content

Citation: La Torre, C.; Fazio, A.; Caputo, P.; Plastina, P.; Caroleo, M.C.; Cannataro, R.; Cione, E. Effects of Long-Term Storage on Radical Scavenging Properties and Phenolic Content of Kombucha from Black Tea. *Molecules* **2021**, *26*, 5474. https://doi.org/10.3390/molecules26185474

Academic Editor: Mirella Nardini

Received: 11 August 2021
Accepted: 7 September 2021
Published: 8 September 2021

Publisher's Note: MDPI stays neutral with regard to jurisdictional claims in published maps and institutional affiliations.

1. Introduction

"Kombucha" is the name of a drink obtained by fermenting tea, mainly black or green, with the addition of sucrose, that acts as a substrate for fermentation, and a symbiotic culture of yeast and bacteria, known as "SCOBY" (Symbiotic Cultures of Bacteria and Yeast). The taste of this drink is slightly acidic and slightly carbonated, which makes it popular and pleasing to consumers [1]. Kombucha was first used in East Asia for its beneficial and curative effects only based on anecdotal evidences, since the Tsin dynasty began consuming it in Manchuria. It spread from China to Russia after World War I and then throughout Europe [1].

The fermentation is due to a symbiotic culture of acetic bacteria of the genus *Acetobacter* and *Gluconobacter* and different osmophilic yeast species, including genera such as *Saccharomycode*, *Schizosaccharomyces*, *Zygosaccharomyces*, *Brettanomyces/Dekkera*, *Candida*, *Torulospora*, *Koleckera*, and *Pichia* e *Mycoderma*. After fermentation, the kombucha tea is filtered through a cheesecloth and is consumed as a healthy drink. It can also be bottled for commercialization [2].

Almost forgotten for decades, kombucha became very popular again in the early 2000s, thanks to its sudden spread in Australia and in the United States. During the last decades, kombucha transitioned from a homemade fermented beverage to a soft drink produced on a large scale for commercial use [2,3]. Chemical analysis of kombucha beverage highlighted the presence of a variety of compounds, such as organic acids, mainly acetic, gluconic, and glucuronic acid, sugars (sucrose, glucose, and fructose), water-soluble vitamins (B1, B2, B6, B12, C), lipids, amino acids, biogenic amines, proteins, ethanol, minerals (manganese, iron, nickel, copper, zinc, plumb, cobalt, chromium, and cadmium), anions (fluoride, chloride, bromide, iodide, nitrate, phosphate, and sulphate), D-saccharic acid-1,4-lactone (DSL), carbon dioxide, and polyphenols [2,3]. Kombucha beverage is a source of bioactive components, such as glucuronic acid and polyphenols displaying antioxidant activity [4–6]. The low pH value of this beverage, especially owing to the presence of acetic acid in particular and a range of other organic acids, makes kombucha a drink with remarkable antimicrobial activity against a broad range of microorganisms [7–10] having also probiotic and symbiotic properties [11].

Many claimed beneficial effects of kombucha may be associated with its antioxidant activities, but when kombucha tea is stored at ordinary temperatures, the biofilm due to the presence of microorganisms continues to form, and might also affect the antioxidant activity. Epigallocatechin-3-gallate (EGCG) and epicatechin-3-gallate are converted into the corresponding epigallocatechin (EGC) and epicatechin (EC), the phenolic concentration in kombucha tea shows a linear increase during the fermentation time [11]. It is worth to note that the beneficial outcomes of the kombucha drink are mainly attributed to the activity of polyphenols, which in turn can act epigenetically [12,13].

Jayabalan et al. studied the effect of temperature (50–90 °C) on biochemical components and free radical scavenging properties of kombucha tea during a storage period of three months [14], concluding that heat treatment was not a suitable method for kombucha tea preservation. Therefore, it is of interest to elucidate the relationship between storage time and the changes of the antioxidant ability of kombucha. In fact, time, temperature, and light can significantly impact the quality and the biological activities of this beverage. To the best of our knowledge, no studies were carried out to determine the effects of storage times for more than three months and at low temperature. Therefore, this study aimed to evaluate the effects of the long-term storage at 4 °C on the pH, total phenolic and flavonoid contents, and free radical scavenging properties of kombucha during nine months with a sampling of 30 days, in order to evaluate the period during which these parameters are stable.

2. Results

2.1. Monitoring of the pH Values of Kombucha during the Stogare Period

The pH values of black tea alone, as well as kombucha tea after one month of fermentation and all the samples over the storage are shown in Figure 1. It was observed that the pH value of sweetened black tea was 5.59, and it dropped to 2.82 in the kombucha beverage obtained after 30 days of fermentation (white bar) decreasing by about 2.77 units. This latter pH was used as the control for all the kombucha samples analyzed during nine months of storage. The value remained constant (2.84), till two months by decreasing significantly compared with the control (**** $p < 0.0001$) of about 0.2 units from months four to six. Then, the pH value of sample significantly increased up to 3.24, and it remained constant at these values at less than 0.05 units for the last three months (from months seven to nine).

2.2. Total Phenolic Content (TPC)

The total phenolic content (TPC) in all kombucha samples is shown in Figure 2. The results were expressed as µg equivalents of gallic acid (µg GAE) per mL of sample.

The results highlighted that the kombucha obtained after one month of fermentation, used as reference, showed the maximum TPC level which was 1.7 times higher (234.1 ± 1.4 µg GAE mL^{-1}) than the value of black tea, (137.5 ± 10.7 µg GAE mL^{-1}), respectively. In the following months (from months two to four), the TPC slowly decreased from 234.1 ± 1.4 to 223.5 ± 0.7 µg GAE mL^{-1} without significant statistic difference. On the contrary, it was decreasing in a time dependent manner, from months five to nine, significantly by about 13% (202.9 ± 2.1 µg GAE mL^{-1}, **** $p < 0.0001$) at month five, and 34% at month nine (80.8 ± 5.4 µg GAE mL^{-1}).

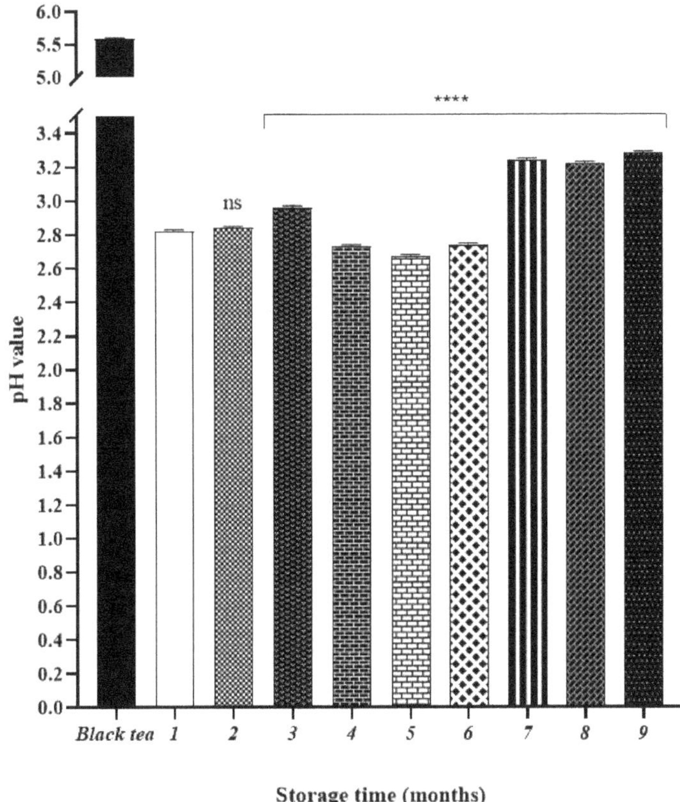

Figure 1. Values of pH determinate in black tea, kombucha tea after 30 days of fermentation and all the samples during storage. Black bar refers to the starting tea. White bar refers to the kombucha sample after 30 days of fermentation (control). Values represent the three-measure mean ± standard deviation. Asterisks on the bars indicate that mean values were statistically different from the control (**** $p < 0.0001$).

2.3. HPLC-DAD Analyses

Five compounds were identified and quantified by HPLC analyses in all samples (Figure 3). Chromatographic evolution at 280 nm is show in Figure 4A–D.

Caffeine was the main compound in all tea samples. Its initial value in black tea was 568.61 ± 0.84 µg mL^{-1} (Table 1), which underwent a reduction of 37.34% after fermentation, of 40.29% after one month and of 45.37% after two months. The minimum value was reached after four months (135.36 ± 1.63 µg mL^{-1}) which corresponded to a reduction of the initial value of 76.19%. After six months the caffeine content increased (674.98 ± 0.49 µg mL^{-1}), reaching its maximum value (702.93 ± 0.02 µg mL^{-1}) after nine months. Chlorogenic acid was the only compound that maintained its content unchanged during fermentation and over time compared to the initial value in black tea (29.60 ± 0.01 µg mL^{-1}). EGCG content in tea was 20.58 ± 0.32 and remained unchanged after 30 days of fermentation and in the next three months of storage, becoming undetectable at by the fourth month. A similar trend was displayed by ferulic acid, present in a smaller amount in black tea (3.15 ± 0.03 µg mL^{-1}). Its content gradually decreased during fermentation in the next five months of storage, becoming undetectable at the sixth month. On the other hand, quercetin, absent in black tea, was identifiable only in the samples after four months from fermentation (23.07 ± 0.01 µg mL^{-1}). Its content remained constant from months five to nine. The results are shown in Table 1.

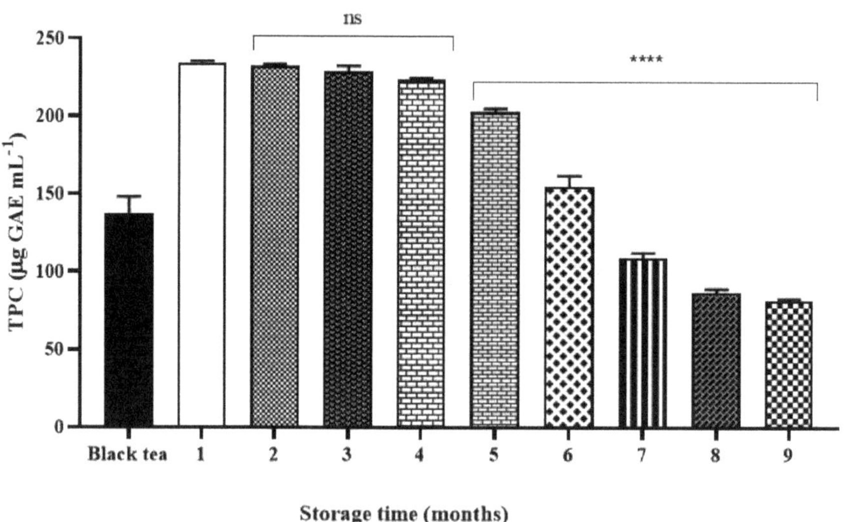

Figure 2. Total phenolic content (TPC) (µg GAE mL^{-1}) of black tea, kombucha tea after 30 days of fermentation and all the samples during storage. Black bar refers to the starting tea. White bar refers to the kombucha sample after 30 days of fermentation (control). Values represent the three-measure mean $\pm$ standard deviation. Asterisks on the bars indicate that mean values were statistically different from the white bar that represents the control (**** $p < 0.0001$).

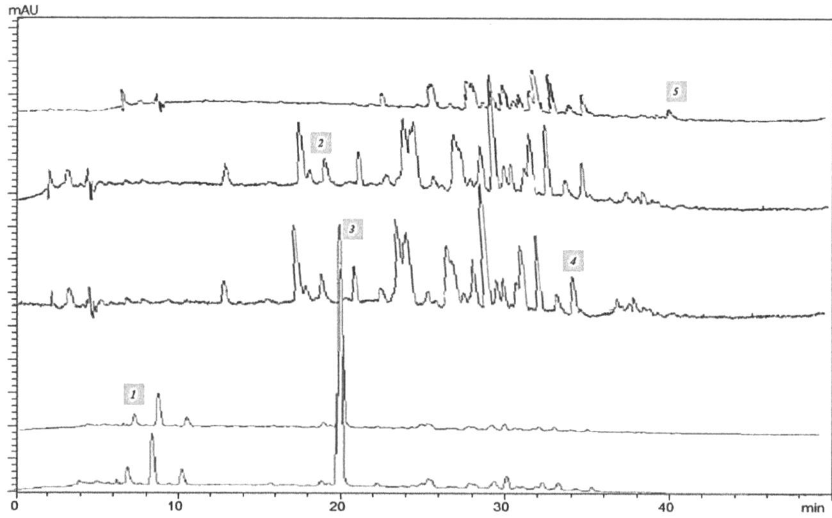

Figure 3. HPLC chromatogram of identified compounds in all the samples at the relative wavelengths to which they have been detected and quantified: 1. EGCG ($\lambda = 280$ nm); 2. Chlorogenic acid ($\lambda = 327$ nm); 3. Caffeine ($\lambda = 273$ nm); 4. Ferulic acid ($\lambda = 325$ nm); 5. Quercetin ($\lambda = 365$ nm).

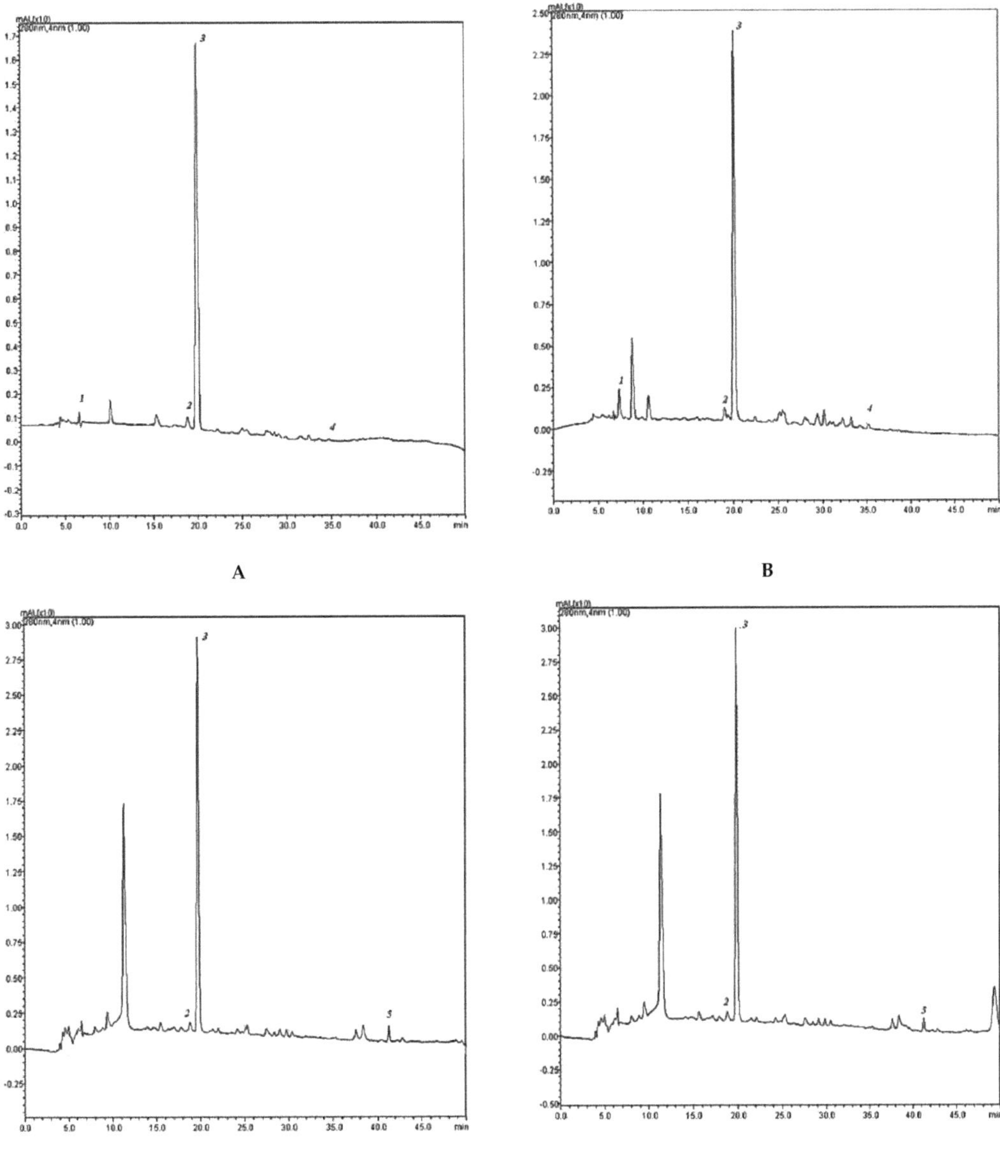

Figure 4. Chromatographic evolution at 280 nm including: (**A**) black tea; (**B**) kombucha after one month of fermentation; (**C**) kombucha tea sample at the fifth month of storage, when the epigallocatechin gallate content was not detected whereas quercetin, absent in black tea, was identifiable; (**D**) kombucha tea sample at the ninth month of storage.

2.4. Total Flavonoid Content (TFC)

Total flavonoid content (TFC) underwent a decrease during fermentation (Figure 5). The initial value of TFC in black tea (2.3 ± 0.2 QE mL^{-1}) was reduced by 50% in kombucha tea (1.1 ± 0.3 µg QE mL^{-1}) and it remained constant over time.

Table 1. Content of HPLC identified compounds, expressed as $\mu g \, mL^{-1}$.

Samples	Caffeine	Chlorogenic Acid	EGCG	Ferulic Acid	Quercetin
Black tea	568.61 ± 0.84	29.70 ± 0.01	20.58 ± 0.32	3.15 ± 0.03	nd [a]
Fermented Kombucha	356.25 ± 7.17	29.60 ± 0.03	20.24 ± 0.53	2.97 ± 0.90	nd
1st storage month	339.50 ± 4.15	29.51 ± 0.06	19.90 ± 0.12	1.21 ± 0.01	nd
2nd storage month	310.63 ± 0.97	29.56 ± 0.01	19.96 ± 0.07	1.30 ± 0.20	nd
3rd storage month	159.20 ± 0.55	29.24 ± 0.38	19.92 ± 0	0.82 ± 0.36	nd
4th storage month	135.36 ± 1.63	29.41 ± 0.41	nd	0.75 ± 0	nd
5th storage month	162.60 ± 0.26	29.64 ± 0.91	nd	0.41 ± 0.04	23.07 ± 0.01
6th storage month	674.98 ± 0.49	29.72 ± 0.01	nd	nd	23.16 ± 0.59
7th storage month	672.27 ± 0.07	29.58 ± 0.03	nd	nd	23.23 ± 0.08
8th storage month	702.91 ± 0.71	29.72 ± 0.01	nd	nd	23.03 ± 0.07
9th storage month	702.93 ± 0.02	29.60 ± 0.01	nd	nd	23.04 ± 0.03

[a] Not-detected.

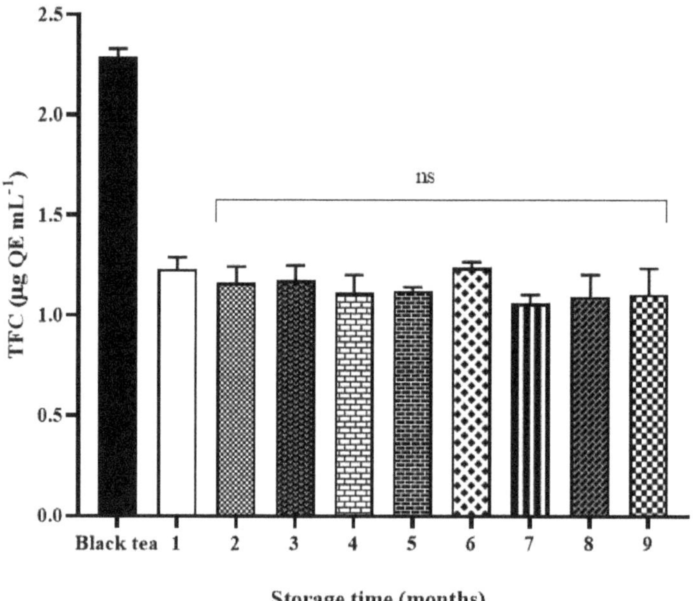

Figure 5. Total flavonoid content (TFC) ($\mu g \, QE \, mL^{-1}$) of black tea, kombucha tea after 30 days of fermentation and all the samples over storage. Black bar refers to the starting tea. White bar refers to the kombucha sample after 30 days of fermentation (control).

2.5. Determination of Antioxidant Activity

Scavenging abilities of black tea, kombucha tea after one month of fermentation and all the samples during the storage were monitored using two common in vitro models, DPPH and ABTS assays.

2.5.1. DPPH Assay

Kombucha exhibited good antioxidant activity against DPPH radical during the storage at all tested concentrations when compared to black tea (Figure 6). The results of the DPPH radical assay were also expressed as the Trolox equivalent antioxidant capacity (TEAC) using Trolox as reference standard. The TEAC values are reported in Table 2. During fermentation, $\%I_{DPPH}$ of kombucha tea at the highest and lowest concentrations (55.1 ± 1.3 at 200 µL, and 10.8 ± 0.8 at 10 µL, respectively) increased by about 70% compared to that of black tea, which was 16.26 ± 0.7 (2.6 ± 0.1 µg TE mL^{-1}) at 200 µL, and

3.1 ± 0.1 (0.8 ± 0.1 µg TE mL^{-1}) at 10 µL. At the highest concentration, a 13% decrease was observed during the first three months of storage with respect to the control value, while it reduced by 30% after four months. At the fifth month it dropped drastically to a value of $23.2 \pm 6.1\%$ (3.0 ± 0.7 µg TE mL^{-1}), corresponding to a reduction of 58% of the control value (**** $p < 0.0001$). The scavenging ability of kombucha declined to $18.9 \pm 0.4\%$ (2.9 ± 0.1 µg TE mL^{-1}) after nine months.

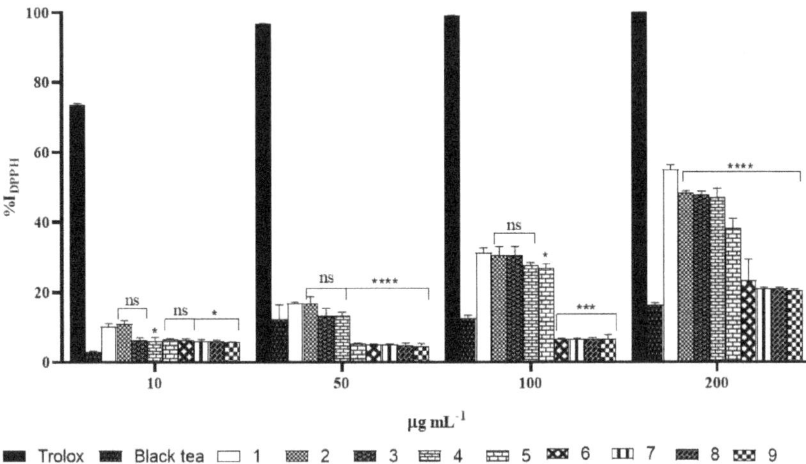

Figure 6. $\%I_{DPPH}$ of Trolox, black tea, kombucha tea after 30 days of fermentation and all the samples during storage. White bar refers to $\%I_{DPPH}$ of kombucha tea after 30 days of fermentation (control), the others refer to kombucha samples over nine months of storage. Asterisks on the bars indicate that mean values were statistically different from the control (* $p < 0.05$, *** $p < 0.001$, **** $p < 0.0001$).

Table 2. Trolox equivalent antioxidant capacity (TEAC) values (µg TE mL^{-1}) of different samples against DPPH.

Samples	Concentrations			
	10 µL	50 µL	100 µL	200 µL
Black tea	0.8 ± 0.1	2.0 ± 0.6	2.1 ± 0.1	2.6 ± 0.1
Fermented Kombucha	1.2 ± 0.1	2.1 ± 0.1	3.0 ± 0.2	7.4 ± 0.2
1st storage month	1.3 ± 0.4	2.1 ± 0.3	4.3 ± 0.2	6.3 ± 0.1
2nd storage month	0.7 ± 0.1	1.5 ± 0.3	4.0 ± 0.2	6.4 ± 0.3
3rd storage month	0.6 ± 1.0	1.7 ± 0.2	3.6 ± 0.1	6.3 ± 0.2
4th storage month	0.4 ± 0.1	0.5 ± 0.1	3.4 ± 0.1	5.1 ± 0.4
5th storage month	0.5 ± 0.1	0.5 ± 0.1	0.8 ± 0.1	3.0 ± 0.7
6th storage month	0.6 ± 0.1	0.7 ± 0.1	0.8 ± 0.1	2.8 ± 0.3
7th storage month	0.7 ± 0.1	1.3 ± 0.1	2.0 ± 0.2	2.7 ± 0.1
8th storage month	0.6 ± 0.1	0.9 ± 0.1	1.0 ± 0.6	3.0 ± 0.3
9th storage month	0.6 ± 0.1	0.9 ± 0.1	1.2 ± 0.1	2.9 ± 0.1

2.5.2. ABTS Assay

The inhibition percentage of ABTS is reported in Figure 7 and the corresponding TEAC values against ABTS are reported in Table 3. Kombucha tea samples showed lower inhibitory abilities on ABTS•$^{+}$ radical cation at the highest concentrations (100 and 200 µL) than those against DPPH radical. After one month of fermentation $\%I_{ABTS}$ of tea Kombucha at the highest concentrations (200 and 100 µL) increased by about 54% (47.4 ± 1.3, 9.1 ± 0.3 µg TE mL^{-1}) and 40.6% (26.6 ± 0.5, 6.1 ± 0.1 µg TE mL^{-1}) respectively, as compared to the black tea at the same concentrations, that is 21.8 ± 0.7 corresponding to

4.8 ± 0.1 µg TE mL^{-1} at 200 µL and 15.8 ± 1.5 or 3.4 ± 0.2 µg TE mL^{-1} at 100 µL. After one month of storage, the abilities of the samples at 200 (42.3 ± 0.5, 9.1 ± 0.3 µg TE mL^{-1}) and 100 µL (25.0 ± 0.8, 5.5 ± 0.1 µg TE mL^{-1}) were lowered by 11% and 6%, respectively, compared to the control. The antioxidant capacity of the control at the highest concentration underwent further but progressive reduction up to 27% in the first three months of storage, but it was reduced by 73% (%I$_{ABTS}$ 12.7 ± 0.9, or 3.4 ± 0.2 µg TE mL^{-1}) at the fifth month of storage and 85% (%I$_{ABTS}$ 7.0 ± 0.4, or 1.6 ± 0.1 µg TE mL^{-1}) at the ninth month.

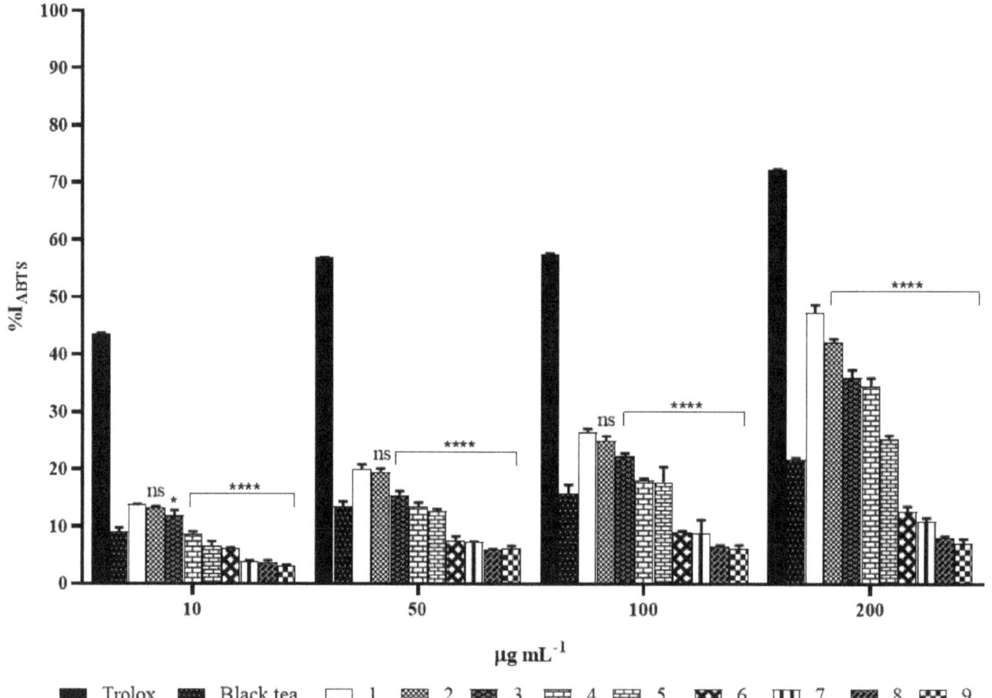

Figure 7. %I$_{ABTS}$ of Trolox, black tea, kombucha tea after 30 days of fermentation and all the samples during storage. White bar (month 1) refers to %I$_{ABTS}$ of kombucha tea after 30 days of fermentation (control), the others refer to kombucha samples over nine months of storage. Asterisks on the bars indicate that mean values were statistically different from the control ($^* p < 0.05$, $^{****} p < 0.0001$).

Table 3. Trolox equivalent antioxidant capacity (TEAC) values (µg TE mL^{-1}) of different samples against ABTS.

Samples	Concentrations			
	10 µL	50 µL	100 µL	200 µL
Black tea	1.8 ± 0.2	3.0 ± 0.2	3.4 ± 0.2	4.8 ± 0.1
Fermented Kombucha	2.8 ± 0.1	4.5 ± 0.1	6.1 ± 0.1	9.6 ± 0.3
1st storage month	2.8 ± 0.1	4.4 ± 0.1	5.5 ± 0.1	9.1 ± 0.3
2nd storage month	2.6 ± 0.1	3.3 ± 0.3	5.1 ± 0.2	7.9 ± 0.2
3rd storage month	1.9 ± 0.1	2.8 ± 0.2	4.1 ± 0.7	7.5 ± 0.3
4th storage month	1.4 ± 0.2	2.8 ± 0.1	4.0 ± 0.5	5.7 ± 0.1
5th storage month	1.2 ± 0.5	1.6 ± 0.2	2.3 ± 0.5	3.4 ± 0.2
6th storage month	0.7 ± 0.1	1.6 ± 0.1	1.9 ± 0.1	3.3 ± 0.1
7th storage month	0.5 ± 0.1	1.3 ± 0.1	1.4 ± 0.1	1.6 ± 0.1
8th storage month	0.4 ± 0.1	1.1 ± 0.1	1.3 ± 0.1	1.5 ± 0.3
9th storage month	0	0	1.3 ± 0.2	1.6 ± 0.1

3. Discussion

The consumption of kombucha has increased over the last decades due to its perceived beneficial effects. For this reason, it has become a highly commercialized drink, industrially produced but is still also a homemade beverage. To evaluate the effects of long-term storage of kombucha on radical scavenging properties and its phenolic content, we kept the tea samples at refrigerator temperature in the dark. Accordingly, to the literature, the pH values of black tea in the kombucha decreased after 30 days of fermentation due to the metabolic activity of tea fungus yeasts and acetic acid bacteria that produce mainly acetic acid [15].

In fact, the pH value of sweetened black tea was 5.59, and it dropped to 2.82 in the kombucha beverage obtained after 30 days. Then, the changes in the first five months of storage were less than 0.2 units, while at the sixth month, the pH value of sample significantly increased up to 3.24, and it remains constant at these values at less than 0.05 units for the last three months. This pH rise is most likely due to the subsequent use of acids by bacteria as a carbon source in the absence of sugar in the tea [16,17]. The final pH of kombucha samples (3.16), after nine months of storage, is still in the safe pH range of 2.5 to 4.2 for human consumption [18,19].

Total phenolic and flavonoid contents were also monitored after fermentation and during storage. The results highlighted that kombucha after one month of fermentation showed the highest total phenolic content level, which was 1.7 times higher than the value of black tea, as a consequence of the action of microbial enzymes from bacteria and yeasts in an acidic environment, which hydrolyzes complex tea polyphenols into smaller molecular weight phenolic compounds causing an increase in polyphenol concentration [19]. Total phenolic content is reduced after four months of storage. A similar trend was seen for the total flavonoid content but earlier than the total phenolic content.

In contrast to Ning et al. [20], HPLC analysis pointed out that chlorogenic acid maintained its content unchanged either during fermentation and over time compared to the initial value in black tea. On the other hand, all the other monitored phenolics dropped at the fifth month. This agrees with the literature in which the strategy to prolong phenolic content is studied [20]. Conversely, total flavonoid content was constantly lower in black tea, most likely due to the microbial activity of the SCOBY during fermentation.

The results of DPPH inhibition properties of kombucha tea directly depend on the tea constituents and the components produced during fermentation time (30 days). The decrease of antioxidant capacity during storage was most likely related to microbial transformation of the compounds responsible for the maximum scavenging ability into less potential scavenging structures. On the other hands, the inhibition percentage of ABTS assay showed lower inhibitory abilities in respect to ABTS$\bullet^+$ radical action.

The antioxidant activities of our tested samples likely depend on the composition and the chemical nature of phenolic compounds [21,22]. Then, during the storage, changes in the composition of antioxidant compounds of kombucha tea might result from the formation of certain compounds as in the case of quercetin in our results, thus leading to a lower antioxidant activity [22].

It is important to mention here, that during the COVID-19 pandemic of 2019–2020 the consumption of fermented food, especially beverages, increased in several countries [17,23,24]. In particular, the consumption of industrial kefir and kombucha increased [15,25] and the latter was reported, in the magazine Forbes, as the drink of 2020 [26]. Although, as source of bioactive components that could be beneficial for human health, there is no evidence about systematic human trials being done using kombucha tea [27] and some toxicity related to kombucha consumption has been reported so far when kept in a ceramic pot for six months or in lead-glazed earthenware at refrigerator temperature [28,29].

4. Materials and Methods

4.1. Standards and Chemicals

The kombucha starter (Figure 8) was obtained from Kefiralia (Burumart Commerce S.L, Arrasate, Spain) and was maintained in sugared black tea. Dimethylsulphoxide (DMSO), absolute ethanol and methanol, formic acid, and acetonitrile HPLC-grade were purchased from Carlo Erba (Milan, Italy), Folin-Ciocâlteu reagents, sodium carbonate, DPPH, ABTS, potassium persulfate ($K_2S_2O_8$), aluminium chloride ($AlCl_3$), potassium acetate, chloroform, and ethyl acetate were purchased from Sigma Aldrich (Milan, Italy).

Figure 8. Kombucha starter, the "SCOBY".

4.2. Preparation of Kombucha Tea and Fermentation Conditions

Black tea (3 g) was immersed into 1 L of boiling water and infused for about 15 min. Then it was filtered through a sterile sieve. This was repeated for three times and 1 L of each preparation was kept into sterilized glass jars. Commercial sucrose (7%) was then added to the hot drink and, after cooling to room temperature, the infusion was inoculated with a commercial kombucha SCOBY (150 g) size (15 × 2 × 10 cm). The jars were covered with a clean cloth. The fermentation was carried out in the dark at 25 ± 2 °C for 30 days and, at the end of this time, the kombucha tea samples were filtered through a cheesecloth and transferred to three amber jars.

4.3. Storage Condition

The jars containing kombucha tea were placed in a refrigerator (T = 4 °C) for nine months. Sampling was performed every 30 days by taking an aliquot (100 mL) which was analyzed. The tea fermented for one month was used as the control for the kombucha tea samples stored at 4 °C for longer times—up to nine months. pH values, content of total polyphenol compounds, qualitative and quantitative profile of the main tea polyphenols, content of total flavonoids, and free radical scavenging activities of each sample were determined.

4.4. pH Values of Tea Kombucha during Storage

The pH values of all the kombucha tea samples were measured using an electronic pH meter (Hanna Instruments, George Washington Hwy, Smithfield, RI, USA) calibrated at pH 4.0 and 7.0.

4.5. Total Phenolic Content (TPC)

The total polyphenolic content (TPC) compounds in the tea samples were quantified by the Folin-Ciocâlteu colorimetric method as previously described [30], with appropriate modifications. The fermented tea sample (0.1 mL) was transferred in an amber glass vial and was added by 2 mL of distilled water, 0.5 mL of the Folin-Ciocâlteu reagent (diluted 1:10 with distilled water), and 0.4 mL of a 7.5% sodium carbonate solution (Na_2CO_3), up to a final volume of 3 mL. The mixture was shaken under constant magnetic stirring for 30 min, at room temperature in the dark. The absorbance was measured at 760 nm using a spectrophotometer Jasco UV-550. Three analyses were carried out for each sample. Gallic acid was used as the standard in order to plot the calibration curve. For the linearity study, an eight-point calibration curve was constructed using different concentrations of gallic acid stock solutions (range 0.5–0.01 mg mL^{-1}). A linear correlation was found between absorbance of the blue complex at 760 nm and concentration of gallic acid in the range 0.5–0.01 mg mL^{-1} (y = 3.6607x − 0.0036). The coefficient (R^2) obtained from the linear regression was 0.9998, indicating an excellent linear correlation between the data. The total phenolic content (TPC) was expressed as µg equivalents of gallic acid (µg GAE) per mL of kombucha.

4.6. HPLC-DAD Analyses

Five compounds in kombucha tea samples were identified and quantified by reversed-phase high performance liquid chromatography coupled with diode array detector (HPLC-DAD) [31]. The samples were filtered through a membrane filter (0.45 µm) into HPLC vials and analyzed as such. An aliquot (10 µL) of each sample was injected into a Shimadzu (Kyoto, Japan) HPLC system equipped with a diode array detector (SPD-M10Avp). The chromatographic separation was performed on a Mediterranea SEA C-18 column (4.6 mm i.d. × 25 cm, 5 µm). The mobile phase was a 0.1% formic acid (A) and acetonitrile (B) mixture. The gradient used was the following: 0 min, 10% B; 20 min, 22% B; 40 min, 40% B; 45 min, 10% B, 51 min, 10% B. The flow rate and column temperature were maintained as 0.6 mL min^{-1} and at room temperature, respectively. Detection was made at the absorption maxima of the pure standard compounds: caffeine was detected at 273 nm, EGCG at 280 nm, ferulic acid at 325 nm, chlorogenic acid at 327 nm, and quercetin at 365 nm, and identification was made by comparison of the retention times and characteristic UV-Vis spectra of pure standard compounds used as references. Individual components were analyzed quantitatively by the external standard method. The calibration curves for standards (caffeine, EGCG, ferulic acid, chlorogenic acid, and quercetin) were prepared with six appropriate concentrations. The limit of detection (LOD) and the limit of quantification (LOQ) for each standard were calculated as follows: $LOD = 3(S_y/S)$ and $LOQ = 10(S_y/S)$, where S_y is the standard deviation of the response of the curve and S is the slope of the calibration curve.

4.7. Total Flavonoid Content (TFC)

Total flavonoid content (TFC) was determined by a colorimetric method as described previously [32]. Briefly, 0.30 mL of the sample solution were diluted with 1.68 mL of distilled water. Then, 0.9 mL of MeOH, 0.06 mL of a 10% $AlCl_3$ solution, and 0.06 mL of 1 M solution of potassium acetate were added to the solution. The mixture was allowed to stand for 30 min at room temperature, under constant magnetic stirring, in the dark, and then the absorbance was measured against the blank at 420 nm using a spectrophotometer Jasco UV-550. Three analyses were carried out for each sample. Quercetin was used as the standard in order to plot an eight-point calibration curve. The linearity range of calibration curve was 10–0.001 µg mL^{-1} (y = 0.084x − 0.0019). The coefficient (R^2) obtained from the linear regression was 0.9984, indicating a good linear correlation between the data. The results were expressed as µg of quercetin equivalents (µg QE) per mL of kombucha [31].

4.8. Radical Scavenging Activity

Two different in vitro assays, DPPH and ABTS, were used to evaluate the changes over time in free-radical scavenging abilities of all kombucha tea samples.

4.8.1. DPPH Assay

The scavenging activity on DPPH radical was determined by the colorimetric method previously described [25] with slight modification. To different volumes of each sample (10, 50, 100, 200 µL), 0.1 mL of DPPH solution (1 mM) and 2.8 mL of MeOH were added. After an incubation time of 30 min, under magnetic stirring, at room temperature and in the dark, the reduction of DPPH free radical was measured by reading the absorbance at 517 nm using a spectrophotometer Jasco UV-550. The experiments were carried out against a blank (3 mL of MeOH) and a control (2.9 mL of MeOH, 0.1 mL DPPH solution). Each sample was tested in triplicate. The antioxidant activity was given as a percentage of free radical inhibition ($\%I_{DPPH}$), according to the formula:

$\%I_{DPPH}$ = [(absorbance of the control − absorbance of the sample)/absorbance of the control)] × 100.

The results were expressed also as µg of Trolox equivalents per mL of tea samples ($\mu g\ TE\ mL^{-1}$).

4.8.2. ABTS Assay

$ABTS\bullet^+$ radical cation is well soluble in both aqueous and organic solvents, so this method can be extensively used to determine antioxidant activity for both hydrophilic and lipophilic compounds [33]. This radical cation was formed by a reaction between 7 mM ABTS solution and 2.45 mM potassium persulfate ($K_2S_2O_8$), and then allowing the mixture to stand for 16 h in darkness at room temperature. It remains stable for the following 48 h, and is characterized by an intense green/blue color. $ABTS\bullet^+$ solution was diluted in methanol until the absorbance reached the value of 0.70 ± 0.02 at 734 nm. Different volumes of each tea broth sample (10, 50, 100, 200 µL) were mixed with ABTS solution (3 mL), and the absorbance was recorded at 734 nm after 10 min of incubation at room temperature in the dark, against a blank and a control. Each sample was tested in triplicate. The radical scavenging activity was given as percentage of $ABTS\bullet^+$ radical inhibition ($\%I_{ABTS}$), according to the following formula:

$\%I_{ABTS}$ = [(absorbance of the control − absorbance of the sample)/absorbance of the control)] × 100.

The results were expressed as µg of Trolox equivalents per mL of tea kombucha samples ($\mu g\ TE\ mL^{-1}$).

4.9. Statistical Analysis

Each tea broth from three different preparations, was analyzed in triplicate and, the results were expressed as mean $\pm$ standard deviation (SD). One-way ANOVA method and a Holm-Sidak comparison method via GraphPad Prism 8 were used. The significance was established at p values < 0.05 (*), p < 0.01 (**), p < 0.001 (***), and p < 0.0001 (****).

5. Conclusions

The polyphenol content of kombucha during long-term storage decreases significantly from the fifth month on and becomes one-third of the initial value after nine months. Therefore, a period of up to four months ensures the preservation of polyphenols of kombucha tea and their antioxidant activities. The results of this pilot study highlighted that the "shelf life" of kombucha stored at refrigerator temperature could be no longer than four months, as only during this period the preservation of polyphenol content and its antioxidant activities are ensured.

Author Contributions: Methodology and writing—original draft preparation, C.L.T.; coordinating the work and writing—review and editing, A.F., E.C.; software, P.C.; validation, P.P., R.C. and M.C.C.; conceptualization and supervision, E.C. All authors have read and agreed to the published version of the manuscript.

Funding: C.L.T. is supported by MIUR (Ministero Istruzione Università e Ricerca) fellow grant for PhD students for Translational Medicine doctorate course at University of Calabria.

Molecules **2021**, *26*, 5474

Institutional Review Board Statement: Not applicable.

Informed Consent Statement: Not applicable.

Data Availability Statement: The data presented in this study are available on request from the corresponding author.

Conflicts of Interest: The authors declare no conflict of interest.

Sample Availability: Samples of the compounds are available from the authors.

Abbreviations

EtOH	Ethanol
MeOH	methanol
DMSO	dimethylsulfoxide
TPC	total phenolic content
TFC	total flavonoid content
DPPH	2,2'-diphenyl-1-picrylhydrazyl
ABTS	2,2'-azino-bis(3-ethylbenzothiazolin-6-sulfonic)
Trolox	6-hydroxy-2,5,7,8-tetramethylcroman-2-carboxylic acid
HPLC	High Performance Liquid Chromatography
QE	quercetin equivalents
GAE	gallic acid equivalents
EGCG	epigallocatechin-3-gallate
ECG	epicatechin-3-gallate
EGC	epigallocatechin
EC	epicatechin

References

1. Watawana, M.I.; Jayawardena, N.; Gunawardhana, C.B.; Waisundara, V.Y. Health, wellness, and safety aspects of the consumption of kombucha. *J. Chem.* **2015**, *2015*, 591869. [CrossRef]
2. Jayabalan, R.; Malbaša, R.V.; Lončar, E.S.; Vitas, J.S.; Sathishkumar, M. A review on kombucha tea—Microbiology, composition, fermentation, beneficial effects, toxicity, and tea fungus. *Compr. Rev. Food Sci.* **2014**, *1*, 538–550. [CrossRef]
3. Jayabalan, R.; Malini, K.; Sathishkumar, M.; Swaminathan, K.; Yun, S.E. Biochemical characteristics of tea fungus produced during kombucha fermentation. *Food Sci. Biotechnol.* **2010**, *19*, 843–847. [CrossRef]
4. Vīna, I.; Linde, R.; Patetko, A.; Semjonovs, P. Glucuronic acid from fermented beverages: Biochemical functions in humans and its role in health protection. *Int. J. Recent Res. Appl. Stud.* **2013**, *14*, 17–25.
5. Martínez Leal, J.; Valenzuela Suárez, L.; Jayabalan, R.; Huerta Oros, J.; Escalante-Aburto, A. A review on health benefits of kombucha nutritional compounds and metabolites. *CYTA J. Food* **2018**, *16*, 390–399. [CrossRef]
6. Fazio, A.; La Torre, C.; Caroleo, M.C.; Caputo, P.; Cannataro, R.; Plastina, P.; Cione, E. Effect of addition of pectins from jujubes (*Ziziphus jujuba* Mill.) on vitamin C production during heterolactic fermentation. *Molecules* **2020**, *25*, 2706. [CrossRef]
7. Chu, S.C.; Chen, C. Effects of origins and fermentation time on the antioxidant activities of kombucha. *Food Chem.* **2006**, *98*, 502–507. [CrossRef]
8. Sreeramulu, G.; Zhu, Y.; Knol, W. Kombucha fermentation and its antimicrobial activity. *J. Agric. Food Chem.* **2000**, *48*, 2589–2594. [CrossRef] [PubMed]
9. Sreeramulu, G.; Zhu, Y.; Knol, W. Characterization of antimicrobial activity in Kombucha fermentation. *Acta Biotechnol.* **2001**, *21*, 49–56. [CrossRef]
10. Santos Junior, R.J.; Batista, R.A.; Rodrigues, S.A.; Filho, L.X.; Silva Lima, A. Antimicrobial activity of broth fermented with Kombucha colonies. *J. Microb. Biochem. Technol.* **2009**, *1*, 72–78. [CrossRef]
11. Kozyrovska, N.O.; Reva, O.M.; Goginyan, V.B.; De Vera, J.P. Kombucha microbiome as a probiotic: A view from the perspective of post-genomics and synthetic ecology. *Biopolym. Cell* **2012**, *28*, 103–113. [CrossRef]
12. Cannataro, R.; Fazio, A.; La Torre, C.; Caroleo, M.C.; Cione, E. Polyphenols in the Mediterranean Diet: From Dietary Sources to microRNA Modulation. *Antioxidants* **2021**, *10*, 328. [CrossRef] [PubMed]
13. Cione, E.; La Torre, C.; Cannataro, R.; Caroleo, M.C.; Plastina, P.; Gallelli, L. Quercetin, Epigallocatechin Gallate, Curcumin, and Resveratrol: From Dietary Sources to Human MicroRNA Modulation. *Molecules* **2019**, *25*, 63. [CrossRef]
14. Jayabalan, R.; Marimuthu, S.; Thangaraj, P.; Sathishkumar, M.; Binupriya, A.R.; Swaminathan, K.; Yun, S.E. Preservation of Kombucha Tea—Effect of Temperature on Tea Components and Free Radical Scavenging Properties. *J. Agric. Food Chem.* **2008**, *56*, 9064–9071. [CrossRef] [PubMed]

15. Plastina, P.; Gabriele, B.; Fazio, A. Characterizing traditional rice varieties grown in temperate region of Italy: Free and Bound phenolic and lipid compounds and in vitro antioxidant properties. *Food Qual. Saf.* **2018**, *2*, 89–95. [CrossRef]
16. Sembiring, L.N.; Elya, B.; Sauriasari, R. Phytochemical screening, Total Flavonoid and Total Phenolic content and Antioxidant Activity of Different Parts of *Caesalpinia bonduc* (L.) Roxb. *Pharmacogn. J.* **2018**, *10*, 123–127. [CrossRef]
17. Essawet, N.A.; Cvetković, D.; Velićanski, A.; Canadanović-brunet, J.; vulić, J.; Maksimović, V.; Markov, S. Polyphenols and antioxidant activities of kombucha beverage enriched with coffeeberry extract. *Chem. Ind. Chem. Eng. Q* **2015**, *21*, 399–409. [CrossRef]
18. Plastina, P.; Fazio, A.; Gabriele, B. Comparison of fatty acid profile and antioxidant potential of extracts of seven Citrus rootstock seeds. *Nat. Prod. Res.* **2012**, *26*, 2182–2187. [CrossRef] [PubMed]
19. Voidarou, C.; Antoniadou, M.; Rozos, G.; Tzora, A.; Skoufos, I.; Varzakas, T.; Lagiou, A.; Bezirtzoglou, E. Fermentative Foods: Microbiology, Biochemistry, Potential Human Health Benefits and Public Health Issues. *Foods* **2021**, *10*, 69. [CrossRef]
20. Skocińska, N.; Sionek, B.; Ścibisz, I.; Kołożyn-Krajewska, D. Acid contents and the effect of fermentation condition of Kombucha tea beverages on physicochemical, microbiological and sensory properties. *CyTA-J. Food* **2017**, *15*, 601–607. [CrossRef]
21. Cardoso, R.R.; Neto, R.O.; Dos Santos D'Almeida, C.T.; do Nascimento, T.P.; Pressete, C.G.; Azevedo, L.; Stampini Duarte Martino, H.; Cameron, L.C.; Ferreira, M.S.L.; de Barros, F.A.R. Kombuchas from green and black teas have different phenolic profile, which impacts their antioxidant capacities, antibacterial and antiproliferative activities. *Food Res. Int.* **2020**, *128*, 108782. [CrossRef]
22. Jakubczyk, K.; Kałduńska, J.; Kochman, J.; Janda, K. Chemical Profile and Antioxidant Activity of the Kombucha Beverage Derived from White, Green, Black and Red Tea. *Antioxidants* **2020**, *9*, 447. [CrossRef]
23. Xu, Y.Q.; Chen, J.X.; Du, Q.Z.; Yin, J.F. Improving the quality of fermented black tea juice with oolong tea infusion. *J. Food Sci. Technol.* **2017**, *54*, 3908–3916. [CrossRef]
24. Ning, Y.; Wu, Z.; Li, Z.; Meng, R.; Xue, Z.; Wang, X.; Lu, X.; Zhang, X. Optimization of Fermentation Process Enhancing Quality of Dandelion Black Tea on the Functional Components, Activity and Sensory Quality. *Open Access Libr. J.* **2020**, *7*, 1–11. [CrossRef]
25. Grumezescu, A.M.; Holban, A.M. The Science of Beverages. In *Fermented Beverages*, 1st ed.; Woodhead publishing-Elsevier: Sawston, UK, 2019; Volume 5, Chapter 10.
26. Jayabalan, R.; Subathradevi, P.; Marimuthu, S.; Sathishkumar, M.; Swaminathan, K. Changes in free-radical scavenging ability of kombucha tea during fermentation. *Food Chem.* **2008**, *109*, 227–234. [CrossRef] [PubMed]
27. Di Renzo, L.; Gualtieri, P.; Pivari, F.; Soldati, L.; Attinà, A.; Cinelli, G.; Leggeri, C.; Caparello, G.; Barrea, L.; Scerbo, F.; et al. Eating habits and lifestyle changes during COVID-19 lockdown: An Italian survey. *J. Transl. Med.* **2020**, *18*, 229. [CrossRef] [PubMed]
28. Janssen, M.; Chang, B.P.I.; Hristov, H.; Pravst, I.; Profeta, A.; Millard, J. Changes in Food Consumption during the COVID-19 Pandemic: Analysis of Consumer Survey Data From the First Lockdown Period in Denmark, Germany, and Slovenia. *Front Nutr.* **2021**, *8*, 635859. [CrossRef] [PubMed]
29. Ganatsios, V.; Nigam, P.; Plessas, S.; Terpou, A. Kefir as a Functional Beverage Gaining Momentum towards Its Health Promoting Attributes. *Beverages* **2021**, *7*, 48. [CrossRef]
30. Available online: https://www.forbes.com/sites/bridgetshirvell/2020/08/13/how-hard-kombucha-became-the-drink-of-2020 (accessed on 6 September 2021).
31. Katarzyna Kapp, J.M.; Sumner, W. Kombucha: A systematic review of the empirical evidence of human health benefit. *Ann. Epidemiol.* **2019**, *30*, 66–70. [CrossRef] [PubMed]
32. Srinivasan, R.; Smolinske, S.; Greenbaum, D. Probable gastrointestinal toxicity of Kombucha tea: Is this beverage healthy or harmful? *J. Gen. Int. Med.* **1997**, *12*, 643–644. [CrossRef] [PubMed]
33. SungHee Kole, A.; Jones, H.D.; Christensen, R.; Gladstein, J. A case of Kombucha tea toxicity. *J. Int. Care Med.* **2009**, *24*, 205–207. [CrossRef] [PubMed]

 *molecules*

Review

Latest Insights on Novel Deep Eutectic Solvents (DES) for Sustainable Extraction of Phenolic Compounds from Natural Sources

Julio Serna-Vázquez [1], Mohd Zamidi Ahmad [2], Grzegorz Boczkaj [3] and Roberto Castro-Muñoz [3,4,*]

1. Tecnologico de Monterrey, Campus Ciudad de México, Calle del Puente 222, Ejidos de Huipulco, Ciudad de México 14380, Mexico; juliosernav@outlook.es
2. Organic Materials Innovation Center (OMIC), Department of Chemistry, University of Manchester, Oxford Road, Manchester M13 9PL, UK; mohdzamidi.ahmad@manchester.ac.uk
3. Department of Process Engineering and Chemical Technology, Faculty of Chemistry, Gdansk University of Technology, 11/12 Narutowicza St., 80-233 Gdansk, Poland; grzegorz.boczkaj@pg.edu.pl
4. Tecnologico de Monterrey, Campus Toluca, Av. Eduardo Monroy Cárdenas 2000 San Antonio Buenavista, Toluca de Lerdo 50110, Mexico
* Correspondence: food.biotechnology88@gmail.com or castromr@tec.mx

Abstract: Phenolic compounds have long been of great importance in the pharmaceutical, food, and cosmetic industries. Unfortunately, conventional extraction procedures have a high cost and are time consuming, and the solvents used can represent a safety risk for operators, consumers, and the environment. Deep eutectic solvents (DESs) are green alternatives for extraction processes, given their low or non-toxicity, biodegradability, and reusability. This review discusses the latest research (in the last two years) employing DESs for phenolic extraction, solvent components, extraction yields, extraction method characteristics, and reviewing the phenolic sources (natural products, by-products, wastes, etc.). This work also analyzes and discusses the most relevant DES-based studies for phenolic extraction from natural sources, their extraction strategies using DESs, their molecular mechanisms, and potential applications.

Keywords: green solvents; biologically active compounds; selective separation; medicinal plants; ultrasonic-assisted extraction; microwave-assisted extraction

Citation: Serna-Vázquez, J.; Ahmad, M.Z.; Boczkaj, G.; Castro-Muñoz, R. Latest Insights on Novel Deep Eutectic Solvents (DES) for Sustainable Extraction of Phenolic Compounds from Natural Sources. *Molecules* **2021**, *26*, 5037. https://doi.org/10.3390/molecules26165037

Academic Editor: Mirella Nardini

Received: 12 July 2021
Accepted: 18 August 2021
Published: 19 August 2021

Publisher's Note: MDPI stays neutral with regard to jurisdictional claims in published maps and institutional affiliations.

1. Introduction

Since the beginnings of humankind, medicines have been a crucial part of our progress as a species. The search for remedies against diseases probably started with plants, as evidenced by archaeological artifacts that demonstrated the use of medicinal plants during the Paleolithic. Furthermore, the oldest written evidence was found on a Sumerian clay slab, dating back to approximately 3000 B.C. Many medicinal plant benefits are attributed to phenolic compounds, given their antioxidant, anticancer, antibiotic, antifungal, and anti-inflammatory activities [1,2]. Phenolic compounds contain at least one aromatic ring with one or more hydroxyl groups. They play an essential role in plant growth, reproduction, and protection against parasitoids, pathogens, and predators [3–5]. There are more than 8000 plant phenolic compounds, and their extraction from natural sources is of high interest in the industry due to their use in pharmaceuticals, beverages, food, and cosmetics [6,7].

Typically, phenolic compounds are extracted using organic solvents (e.g., methanol, ethanol, acetone, ethyl acetate, hexane, benzene, etc.) or petroleum-based solvents [8,9]. Nevertheless, due to environmental, health, and safety concerns, a large amount of research has been done in order to synthesize safer extraction solvents from renewable sources [10]. Ionic liquids (ILs) are presented as an alternative to the solvents described above, given their notable attributes, such as high thermal and chemical stability, non-flammability, and low vapor pressure [11]. However, their high cost, toxicity, dangerous decomposition by-products, and poor biodegradation levels are matters of concern, and given that some of the most used ILs constituents (i.e., imidazolium, pyridinium) are petroleum-derived, further research on alternative solvents has been carried out, leading to the creation of deep eutectic solvents (DES) [12].

The term DES was first introduced in 2003 by Andrew P. Abbott and his group when creating eutectic mixtures of urea and quaternary ammonium salts that could be used as solvents at room temperature as a result of hydrogen bonding between the species that make up these solvents [13]. Throughout the years, DESs have been catalogued as great substitutes for conventional solvents considering their similar physical properties to ILs, presenting advantageous characteristics such as low cost, biodegradability, non-toxicity, low volatility, easy preparation, and even biocompatibility [14,15]. Additionally, phenolic stability in DESs has been reported [16,17]; hydrogen bonding between DESs and phenolic compounds prevents degradation by reducing the phenolic molecules' movement, thus decreasing their contact time with air and simultaneously avoiding oxidative degradation [18]. Furthermore, DESs can be designed for highly efficient extraction of specific compounds improving their bioactivity and stability compared to traditional solvents [19].

Given the high value of several phenolic compounds for pharmaceutical and food industries, particular interest is shown in DES-based extraction methods since various DES-constituting compounds are approved for human consumption as food additives or supplements. This is the case of choline chloride (ChCl), the most widely studied hydrogen bond acceptor (HBA) in DESs [20,21], as well as other components including 1,2-propanediol [22], organic acids, such as acetic, citric, lactic, malic, and tartaric acids [23]; sugars like fructose, glucose, sucrose, ribose, and mannose [24–26]; and amino acids like L-proline and L-alanine [27–29].

DES importance also relies on their wide application range, including analytical chemistry, biomolecule extraction, biocatalysis, biomass processing, biodiesel and gas separation, metallurgy, nanomaterials synthesis, therapeutics, food and water sample analysis [10,15,30–36]. Their high potential results from almost unlimited possible chemical combinations for their synthesis. This way, a fit-to-purpose DES with unique selectivity can be synthesized [37–39]. DES research has had substantial growth throughout the years (illustrated in Figure 1), owing to its remarkable properties and its large application field, including phenolic compounds extraction from natural sources.

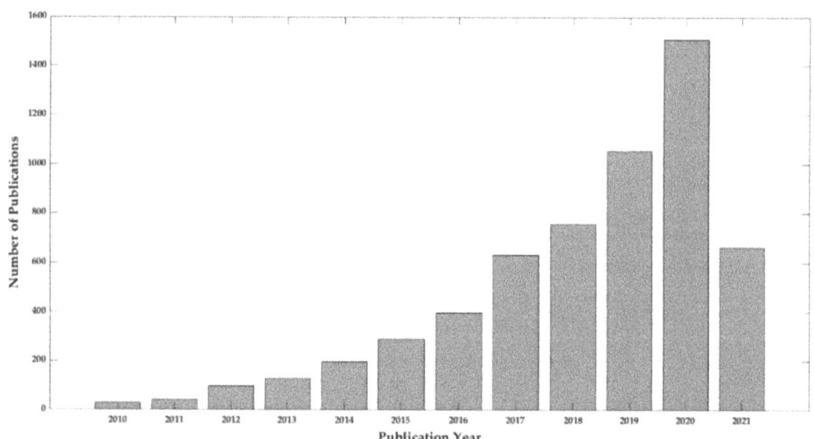

Figure 1. Number of publications that include the term "deep eutectic solvent" in their title since 2010. Source: Web of Science, 12 June 2021 [40].

This review summarizes the most relevant studies in the last year involving phenolic compounds extraction from natural sources through different DES-based extraction methods, where molecular mechanisms involved in each extraction procedure are discussed. This work also aims to exhibit the current DES research and its potential use in various industries.

2. Principles of DES

DESs are mixtures of two or more compounds that present a significantly lower eutectic point than that of the ideal liquid mixture, as well as being liquid at ambient temperatures [41]. These solvents consist of a compound acting as a hydrogen bond donor (HBD) and a compound that acts as HBA [30,42]. DESs, similar to other eutectic mixture, possess a lower melting point than their components in their pure form [43,44]. The two most studied DES subgroups are natural deep eutectic solvents (NADES), which are based on sugars, alcohols, organic acids, amino acids, or amines [45], and therapeutic deep eutectic solvents (THEDES), where one or more of their components is an active pharmaceutical ingredient [46]. NADESs have been indicated as possible solvents present in living cells, thus explaining the presence of compounds at much higher concentrations than what is soluble in aqueous solutions [47]. Furthermore, NADESs reduce physicochemical constraints of metabolites transport and cellular processes through the formation of liquid microenvironments [14].

DES composition can be described with the general formula, as follows:

$$Cat^+X^-zY, \tag{1}$$

Cat^+ is an ammonium, phosphonium, or sulfonium cation; X^- is a Lewis base, commonly a halide anion; Y is a Lewis or Brønsted acid, and z represents the number of Y molecules interacting with X^-, the anion [48,49].

As reported in Table 1, Abbot's group have classified DES into four different types [48,50].

Table 1. Classification of DES. Adapted from [48].

Type	Formula	Terms
Type I	$Cat^+X^-zMCl_x$	M = Zn, Sn, Fe, Al, Ga, In
Type II	$Cat^+X^-zMCl_x \cdot yH_2O$	M = Cr, Co, Cu, Ni, Fe
Type III	Cat^+X^-zRZ	Z = $CONH_2$, COOH, OH
Type IV	$MCl_x + RZ = MCl_{x-1}^+ \cdot RZ + MCl_{x+1}^-$	M = Al, Zn Z = $CONH_2$, OH

Type III DESs are the most studied and are commonly based on ChCl, an extensively used HBA for its low cost, non-toxicity, and biodegradability [49]. Type III HBDs are generally alcohols, amides, amino acids, carboxylic acids, and sugars [49,51].

The two most used methods in DES preparation are (1) grinding and (2) mixing and heating, preferred due to their simplicity. Grinding involves the use of a mortar and pestle where the DES components are triturated at room temperature until a homogeneous solution is formed. Mixing and heating is carried out in a container under constant stirring at a set temperature (usually between 50 and 100 °C). The DES precursors are mixed until the eutectic solution is formed [49]. Other preparation methods are freeze-drying and evaporation [11,52]. Several DES-based extraction methods have been reported; in fact, many of them are well-established methods that have been operated with conventional solvents, now adapted for DES uses. Several of the most reported methods include ultrasound-assisted extraction (UAE) and microwave-assisted extraction (MAE); however, traditional heating and stirring continues to be employed because of its simplicity and low cost.

DES viscosity is an important property that affects extraction efficiency as lower viscosities are correlated with a higher mass transfer and extraction yield [53]. Essentially, the two options to reduce viscosity are water addition or changing one or more of the solvent components [20]. Through the hole theory, Abbot et al. [54] concluded that high viscosity is attributed to solvent species having a larger molecular radius compared to the average void radius. This hindrance can be overcome by using smaller cations or fluorinated HBDs [54]. Viscosity reduction through co-solvents addition (e.g., water) is commonly used for mass transfer enhancement. Water increases solvent polarity, which might be beneficial for certain compound extraction, depending on their polarities. However, water addition can result in loss of the DES molecular structure caused by the weakening of intramolecular interactions between the DES components [43].

Hydrogen bond strength between the DES components and cation symmetry are key parameters in DES synthesis. The most significant freezing point depressions are observed in those mixtures where the DES components have a higher ability to form hydrogen bonds and a lower cation symmetry [55]. Other factors affecting DESs physicochemical characteristics are the charge delocalization process from hydrogen bonding formation, the electron density in hydrogen bond networks, and the presence or absence of functional groups, leading to different supramolecular structures with different melting points [56]. A schematic diagram of hydrogen bonding in a ChCl:urea DES is presented in Figure 2.

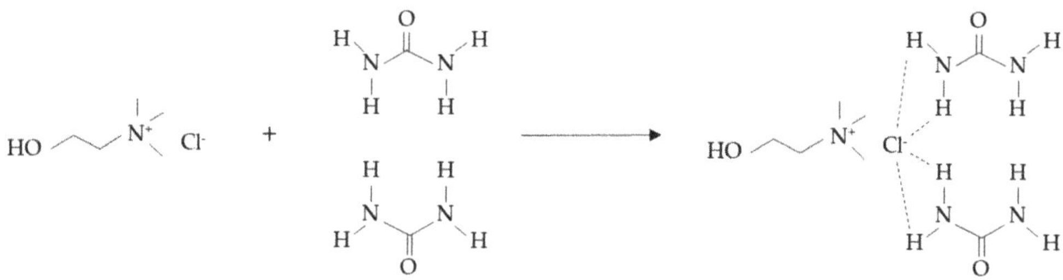

Figure 2. Hydrogen bonding in a ChCl:urea DES.

3. Phenolic Compound Extraction from Natural Sources

The section presents the studies reported on DES-based phenolic extraction from natural sources, emphasizing their molecular mechanisms, characteristics and advantages of each extraction process. Table 2 summarizes the sources, DES components and their molar ratio, co-solvent addition, extraction method, and extraction yield of each study.

Table 2. Phenolic compound extraction processes presented in this review.

Source	DES Used (Molar Ratio)	Co-Solvent	Extraction Method and Conditions	Yield	Reference
Manuka leaves (*Leptospermum scoparium*)	ChCl:Lactic acid (1:2)	-	Heating and stirring Temperature: 50 °C Extraction time: 67 min Liquid-solid ratio: 100:5.07	TPC: 59.82 mg GAE g^{-1}	[57]
Polygonum cuspidatum root	TBAC:Ethylene glycol (1:3)	40% water 4.5% HCl	Heating and stirring Temperature: 80 °C Extraction time: 80 min Rotational speed: 500 rpm Liquid–solid ratio: 40:1	Resveratrol: 12.26 ± 0.14 mg g^{-1}	[58]

Table 2. *Cont.*

Source	DES Used (Molar Ratio)	Co-Solvent	Extraction Method and Conditions	Yield	Reference
Buddleja globosa Hope leaves	ChCl:Lactic acid (1:1)	20% water	Heating and stirring Temperature: 60 °C Extraction time: 50 min Rotational speed: 340 rpm Liquid–solid ratio: 100:1	TPC: ca. 175 mg GAE g^{-1} Luteolin 7-O-glucoside: ca 4.1 mg g^{-1} Verbascoside: ca. 39 mg g^{-1}	[59]
	L-proline:Citric acid (1:1)	20% water		TPC: ca. 175 mg GAE g^{-1} Luteolin 7-O-glucoside: ca. 4 mg g^{-1} Verbascoside: 51.045 mg g^{-1}	
Rosemary leaves (*Rosmarinus officinalis* L.)	ChCl:1,2-propanediol (1:2)	50% water	Heating and stirring Temperature: 65 °C Extraction time: 15 min Rotational speed: 600 rpm Liquid–solid ratio: 40:1	TPC: 78 mg GAE g^{-1}	[60]
Black mustard (*Brassica nigra* L.) seed	TEOA:Glycerol (1:2)	25% ethanol	Heating and stirring Temperature: 65 °C Extraction time: 180 min Rotational speed:1000 rpm Liquid–solid ratio: 10:1	TPC: 32.2 ± 0.2 mg GAE g^{-1} TFC: 7.2 ± 0.2 mg QE g^{-1}	[61]
	TEOA:Propylene glycol (1:2)	25% ethanol		TPC: 29.3 ± 0.3 mg GAE g^{-1} TFC: 7.4 ± 0.3 mg QE g^{-1}	
Coffee	L-serine:Lactic acid (1:4)	30% methanol	Two-phase hollow fiber-liquid microextraction (HF-LPME) Fiber material: polypropylene HF DES immersion time: 30 s Rotational speed: 840 rpm	Caffeic acid: 0.032 µg g^{-1}	[62]
Green Tea				Caffeic acid: 0.022 µg g^{-1}	
Tomato				Caffeic acid: 0.012 µg g^{-1}	
Green tea	ChCl:ethylene glycol (1:2)	29% water	Heating and stirring Temperature: 84 °C Extraction time: 39 min Liquid–solid ratio: 44:1	Phenolic extraction yield: 20.12%	[63]
Radix et Rhixoma Rhei	1-tetradecanol:10-undecenoic acid (1:4)	10% HCl	Heating and stirring Temperature: 67 °C Extraction time: 20 min Rotational speed: 500 rpm Liquid–solid ratio: 12:1	Anthraquinones: 21.52 mg g^{-1}	[64]
Turmeric (*Curcuma longa*)	Sucrose:Lactic acid:Water (1:5:7)	-	Microwave-assisted extraction Temperature: 68.2 °C Extraction time: 15.4 min Maximum power: 1500 W Liquid–solid ratio: 14.5:2	Curcumin: 32.00 mg g^{-1}	[65]
Lour (*Osmanthus fragrans*) flower	Lactic acid:glucose (1:5)	85% water	Microwave-assisted extraction Extraction time: 0.5 s Power: 500 W Liquid–solid ratio: 30:1 Microwave-assisted extraction Extraction time: 60.00 s Power: 500 W Liquid–solid ratio: 38.41:1	TPC: 192.20 ± 1.08 mg GAE g^{-1} TFC: 560.13 ± 0.51 mg RE g^{-1} TPC: 238.09 ± 0.81 mg GAE g^{-1} TFC: 321.07 ± 1.65 mg RE g^{-1}	[66]

Table 2. *Cont.*

Source	DES Used (Molar Ratio)	Co-Solvent	Extraction Method and Conditions	Yield	Reference
Orange peel waste	ChCl:Ethylene glycol (1:2)	50% water	Stirring Extraction time: 7 days Rotational speed: 850 rpm Liquid–solid ratio: 5:1	Polyphenols: 45.7 mg g^{-1}	[67]
Brewer's spent grain	ChCl:Glycerol (1:2)	37.46% water	Microwave-assisted extraction Temperature: 100 °C Extraction time: 13.3 min Maximum power: 700 W	TPC: 2.89 mg GAE g^{-1}	[68]
Maritime Pine (*Pinus pinaster*) wood	Lactic acid:Tartaric acid:ChCl (4:1:1)	-	Heating and stirring Temperature: 175 °C Extraction time: 90 min Liquid–solid ratio: 10:1	Lignin removal: 94.9 ± 3.3%	[69]
Poplar (*Populus*) sawdust	ChCl:Lactic acid (1:10)	-	Heating and stirring Temperature: 130 °C Extraction time: 90 min Liquid–solid ratio: 10:1	Lignin removal: 89.3%	[70]
Oil palm (Elaeis guineensis) frond	ChCl:Urea (1:2)	20% water 2.5% NaOH	Ultrasound-assisted extraction Extraction time: 30 min Ultrasound amplitude: 60%	Lignin removal: 47.00 ± 0.16%	[71]
Blueberry (*Vaccinium* spp.) pomace	ChCl:Oxalic acid (1:1)	30% water	Pulse- Ultrasound-assisted extraction Temperature: 76 °C Extraction time: 3.2 min Power: 325 W Liquid–solid ratio: 60:1	Total anthocyanin yield: 24.27 ± 0.05 mg C3GE g^{-1}	[72]
Lily (*Lilium lancifolium* Thunb.)	ChCl:Ethylene glycol (1:2)	20% water	Ultrasound-assisted extraction Temperature: 50 °C Extraction time: 40 min Liquid–solid ratio: 25:1	Regaloside B: 0.31 ± 0.06 mg g^{-1} Regaloside C: 0.29 ± 0.03 mg g^{-1} Regaloside E: 3.04 ± 0.38 mg g^{-1}	[73]
Lavandula pedunculata subsp. *lusitanica*	Proline:Lactic acid (1:1)	30% water	Ultrasound-assisted extraction Extraction time: 60 min Frequency: 37 kHz	TPC: 56.00 ± 0.77 mg GAE g^{-1}	[74]
Sage (*Salvia officinalis*)	ChCl:Lactic acid (1:2)	10% water	Stirring and heating Temperature: 50 °C Extraction time: 90 min	Carnosic acid: 14.43 ± 0.28 µg mg^{-1} Carnosol: 4.83 ± 0.09 µg mg^{-1}	[75]
	ChCl:Lactic acid (1:2)	10% water	Ultrasound-assisted extraction Temperature: 70 °C Extraction time: 60 min	Carnosic acid: 14.00 ± 0.02 µg mg^{-1} Carnosol: 4.18 ± 0.05 µg mg^{-1}	
	ChCl:Lactic acid (1:2)	10% water	MCE Extraction time: 2 min Vibration speed: 5 m s^{-1}	Carnosic acid: 8.26 ± 0.45 µg mg^{-1} Carnosol: 1.87 ± 0.33 µg mg^{-1}	

Table 2. *Cont.*

Source	DES Used (Molar Ratio)	Co-Solvent	Extraction Method and Conditions	Yield	Reference
Turmeric (*Curcuma longa*)	ChCl:Lactic acid (1:1)	20% water	Ultrasound-assisted extraction—direct sonication Temperature: 30 °C Extraction time: 20 min Frequency: 22 kHz Intensity: 70.8 W cm^{-2} Pulse mode: 60% Particle size: 0.355 mm Liquid–solid ratio: 20:1	Curcuminoids: 77.13 mg g^{-1}	[76]
Olive leaf	ChCl:Fructose:Water (5:2:5)	42.69% water	Ultrasound-assisted extraction Temperature: 75 °C Extraction time: 60 min Power: 140 W Frequency: 37 kHz Liquid–solid ratio: 40.66:1	TPC: 187.31 ± 10.3 mg GAE g^{-1} TFC: 12.75 ± 0.6 mg ApE g^{-1}	[77]
Helichrysum arenarium L. inflorescences	ChCl:1,2-propanediol (1:2) ChCl:1,4-butanediol (1:6)	25% water	Ultrasound-assisted extraction Temperature: 50 °C Extraction time: 60 min	Phenolics: 30.15 ± 0.42 mg g^{-1} Phenolics: 29.75 ± 0.49 mg g^{-1}	[78]
Olive oil	ChCl:Xylitol:Water (2:1:3)	-	Heating and stirring Temperature: 40 °C Extraction time: 60 min Liquid:oil ratio: 1:1	Phenolics: 555.36 ± 21.93 mg kg^{-1}	[79]
Olive oil	ChCl:Extracted phenolics	-	Rotating disk sorptive extraction Extraction time: 15 min Disk material: poly(vinylidene fluoride-co-tetrafluoroethylene) Rotational speed: 150 rpm ChCl concentration: 300 g L^{-1}	Absolute extraction recovery Gallic acid: 87% Protocatechuic acid: 79% Tyrosol: 76% Vanillic acid: 75% *p*-coumarinic acid: 82% Syringaldehyde: 81% Thymol: 66%	[80]

Phenolic compound extraction from natural sources is necessary to obtain high-added value compounds that are crucial feedstock in the manufacture of several pharmaceutical, cosmetic, food, and nutraceutical products [81]. Manuka, the most widespread native shrub in New Zealand, possesses great importance in the country since it has long been used as a traditional medicine with antibacterial and antioxidant properties associated with its high phenolic content, and nowadays, it is a notable source of honey and essential oils. Alsaud et al. [57] studied phenolic extraction from manuka (*Leptospermum scoparium*) leaves using eight different DESs, constituted by either ChCl or tetrabutylammonium chloride (TBAC) as the HBA, combined with either ethylene glycol, 1,3-propanediol, acetic acid, or lactic acid as the HBD. The DESs were synthesized by stirring at 80 °C. Extraction was performed in a stirred vessel at room temperature for 1 h, with 5% biomass. The obtained extraction yields with different DESs were compared with the one achieved with ethanol as the extraction solvent; TPC (Total phenolic content) of 45.04 mg GAE g^{-1} (GAE = Gallic Acid Equivalents). Three of the DESs, ChCl:ethylene glycol (molar ratio of 1:2), ChCl:lactic acid (1:2), and ChCl:1,3-propanediol (1:3), showed higher extraction efficiencies than ethanol by presenting a final yield of 56.87, 52.51, and 50.67 mg GAE g^{-1}, respectively. The high efficiency of ChCl:ethylene glycol and ChCl:1,3-propanediol DESs was attributed to their lower viscosity. However, the superior outcome of ethylene glycol-based DES was ascribed to its ability to form multiple hydrogen bonds with phenolic compounds and be only a two-carbon chain-length, which is believed to interact more easily than other polyalcohols [82]. Regarding the carboxylic acid-based DESs, the more viscous lactic acid DES presented a higher extraction efficiency than using acetic acid solvent, accredited to its higher number of OH groups. Using the ChCl:lactic acid DES, the authors reported the optimal extraction conditions at 50 °C, extraction duration of 1.07 h using 5.07% biomass, where they successfully extracted 59.82 mg GAE g^{-1}. Additionally, the extracted phenolic compounds presented high stability in DES, even after eight days in storage.

Resveratrol is a phenolic compound widely used in the pharmaceutical and food industry due to its anti-inflammatory, anti-oxidative and anti-viral properties [83]. A high level of resveratrol is found in *Polygonum cuspidatum*, particularly in the root. However, it is majorly present as polydatin, its glycoside form [84], consequently, more efficacious methods of extraction and conversion of polydatin to resveratrol have been searched to upgrade or replace traditional methods, which are either costly, time-consuming, non-efficient, or harmful to the environment. Recently, Sun et al. [58] developed a one-pot polydatin-to-resveratrol extraction and conversion method from *Polygonum cuspidatum* root using eleven DESs. The method consists of stirring (at 500 rpm) the plant-DES mixture tube (with 4.5% HCl) in an oil bath at 80 °C for 80 min. Five out of the eleven DESs (tetraethylammonium chloride(TEA):ethylene glycol (1:2), TBAC:ethylene glycol (1:3), TEA:glycerin (1:2), TEA:1,4-butanediol (1:2), and triethyl benzyl ammonium chloride (TEBAC):1,4-butanediol (1:2)) presented higher resveratrol extraction yields than methanol and ethanol, tested at similar test conditions. The highest extraction efficiency was obtained with TBAC:ethylene glycol DES (9.00 ± 0.28 mg/g), attributed to its low viscosity, low pH, and a stronger hydrophobic interaction between the solvent and resveratrol. The optimized extraction parameters, determined using the TBAC:ethylene glycol solvent, are: stirring at 80 °C for 80 min in a 40% (*v/v*) water in DES, 4.5% (*m/v*) HCl, 40:1 liquid–solid ratio. The suggested method was compared against UAE and produced a higher resveratrol yield even though the presence of cavitation facilitates interaction between the solvent and the plant, thus proving the effectiveness of the stirring step. Contrarily, UAE could not penetrate the plant powder deposited at the bottom of the tube, leading to unextracted or unconverted compounds. With these optimum conditions, the extraction–conversion process reached a resveratrol yield of 12.26 ± 0.14 mg g^{-1}.

Torres-Vega et al. [59] reported polyphenols extraction from *Buddleja globosa* Hope leaves through heating and mixing with DESs. *Buddleja globosa* is a medicinal plant traditionally used by the Mapuche for the treatment of wounds and dermatological, gastrointestinal, and hepatic disorders [85,86]. Water was added to all the prepared DESs, producing 20% water aqueous solvents, and the extractions were carried out at 60 °C and 340 rpm stirred for 50 min with 0.1 g plant in 10 mL solvent. Eight DESs were used; however, only five were suitable for individual phenolic quantification given the poor separation of forsythoside B and verbascoside in the other three. As for total phenolic extraction, only the ChCl:1,2-propanediol (1:3) solvent presented a significantly lower yield than 80% compared with methanol under the same extraction conditions. Whereas, the lactic acid:glycerol:water (3:1:3) and ChCl:glycerol (1:2) solvents did not show significant differences with the methanol extraction yield. Nevertheless, the other two DESs, ChCl:lactic acid (1:1) and L-proline:citric acid (1:1), displayed a significantly better total phenolic recovery than methanol. All five solvents showed significantly higher luteolin 7-*O*-glucoside extraction yields than methanol extraction, similarly for verbascoside yields except for the ChCl:1,2-propanediol solvent. This solvent exhibited the highest luteolin 7-*O*-glucoside extraction yield but the lowest for verbacoside and total phenolic yields. The trend is due to ChCl:1,2-propanediol's higher extraction affinity towards flavonoids, similarly reported in other studies [87]. The poor extraction of the other phenolic compounds is probably due to its polarity. Using this new procedure, it was concluded that the ChCl:lactic acid and L-proline:citric acid DES solvents presented the highest phenolic extraction capacities from *Buddleja globosa* and are promising alternatives to traditional solvents.

Wojeicchowski et al. [60] screened ten DESs to extract phenolic compounds from rosemary (*Rosmarinus officinalis* L.) leaves. The extraction was done by heating and stirring at 30 °C and 600 rpm for 15 min with a liquid-solid ratio of 20:1. All DESs were prepared with ChCl as the HBA with various HBDs including acetic acid (1:2), lactic acid (1:2), oxalic acid (1:1), 1,2-propanediol (1:2), ethylene glycol (1:2), glycerol (1:2), xylitol (2:1), sorbitol (2:1), xylose (2:1), and zinc chloride (1:2). The extraction performances of pure DESs and aqueous DESs (30% water) were compared to pure ethanol and 70% ethanol. From the ten pure DESs, phenolic content was not observed in six of them, associated with their high viscosities. Meanwhile, the ten aqueous DESs showed higher extraction yields than 70% ethanol and significantly lower extraction yields than pure ethanol. The extraction optimization was conducted using the best DES (ChCl:1,2-propanediol) and was determined to be at 65 °C, with a 40:1 liquid-solid ratio with a 50% aqueous DES, resulting in the highest total phenolic yield of 78 mg GAE g^{-1}.

Đorđević et al. [61] reported total phenolic and total flavonoid extractions from ground black mustard (*Brassica nigra* L.) seed using triethanolamine (TEOA):glycerol, TEOA:propylene glycol, and ChCl:urea DESs. The DESs were prepared with a molar ratio of 1:2, along with DES mixtures with 25% water or ethanol. Extraction was performed through heating at 65 °C for 3 h and stirring at 1000 rpm, and a liquid–solid ratio of 10:1 (v/w). Out of the three pure DESs, the highest total phenolic extraction was obtained with the TEOA:propylene glycol solvent, providing 23.5 ± 0.7 mg of GAE g^{-1}. Nevertheless, the highest yield was obtained with a mixture of TEOA:glycerol with 25% ethanol, producing 32.2 ± 0.2 mg of GAE g^{-1}. Overall, it was observed that the addition of ethanol or water to each DES increased the extracted phenolic content, owing to the reduced solvent viscosity that simultaneously enhanced the phenolic transfer between the seeds and the DES [12,88]. However, the TEOA:propylene glycol-water mixture behaved differently. It displayed a reduced extraction yield, which is attributed to DES nanostructure debilitation or disintegration due to water-DES interaction, resulting in a decline in solvent extraction capability [3,43,89]. Regarding the total flavonoids content (TFC), the TEOA:propylene glycol with 25% ethanol showed the highest yield at 7.4 ± 0.3 mg of QE (quercetin equivalents) g^{-1}. The higher phenolic and flavonoid recoveries using ethanol-DES mixtures were attributed to an increase in selectivity due to their polarity.

Nia and Hadjmohammadi [62] performed caffeic acid extraction from coffee, green tea, and tomato through a two-phase hollow fiber-liquid microextraction (HF-LPME) using a hydrophobic DES as an acceptor phase. The two-phase HF-LPME consisted of a porous polypropylene hollow fiber filled with an acceptor solution, immersed in an immiscible solution of an aqueous sample and donor phase (see Figure 3). The target analytes in the aqueous sample were extracted into the liquid-supported membrane. The solvent selection for acceptor solution is crucial for an adequate extraction performance, as the solvent must be immiscible with the donor phase and have a high affinity with both the hollow fiber and the target compound. Conventional organic solvents are typically used for extraction; nevertheless, they present significant drawbacks such as instability, toxicity and volatility. More advanced solvents such as ionic liquids have also been used as the acceptor phase; however, their anions can hydrolyze into toxic species, in addition to their readily high cost and time-consuming synthesis. Hence, DESs are the obvious solvents of choice and have been proposed for HF-LPME [90–92]. For this study, three DESs were prepared: L-leucine:lactic acid, L-arginine:lactic acid, and L-serine:lactic acid, each at 1:3, 1:4, and 1:5 molar ratios. Based on the physicochemical properties of the produced DESs, the 1:4 molar ratio L-serine:lactic acid solvent was selected due to its high capacity to form π-type hydrogen bonds between the caffeic acid's conjugated aromatic rings and the solvent's polar groups. Moreover, to further reduce the DES viscosity, 30% methanol was added as a co-solvent. In the experiment, the hollow fiber was first immersed into the DES solvent for

30 s, followed by a 50 µL DES addition into the hollow fiber lumen side. Subsequently, the DES-supported hollow fiber was placed into the sample solution and stirred. The caffeic acid extraction yields from tomato, green tea, and coffee sample obtained were 0.032, 0.022, and 0.012 µg g^{-1}, respectively.

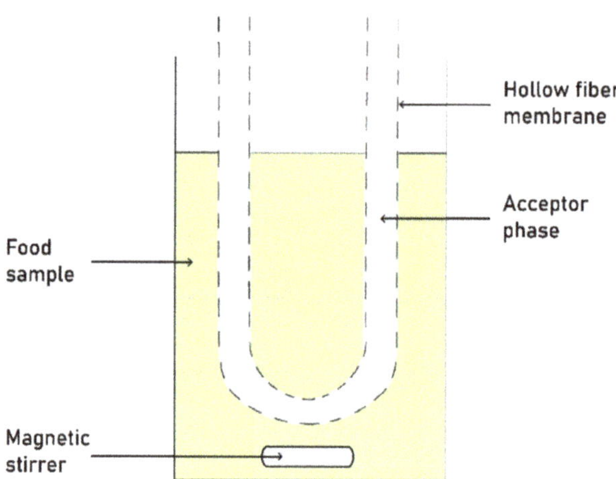

Figure 3. Two-phase hollow fiber-liquid microextraction setup. Adapted from Esrafili et al. [93].

Cui et al. [63] reported a phenolic extraction from green tea by heating and stirring with five different DESs: ChCl-based DES with ethylene glycol, glycerol, glucose oxalic acid, and citric acid as the HBDs. The ChCl:ethylene glycol (1:2) was the best performing DES, attributed to its strongest affinity to phenolic compared to the other DESs, due to the good combination of a strong HBA and a relatively weaker HBD. The extraction produced a maximum yield of 20.12%, conducted with a liquid-solid ratio of 44:1 at 84 °C for 39 min.

Huang et al. [64] extracted anthraquinones from Radix et Rhizoma Rhei, a Chinese medicinal plant from the *Polygonaceae* family, using novel hydrophobic DESs. The plant was used because hydrophobic anthraquinones (e.g., aloe-emodin, emodin, chrysophanol, physcion, and rehin) are one of its main chemical components [94]. Different alkanols (n-octanol, n-decanol, 1-dodecanol, and 1-tetradecanol) and alkyl carboxylic acids (n-octanoic acid, decanoic acid, and 10-undecenoic acid) were used for the synthesis of nine green DES solvents. In general, these DESs presented low viscosities ranging from 4.03 to 12.26 mPa·s, suitable for a high mass transfer between the plant and the solvent due to the absence of coulombic forces [51]. Additionally, these DESs were mixed with water at different volumetric ratios, confirming their immiscibility and thus their hydrophobicity. Based on the physiochemical analysis and protocol optimization, the 1-tetradecanol:10-undecenoic acid (1:4) presented the best dissolution performances and was selected for the extraction process. Before the extraction, a pectinase solution was used for plant cell walls impairment and then deactivated by heat. Later, 385 µL of concentrated HCl were added to obtain a 10% *w/v* HCl solution. For the extraction, DES and plant material were mixed in the aqueous solution at a 12:1 volume-to-mass ratio (mL g^{-1}) at 67 °C for 20 min under 500 rpm stirring. After that, the solution was centrifuged to collect the anthraquinones in the DES phase and quantified by HPLC with diode array detection. Finally, the yields for four different samples from different Chinese provinces were obtained, and the sample from Sichuan presented the highest total anthraquinones extraction yield (ca. 21.52 mg g^{-1}). The yields were comparable to values given for anthraquinones extraction from Chinese pharmacopoeia using a more complex traditional process (heated-reflux extraction process with methanol).

Doldolova et al. [65] studied the curcumin and antioxidant extraction capacity of five different DESs from turmeric (*Curcuma longa*) utilizing microwave-assisted extraction. It is important to note that polyphenols are not the only antioxidants present in turmeric, as several other non-phenolic compounds also exhibit antioxidant activity [95]. Herein, five DESs were prepared; fructose:ChCl:water (molar ratio of 2:5:5), sucrose:ChCl:water (1:4:4), fructose:lactic acid:water (1:5:5), sucrose:lactic acid:water (1:5:7), and lactic acid:ChCl:water (1:1:2), and referred as DES 1, 2, 3, 4, and 5. Different extraction parameters (e.g., extraction time, temperature, and solvent:solid ratio) were studied. The optimal extraction duration was 15–21 min at 64–71 °C, using 14–16 mL/0.2 g of dried sample. The differences in duration is due to different times being needed for different DES to penetrate and dissolve the curcumin in the turmeric. In this study, it is important to note that (1) despite DESs may degrade at high temperatures, there was no degradation observed, even at the harshest experiment conditions (75 °C for 30 min), and (2) higher solvent-solid ratios only produced a small increase in extraction efficiencies, even though it is known that higher ratios often enhance the mass transfer due to the higher concentration gradients. At the optimal conditions, all DESs except DES 1, presented a higher curcumin extraction yield and total antioxidants content than those obtained using 80% methanol:water solvent. DES 1, 2, 3, 4, 5, and 80% methanol:water curcumin yields were 13.95, 21.41, 24.81, 32.00, 23.06, and ca. 15 mg/g of dried sample, respectively.

In a DES-based MAE recovery study, Pan et al. [66] obtained total phenolic and total flavonoid content from Lour (*Osmanthus fragrans*) flowers using a lactic acid:glucose (1:5) DES. An interesting aspect of this study is that optimal conditions for total phenolic and total flavonoid content extraction differ greatly. For example, the highest yield of flavonoids was obtained using a solid–liquid ratio of 30 mL g^{-1} under a microwave power of 500 W for 34.77 s. In contrast, the phenolic yield was only 80.7% at the same condition. The finding highlights that a specific process condition is needed to extract the desired compound effectively. Overall, the highest total phenolic and flavonoid yields were 192.20 $\pm$ 1.08 mg GAE g^{-1} and 560.13 $\pm$ 0.51 mg RE g^{-1}, respectively.

Use of DESs in biomass processing is a growing research field that aims to develop green and efficient methods for biomass valorization. Food industry waste represents a great opportunity for biomass valorization due to the large volumes of residues and by-products, i.e., ~28–35 Mt of orange peel waste (by-product) are produced worldwide, where only a small amount is reused as animal feed, fertilizer, biofuel, and bioactive compounds extraction and the majority is disposed of as land-filling or compost [96]. Concerning this, Panić et al. [67] proposed the use of DESs for valorization of orange peel waste by (*R,S*)-1-phenylethyl acetate hydrolyzation, as well as D-limonene, protein and polyphenol extraction. From preliminary experiments in an integrated bio-refinery protocol, the authors concluded that the ChCl:ethylene glycol (1:2) solvent with 50% water (w/w) was the most efficient out of seven prepared DESs. In the experiment, 10 mL of the selected DES, 0.015 g L^{-1} (*R,S*)-1-phenylethyl acetate, and 0.2 g mL^{-1} orange peel were mixed and placed in a shaker at 850 rpm for seven days. Later, using a glass column with a Sepabeads SP825L porous resin, the polyphenols were separated from the solvent. Using the best DES, a polyphenol yield of 45.7 mg g^{-1} was obtained with good limonene (0.5 mg g^{-1}) and protein (7.7 mg g^{-1}) extraction yields. The DES also showed high enantio-selectivity in the hydrolysis of (*R,S*)-1-phenylethyl acetate.

Additionally, aiming at bio-refinery processing in the food industry, López-Linares et al. [68] evaluated the phenolic extraction from brewer's spent grain using a DES-based MAE. First, four DESs (ChCl:ethylene glycol, ChCl:lactic acid, ChCl:glycerol, and ChCl:1,2-propanediol with a 1:2 molar ratio and 25% water (*v/v*)) were screened in experiments with a maximum microwave power of 700 W, at 65 °C for 20 min. From the initial screening, the ChCl:glycerol solvent proved to be the most efficient, obtaining 2.3 mg GAE g^{-1}, higher than the obtained value of 1.2 mg GAE g^{-1} using methanol:water (80:20 *v/v*) solvent. Additionally, the process extracted not only phenolic compounds but also lignin, with delignification ranging from 0.13 to 20.75%, adding to its advantages for bio-refinery use. Later, variations of temperature (50–100 °C), time (5–25 min), and water percentage in the ChCl:glycerol solvent (20–70%) were studied for process optimization. As expected, an increase in temperature raised the phenolic extraction yield due to the improved phenolic solubility and diffusivity in the solvent. However, low extraction yields were obtained at high temperatures using high-water percentage solvents, believed to be caused by a loss of eutectic properties in the solvent due to the rupture of hydrogen bonds. The highest yield obtained in this MAE process was with 37.46% (*v/v*) water in the ChCl:glycerol DES at 100 °C for 13.3 min, generating a total phenolic yield of 2.89 mg GAE g^{-1}. This study confirms that a fast, greener and more efficient phenolic extraction method from brewer's spent grain is possible using this DES. Furthermore, this process presents a delignification capacity and a low sugar concentration, displaying its potential in bio-refinery processing.

Lignin, an underutilized high-molecular mass phenolic biopolymer [97], is the most abundant aromatic polymer and the biggest source in the world [98,99]. Currently, extensive research is being done to replace fossil fuels with renewable energy sources, such as lignin [100,101]. Extraction of lignin from maritime pine (*Pinus pinaster*) wood using several pure and aqueous DES mixtures was reported by Fernandes et al. [69]. The experiments were performed at an extraction temperature of 150 °C for 2 h with a solid-solvent ratio of 1:10. It was observed that ChCl-based DESs presented a higher lignin extraction yield than that betaine or urea-based DESs. For binary DESs, lactic acid:ChCl with a 5:1 molar ratio was the most efficient, extracting a lignin yield of ca. 40% with a purity of 68.3%; nevertheless, the tartaric acid:ChCl (1:2) DES produced the best lignin purity of ca. 83%. It was observed that carboxylic acids containing alpha-hydroxy acids presented a higher affinity for lignin than linear carboxylic acids. The tertiary lactic acid:tartaric acid:ChCl (4:1:1) DES showed the best performance, surpassing all pure, binary and tertiary DESs, exhibiting a 92.7% and ca. 27% of lignin purity and extraction yield, respectively. Additionally, Fourier-transform infrared spectroscopy (FTIR) showed no significant differences in lignin structures regardless of the DES used. Furthermore, the extraction temperature and time were optimized using this particular solvent, showing the typical trend where the extraction yield increases with temperature. However, lignin purity decreased in extractions at over 150 °C (extractions carried out for 2 h). Concerning time, 1 h extraction time was sufficient for an adequate extraction at 175 °C, as there were no significant differences in extraction yield and lignin purity for 1, 2, and 4 h experiments. At optimal conditions with this DES, a 95 wt.% lignin recovery with an 89 wt.% purity was achieved. The DES was reused in 3 consecutive extractions, producing no significant differences in each extraction yield, thus proving its reusability.

Biomass pre-treatment is an essential step in bio-refineries, given the need to disrupt the lignocellulosic material before further processing [102]. In this aspect, DESs have risen as conventional solvent substitutes. Su et al. [70] studied a DES pre-treatment for poplar (*Populus*) sawdust where lignin removal was evaluated. ChCl:lactic acid DESs with different molar ratios (1:2, 1:4, 1:6, 1:8, and 1:10) were prepared, and the processes were carried out in a reaction vessel at either 110 or 130 °C, for 90 min with 10 g of poplar powder and 100 g of DES. The extracted lignin amount was significantly higher in the process at 130 °C, a phenomenon that was attributed to a viscosity and surface tension reduction in the solvent, which eases the solvent permeation into the poplar's cell wall [103]. There was also a clear relationship between lignin removal and DES molar ratio, where varying the molar ratio from 1:2 to 1:10 increased lignin recoveries accordingly. The optimal extraction was observed at 130 °C with a 1:10 molar ratio DES, reaching a lignin removal up to 89.3%. This process provided selective lignin and xylan removal with high cellulose preservation, making it a very efficient alternative to the conventional solvent pre-treatment of lignocellulosic biomass.

Ong et al. [71] explored delignification of oil palm (*Elaeis guineensis*) frond using an ultrasound-assisted NaOH-aqueous DES pre-treatment. Currently, the pre-treatment processes represent ~20–48% of the total operational cost in bio-refineries, and this is mainly due to the higher fractionation energy required due to the recalcitrant nature of lignocellulosic biomass. UAE is recognized as a powerful technology for phytochemical extraction and easily applied on small and large scales [81]. A ChCl:urea (1:2 molar ratio) DES was mixed with 4-parts of water, obtaining a 1:4 water:DES aqueous mixture, followed by the addition of 0.1, 0.5, 1.5, 2.5, and 3.5% (*w/v*) NaOH to produce the final NaOH-aqueous DESs. Pure and aqueous DESs with NaOH were used as references. The optimum conditions; NaOH concentration of 2.5%, an ultrasound amplitude of 60%, and extraction duration of 30 min, produced a lignin removal of 47.00 ± 0.16%. The authors proposed a working mechanism where the ultrasound-produced microbubbles in the low viscosity NaOH-DES solvent forming micro-jet streams directed towards the biomass surfaces, disintegrating and concurrently removing the hydrophobic wax layer from the frond's surfaces. The cavitation approach as a destructive force is already proven in several applications, confirming the above-mentioned proposed mechanism [104]. These micro-jets also improved the solvent penetration into the biomass, by inducing the frond fibers swelling and facilitating the solvent penetration. On the other hand, the high NaOH basicity increased the lignin solubility to achieve a high lignin removal. The process mechanism is depicted in Figure 4.

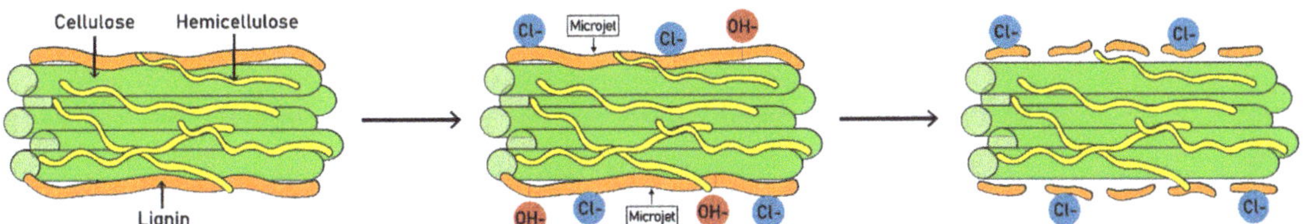

Figure 4. Process mechanism of ultrasound-assisted NaOH-aqueous DES pre-treatment. Adapted from Ong et al. [71].

Several other studies also showed the DES-based UAE extraction potential. Fu et al. [72] developed a DES-coupled pulse-UAE for anthocyanin recovery from blueberry (*Vaccinium* spp.) pomace. Anthocyanins, the biggest group of phenolic pigments with relevant antioxidant activity, are compounds of interest in the food industry due to their biological activities and natural colorants [105]. Total anthocyanins content in blueberry pomace was determined by Giusti and Wrolstad [106] and expressed in mg of Cyanidin-3-*O*-glucoside equivalent (C3GE) per gram of sample. First, seven ChCl-based DESs with different carboxylic acids and polyalcohols as the HBDs. Given the high pure DES viscosities, 10, 30, and 50% DES-water mixtures were produced. The selection for the highest extraction efficiency was conducted using the conventional heat-assisted extractions, and the total extracted anthocyanin contents was compared with the obtained values using water and ethanol. The 30% DES-water mixtures obtained the highest yields. As discussed earlier, water addition into DESs reduces the solvent viscosity, easing the mass transfer of the targeted compound from the source to the solvent; however, when excessive water is added, the DES structure is dismantled causing a reduction in the extraction yield [3,43,89]. It was noted that DESs with carboxylic acids as HBD showed higher extraction efficiencies than those with polyalcohols. This is ascribed to the higher polarity and lower pH value of the carboxylic acid-based DESs, as anthocyanins are highly polar compounds found frequently in their flavylium cation form and stable at pH values below 2 [107]. It was also observed that higher density and lower viscosity in the DES corresponded to a higher extraction of total anthocyanin content. The highest extraction yield was obtained with the ChCl:oxalyc acid DES (1:1 molar ratio) with 30% water, outperforming extractions using ethanol and water. Using the best DES, four different UAE parameters were then assessed, including ultrasonic time, ultrasonic power, temperature, and solvent to solid ratio. The optimum process conditions were: an ultrasonic time of 3.2 min, with 325 W power at 76 °C, and a solvent to solid ratio of 60 mL g^{-1}, acquiring a total anthocyanin yield of 24.27 ± 0.05 mg C3GE g^{-1}.

Chen et al. [73] documented phenolic extraction from lily (*Lilium lancifolium* Thunb.) through DES-based UAE. In this study, four DESs were synthesized using ChCl as the HBA and four different HBDs: ethylene glycol, lactic acid, glycerol, and formic acid, each at a molar ratio of 1:2. The authors utilized 1 g of *Lilium lancifolium* in each extraction, obtaining the ChCl:ethylene glycol DES as the most efficient. Additionally, the effects of water content in the solvent, solid-solvent ratio, temperature, and extraction time were also studied. It was observed that water addition from 10 to 30% increased the extraction rate, with 20% water being the best, whereas 40 to 50% water addition was detrimental to the extraction. The optimum conditions were: 50 °C, a 40 min extraction with a solid-solvent ratio of 1:25. At these optimal conditions, yields of 0.31 ± 0.06, 0.29 ± 0.03, and 3.04 ± 0.38 mg g^{-1} for regaloside B, regaloside C, and regaloside E were achieved, respectively. Compared to the conventional hot water extraction method, higher yields and shorter extraction times were obtained with DES.

Mansinhos et al. [74] carried out phenolic compounds separation from *Lavandula pedunculata* subsp. *lusitanica* with ten different DESs using UAE. *Lavandula pedunculata* is known to have medicinal properties, credited partly to its phenolic content [108]. Mansinhos et al. used four HBAs (glycerol, glucose, ChCl, and proline) with different HBDs (citric acid, urea, lactic acid, xylitol, in their work and malic acid). After a preliminary evaluation, proline:lactic acid (1:1) DES presented the highest total phenolic amount (56.00 ± 0.77 mg GAE g^{-1} DW), in which the most abundant extracted compounds were rosmarinic acid, ferulic acid, and salvianolic acid. The results were in agreement with the previously reported phenolic compositions in *Lavandula* species [109]. It is important to mention that the individual phenolic yield was different for each DES, indicating an opportunity for selective phenolic extraction.

In a study by Jakovljević et al. [75], carnosic acid and carnosol extraction from sage (*Salvia officinalis*) was conducted through three extraction methods, i.e., conventional stirring and heating, UAE, and mechanochemical extraction (MCE), using DESs. Firstly, seventeen DESs with ChCl as the HBA were prepared with their respective 10, 30 and 50% aqueous mixtures and studied through stirring and heating extraction to obtain the best performing solvent. For carnosol extraction, the ChCl:1,4-butanediol (1:2) DES was the most effective, whereas the acidic ChCl:lactic acid (1:2) DES was the most efficient for carnosic acid extraction. This is attributed to the higher acidic DES polarity compared with the DESs with sugar HBDs. Further experiments were carried out with the ChCl:lactic acid (1:2) DES. It was found that the stirring and heating and UAE methods were the most efficient, believed to be due to the continuous stirring in the earlier method and the faster compound diffusion and increased mass transfer due to the cavitation effect in the latter. Nonetheless, MCE reduces the extraction time and solvent amount. Similar amounts of carnosol and carnosic acid were obtained in 90 min stirring and heating and UAE extraction processes, and only 3 min of the MCE process, with a minimal DES amount. This is caused by the glass beads that enhance the plant-DES mixing. The highest carnosic acid extraction yields were 14.43 ± 0.28 µg mg^{-1} for stirring and heating (10% of H_2O, 90 min, 50 °C); 14.00 ± 0.02 µg mg^{-1} for UAE (10% of H_2O, 60 min, 70 °C); and 8.26 ± 0.45 µg mg^{-1} for MCE (10% of H_2O, 2 min, 5 m s^{-1} vibration speed). Moreover, carnosol and carnosic acid extraction with the ChCl:lactic acid (1:2) DES proved to be more efficient than conventional solvents such as water, ethanol, and 30–70% (*v/v*) water-ethanol aqueous mixtures.

Patil et al. [76] described a DES-based UAE of curcuminoids from turmeric. Herein, nine ChCl-based DESs were prepared and screened for curcuminoids extraction using glycerol (1:2), ethylene glycol (1:1), 1,4-butanediol (1:3), lactic acid (1:1), malic acid (1:1), citric acid (2:1), xylose (1:1), glucose (2:1), and fructose (2:1) as the HBDs. The UAE was carried out with sonication at 22 kHz frequency and 200 W of power. The highest curcuminoids extraction yield (ca. 58.87 mg g^{-1}) was obtained with the ChCl:Lactic acid DES, which showed the highest curcuminoid solubility (13.7 mg mL^{-1}) during a solubility test. Hence, process optimization was performed with this DES. The highest extraction yields were obtained using the acid-based DESs, followed by sugar-based DESs, while the alcohol-based DESs presented the lowest yields. In the optimization, increasing the DES molar ratio from 1:1 to 1:4 provided a higher curcuminoid yield; however, there were no significant differences. Thus, the 1:1 ratio was maintained as a larger HBD amount is not cost-effective. Water addition to the DESs was also studied, and 20% water content presented the highest yield, attributed to a better dissipation of ultrasonic energy and cavitation enhancement. The authors also examined the turmeric loading and particle size, where 5% turmeric loading presented the highest yield, at higher loadings a constraint in mass transfer within the cluster of particles was observed [110]. The particle size of 0.355 mm was chosen due to higher aspect-ratio in smaller particles (higher particle cells in contact with solvent), but a further size reduction promoted particle agglomeration and sedimentation. Using the optimal conditions (intensity of 70.8 W cm^{-2}, pulse mode of 60% duty cycle at 30 °C for 20 min), a 77.13 mg g^{-1} curcuminoid yield was obtained. In a comparative study between ethanol-based Soxhlet extraction and DES-based batch extraction, UAE presented a much lower energy requirement and a yield higher than the batch extraction. Interestingly, the yield is similar to the Soxhlet extraction but only consumed 0.67% of the energy required in the Soxhlet process.

Very recently, Ünlü [77] developed a phenolic extraction from the olive leaf through UAE with DESs, expressed as total polyphenol and flavonoid yields. Like many other natural sources presented in this review, olive leaf has been utilized for medicinal purposes since ancient times due to its beneficial antioxidant, antibacterial, and antifungal properties [111]. Eight DESs were prepared and evaluated for total polyphenol and flavonoid yields. Before optimization, the glucose:sucrose:water (1:1:11) solvent was screened to be the most efficient for total polyphenol extraction, yielding around 20.49 ± 0.50 mg GAE g^{-1}, ascribed to the low pH compared to the other sugar-based DESs prepared. For the total flavonoid content, the highest yield was obtained with the ChCl:lactic acid (1:2) DES, achieving 8.44 ± 0.30 mg ApE g^{-1}, thanks to its low viscosity. Afterwards, the experimental conditions were optimized, and both polyphenol and flavonoid content were extracted using a ChCl:fructose:water (5:2:5) DES (42.69% water content, liquid-to-solid ratio of 40.66 mL g^{-1}, 75 °C, 1 h UAE at 140 W and 37 kHz) obtaining 187.31 ± 10.3 mg GAE g^{-1} and 12.75 ± 0.6 mg ApE g^{-1} for polyphenols and flavonoids, respectively.

Using eleven different DESs in a UAE study, Ivanović et al. [78] extracted phenolics from *Helichrysum arenarium* L. inflorescences. The maximum phenolic contents (determined by spectrophotometry) of 30.15 ± 0.42, and 29.75 ± 0.49 mg g^{-1} were achieved using a ChCl:1,2-propanediol (1:2) and a ChCl:1,4-butanediol (1:6) 25% water DESs, respectively. These yields were higher than those obtained with processes carried out with 80% methanol and water as solvents under the same conditions. Furthermore, since the phenolic compounds were quantified individually, the authors determined that the high yields were attributed to the less polar phenolic compounds, such as secoiridoids and flavonoids. Similar findings were also previously reported by Garcia et al. [87].

Olive oil presents an important amount of phenolic compounds. It contains at least 36 structurally different phenolic compounds, and studies have proven that olive oil's phenolic compounds display positive effects on human health [112]. Rodríguez-Juan et al. [79] extracted phenolic compounds from virgin olive oil with a ChCl:xylitol aqueous DES through agitation in a water bath, followed by solvent removal using a XAD-16 resin column. The DES was washed with water in the column, and the phenolics were eluted with ethanol, thus recycling the DES. The reported phenolic content was the sum of specific compounds quantified by HPLC (e.g., hydroxytyrosol, tyrosol, 1-acetoxypinoresinol, 3,4-DHPEA-EDA, etc.). Different test conditions were explored; a temperature of 30–90 °C, 0.5–6 h of extraction time, and different virgin olive oil:DES ratios (1:1, 1:3, 1:4, 1:7). It was observed that the most efficient phenolic extraction conditions were at 40 °C for 1 h using an oil:DES ratio of 1:1, obtaining a yield of 555.36 ± 21.93 mg of phenolic per kg of oil.

Shishov et al. [80] investigated phenolic extraction from olive oil using a rotating disk sorptive extraction method, where the extraction mechanism relies on in situ DES formation between a ChCl-coated rotating disk and the phenolic compounds in the oil sample. The rotating disk was made of two polymer films, sandwiching a Parafilm M-wrapped iron wire for magnetic stirring (see the configuration in Figure 5a). After the assembly, the disk was clamped with glass and heated at 60 °C for 15 min, followed by a coating with 10 μL of

ChCl solution. The rotation disk was placed inside a vial consisting of 1 mL of olive oil and 1 mL of n-hexane (used to reduce the oil's viscosity). After a period of stirring, the disk was removed, and the n-hexane was evaporated before placing the disk into a vial with 200 μL of a water-ethanol mixture (2:1 *v/v*). In this stage, the phenolic compounds were separated from the DES due to DES decomposition (ChCl dissolves in water and disintegrates from the DES formation, and methanol enhances phenolic solubility). The extraction process is illustrated in Figure 5b. The optimal extraction and elution conditions were studied, and the authors found that poly(vinylidene fluoride-co-tetrafluoroethylene) (PVDF-co-PTFE) films presented a higher absolute extraction recovery than cellulose acetate, nylon and polyethersulfone films. This is due to its high hydrophobicity, a crucial characteristic that allows PVDF-co-PTFE to be more oil-selective. As for the oil dilution, no significant differences were found between n-hexane, n-heptane, and isooctane; however, n-hexane was chosen for its high volatility, thus reducing the evaporation time. A concentration of 300 g L^{-1} ChCl was chosen as it provided the highest absolute extraction recovery. As for the extraction conditions, chemical equilibrium was reached at 15 min at 150 rpm rotation. The optimal elution volume and time were 0.2 mL and 2 min, respectively. Absolute extraction recoveries of gallic acid, protocatechuic acid, tyrosol, vanillic acid, *p*-coumaric acid, syringaldehyde and thymol were 87, 79, 76, 75, 82, 81, and 66%, respectively. The extraction performances were comparable to the obtained values in a methanol-water (60:40 *v/v*) extraction method. Using specific techniques, such as FTIR and scanning electron microscope (SEM), the authors confirmed ChCl disk-impregnation, as well as DES formation and retention. The FTIR analysis confirmed that the O-H absorption peak shifted after ChCl coating and after ChCl-phenolic compound interaction, a change believed to be due to hydrogen bond formation. The spectra also indicated several peaks associated with the phenolic compounds' functional groups, confirming its extraction into the rotating disk. The SEM images showed a fibrous pristine disk structure, which smoothed out after ChCl impregnation and later presented a different structure after phenolic uptake. This study offers a novel, rapid extraction and elution method, utilizing an in situ synthesized DES with the phenolic compounds from the sample source.

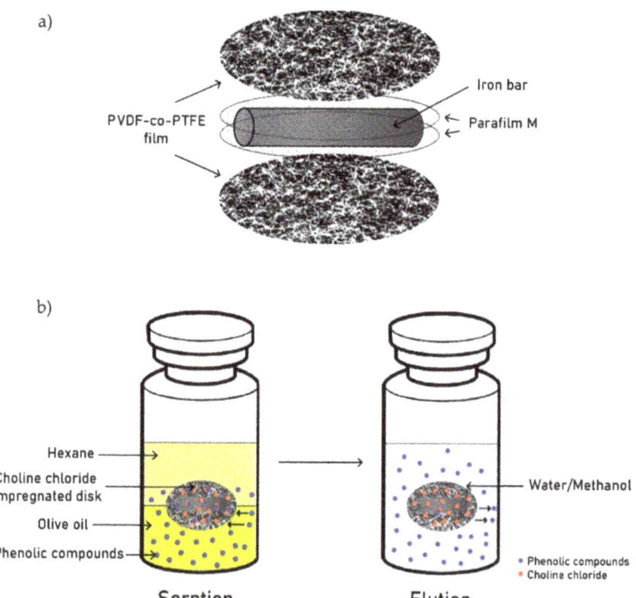

Figure 5. Rotating disk sorptive extraction: disk composition (**a**), sorptive extraction and elution process (**b**). Adapted from Shishov et al. [80].

4. Concluding Remarks

The DES-based phenolic extraction research has increased exponentially due to its apparent advantages over conventional solvents. As presented, DESs are catalogued as green alternatives, given their biodegradability and reusability; moreover, numerous DES components are non-toxic, and some are even safe for human consumption, further broadening their industrial applicability. DESs have shown the ability to preserve phenolic compounds in solutions. These properties make DESs excellent candidates in the industry not only for extractive processes but also as their final carrier eliminating the need for further processing, as suggested by Skarpalezos and Detsi [113]. In addition to the intrinsic characteristics such as low volatility and melting point, good thermal and chemical stability, and low cost, their advantages are very attractive not only in biomolecule extraction or separation but also in biomass waste revalorization through DES-based bio-refinery processes.

DES selection and optimum process conditions are vital for an adequate extraction. It is important to note that there are limitless DES components, leading to a whole new variety of DESs (with different physicochemical properties) that may present enormous opportunities for more selective extractions. Likewise, extraction method selection will vary depending on the compound of interest and its source. More importantly, DESs have demonstrated their ability to assist the emerging techniques and protocols in the phenolic compound separation. Many studies have shown the great DES performances in biomolecule extractions, displaying comparable or higher yields than conventional solvents. More industrial applicability research is encouraged and required, given its potential to reduce operation costs and decrease operational risks for workers, consumers, and the environment. However, as it can be observed in this review, the screening for an adequate DES is a time-consuming process, hence, several authors encourage the use of computational tools for DES extraction efficiency modeling [114], which is increasingly applied by researchers along with the extraction experiments.

DES components are often non-volatile. Thus, phenolic separation from the extract can be more challenging than the classic organic solvents (simple solvent evaporation will not be effective in DESs). In the reviewed studies, these aspects are often not being addressed accordingly. In our opinion, (1) new DES application study (for precious compound extraction) should include the extracted component separation, and (2) the DES advantages compared to the conventional solvents should be proved appropriately and may include a brief economic assessment. The whole process cost can be strongly affected by the DES components and the final phenolic separation from DES. Finally, the DES recycling possibility should also be addressed.

Author Contributions: Conceptualization, investigation, writing—original draft preparation, editing, and writing—review, J.S.-V. and R.C.-M.; editing—review, M.Z.A. and G.B. All authors have read and agreed to the published version of the manuscript.

Funding: This research received no external funding.

Institutional Review Board Statement: Not applicable.

Informed Consent Statement: Not applicable.

Data Availability Statement: Data are contained within the article.

Acknowledgments: R. Castro-Muñoz acknowledges the School of Science and Engineering and the FEMSA-Biotechnology Center at Tecnológico de Monterrey for their support through the Bioprocess (0020209I13) Focus Group. Financial support from the Polish National Agency for Academic Exchange (NAWA) under the Ulam Programme (Agreement No. PPN/ULM/2020/1/00005/U/00001) is also gratefully acknowledged.

Conflicts of Interest: The authors declare no conflict of interest.

Sample Availability: Samples of the compounds are not available from the authors.

References

1. Castro-Muñoz, R.; Yáñez-Fernández, J.; Fíla, V. Phenolic compounds recovered from agro-food by-products using membrane technologies: An overview. *Food Chem.* **2016**, *213*, 753–762. [CrossRef] [PubMed]
2. Tungmunnithum, D.; Thongboonyou, A.; Pholboon, A.; Yangsabai, A. Flavonoids and other phenolic compounds from medicinal plants for pharmaceutical and medical aspects: An overview. *Medicines* **2018**, *5*, 93. [CrossRef] [PubMed]
3. Ruesgas-Ramón, M.; Figueroa-Espinoza, M.C.; Durand, E. Application of deep eutectic solvents (DES) for phenolic compounds extraction: Overview, challenges, and opportunities. *J. Agric. Food Chem.* **2017**, *65*, 3591–3601. [CrossRef] [PubMed]
4. Castro-Muñoz, R. Pervaporation: The emerging technique for extracting aroma compounds from food systems. *J. Food Eng.* **2019**, *253*, 27–39. [CrossRef]
5. Galanakis, C.M.; Castro-Muñoz, R.; Cassano, A.; Conidi, C. *Recovery of High-Added-Value Compounds from Food Waste by Membrane Technology*; Woodhead Publishing: Sawston, UK, 2016; ISBN 9780081004524/9780081004517.
6. Xia, E.-Q.; Deng, G.-F.; Guo, Y.-J.; Li, H.-B. Biological activities of polyphenols from grapes. *Int. J. Mol. Sci.* **2010**, *11*, 622–646. [CrossRef]
7. de la Rosa, L.A.; Moreno-Escamilla, J.O.; Rodrigo-García, J.; Alvarez-Parrilla, E. Chapter 12-phenolic compounds. In *Postharvest Physiology and Biochemistry of Fruits and Vegetables*; Yahia, E., Carrillo-Lopez, A., Eds.; Woodhead Publishing: Sawston, UK, 2019; pp. 253–271, ISBN 978-0-12-813278-4.
8. Stalikas, C.D. Extraction, separation, and detection methods for phenolic acids and flavonoids. *J. Sep. Sci.* **2007**, *30*, 3268–3295. [CrossRef]
9. Hidalgo, G.I.; Almajano, M.P. Red fruits: Extraction of antioxidants, phenolic content, and radical scavenging determination: A review. *Antioxidants* **2017**, *6*, 7. [CrossRef] [PubMed]
10. Santos, F.; Duarte, A.R.C. Therapeutic deep eutectic systems for the enhancement of drug bioavailability. In *Deep Eutectic Solvents for Medicine, Gas Solubilization and Extraction of Natural Substances*; Fourmentin, S., Costa Gomes, M., Lichtfouse, E., Eds.; Springer International Publishing: New York, NY, USA, 2021; pp. 103–129, ISBN 978-3-030-53068-6.
11. Şahin, S. Tailor-designed deep eutectic liquids as a sustainable extraction media: An alternative to ionic liquids. *J. Pharm. Biomed. Anal.* **2019**, *174*, 324–329. [CrossRef]
12. Pena-Pereira, F.; Kloskowski, A.; Namieśnik, J. Perspectives on the replacement of harmful organic solvents in analytical methodologies: A framework toward the implementation of a generation of eco-friendly alternatives. *Green Chem.* **2015**, *17*, 3687–3705. [CrossRef]
13. Van Osch, D.J.G.P.; Dietz, C.H.J.T.; Van Spronsen, J.; Kroon, M.C.; Gallucci, F.; Van Sint Annaland, M.; Tuinier, R. A search for natural hydrophobic deep eutectic solvents based on natural components. *ACS Sustain. Chem. Eng.* **2019**, *7*, 2933–2942. [CrossRef]
14. Durand, E.; Villeneuve, P.; Bourlieu-lacanal, C.; Carrière, F. Natural deep eutectic solvents: Hypothesis for their possible roles in cellular functions and interaction with membranes and other organized biological systems. In *Advances in Botanical Research*; Verpoorte, R., Witkamp, G.-J., Choi, Y.H.B.T., Eds.; Academic Press: Cambridge, MA, USA, 2021; Volume 97, pp. 133–158, ISBN 0065-2296.
15. Faraz, N.; Ul Haq, H.; Balal, M.; Castro-Muñoz, R.; Boczkaj, G.; Khan, A. Deep eutectic solvent based method for analysis of Niclosamide in pharmaceutical and wastewater samples—A green analytical chemistry approach. *J. Mol. Liq.* **2021**, *335*, 116142. [CrossRef]
16. Barbieri, J.B.; Goltz, C.; Batistão Cavalheiro, F.; Theodoro Toci, A.; Igarashi-Mafra, L.; Mafra, M.R. Deep eutectic solvents applied in the extraction and stabilization of rosemary (*Rosmarinus officinalis* L.) phenolic compounds. *Ind. Crops Prod.* **2020**, *144*, 112049. [CrossRef]
17. Zannou, O.; Koca, I.; Aldawoud, T.M.S.; Galanakis, C.M. Recovery and stabilization of anthocyanins and phenolic antioxidants of roselle (*Hibiscus sabdariffa* L.) with hydrophilic deep eutectic solvents. *Molecules* **2020**, *25*, 3715. [CrossRef] [PubMed]
18. Dai, Y.; Rozema, E.; Verpoorte, R.; Choi, Y.H. Application of natural deep eutectic solvents to the extraction of anthocyanins from Catharanthus roseus with high extractability and stability replacing conventional organic solvents. *J. Chromatogr. A* **2016**, *1434*, 50–56. [CrossRef] [PubMed]
19. Gullón, P.; Gullón, B.; Romaní, A.; Rocchetti, G.; Lorenzo, J.M. Smart advanced solvents for bioactive compounds recovery from agri-food by-products: A review. *Trends Food Sci. Technol.* **2020**, *101*, 182–197. [CrossRef]
20. Patyar, P.; Ali, A.; Malek, N.I. Experimental and theoretical excess molar properties of aqueous choline chloride based deep eutectic solvents. *J. Mol. Liq.* **2021**, *324*, 114340.
21. Phillips, M.M. Choline: Properties and determination. In *Encyclopedia of Food and Health*; Caballero, B., Finglas, P.M., Toldrá, F., Eds.; Elsevier Inc.: Amsterdam, The Netherlands, 2015; pp. 73–78, ISBN 9780123849533.
22. Younes, M.; Aggett, P.; Aguilar, F.; Crebelli, R.; Dusemund, B.; Filipič, M.; Frutos, M.J.; Galtier, P.; Gott, D.; Gundert-Remy, U.; et al. Re-evaluation of propane-1,2-diol (E 1520) as a food additive. *EFSA J.* **2018**, *16*, 1–40. [CrossRef]
23. Dziezak, J.D. Acids: Natural acids and acidulants. In *Encyclopedia of Food and Health*; Caballero, B., Finglas, P.M., Toldrá, F., Eds.; Elsevier Inc.: Amsterdam, The Netherlands, 2015; pp. 15–18, ISBN 9780123849533.
24. Hu, X.; Shi, Y.; Zhang, P.; Miao, M.; Zhang, T.; Jiang, B. d-Mannose: Properties, production, and applications: An overview. *Compr. Rev. Food Sci. Food Saf.* **2016**, *15*, 773–785. [CrossRef] [PubMed]
25. Turck, D.; Bresson, J.L.; Burlingame, B.; Dean, T.; Fairweather-Tait, S.; Heinonen, M.; Hirsch-Ernst, K.I.; Mangelsdorf, I.; McArdle, H.; Naska, A.; et al. Safety of d-ribose as a novel food pursuant to Regulation (EU) 2015/2283. *EFSA J.* **2018**, *16*, e05265. [CrossRef]

26. Silva, L.P.; Fernandez, L.; Conceiçao, J.H.F.; Martins, M.A.R.; Sosa, A.; Ortega, J.; Pinho, S.P.; Coutinho, J.A.P. Design and characterization of sugar-based deep eutectic solvents using conductor-like screening model for real solvents. *ACS Sustain. Chem. Eng.* **2018**, *6*, 10724–10734. [CrossRef]

27. Wu, L.; Chen, Z.; Li, S.; Wang, L.; Zhang, J. Eco-friendly and high-efficient extraction of natural antioxidants from Polygonum aviculare leaves using tailor-made deep eutectic solvents as extractants. *Sep. Purif. Technol.* **2021**, *262*, 118339. [CrossRef]

28. Lakka, A.; Karageorgou, I.; Kaltsa, O.; Batra, G.; Bozinou, E.; Lalas, S.; Makris, D. Polyphenol extraction from *Humulus Lupulus* (hop) using a neoteric glycerol/l-alanine deep eutectic solvent: Optimisation, kinetics and the effect of ultrasound-assisted pretreatment. *AgriEngineering* **2019**, *1*, 30. [CrossRef]

29. Janicka, P.; Przyjazny, A.; Boczkaj, G. Novel "acid tuned" deep eutectic solvents based on protonated L-proline. *J. Mol. Liq.* **2021**, *333*, 115965. [CrossRef]

30. Zainal-Abidin, M.H.; Hayyan, M.; Hayyan, A.; Jayakumar, N.S. New horizons in the extraction of bioactive compounds using deep eutectic solvents: A review. *Anal. Chim. Acta* **2017**, *979*, 1–23. [CrossRef] [PubMed]

31. Loow, Y.-L.; New, E.; Yang, G.; Ang, L.; Foo Yang Wei, L.; Wu, T. Potential use of deep eutectic solvents to facilitate lignocellulosic biomass utilization and conversion. *Cellulose* **2017**, *24*, 3591–3618. [CrossRef]

32. Tang, B.; Zhang, H.; Row, K.H. Application of deep eutectic solvents in the extraction and separation of target compounds from various samples. *J. Sep. Sci.* **2015**, *38*, 1053–1064. [CrossRef] [PubMed]

33. Shishov, A.; Bulatov, A.; Locatelli, M.; Carradori, S.; Andruch, V. Application of deep eutectic solvents in analytical chemistry. A review. *Microchem. J.* **2017**, *135*, 33–38. [CrossRef]

34. Hansen, B.B.; Spittle, S.; Chen, B.; Poe, D.; Zhang, Y.; Klein, J.M.; Horton, A.; Adhikari, L.; Zelovich, T.; Doherty, B.W.; et al. Deep eutectic solvents: A review of fundamentals and applications. *Chem. Rev.* **2021**, *121*, 1232–1285. [CrossRef]

35. Haq, H.U.; Balal, M.; Castro-Muñoz, R.; Hussain, Z.; Safi, F.; Ullah, S.; Boczkaj, G. Deep eutectic solvents based assay for extraction and determination of zinc in fish and eel samples using FAAS. *J. Mol. Liq.* **2021**, *333*, 115930. [CrossRef]

36. Jia, L.; Huang, X.; Zhao, W.; Wang, H.; Jing, X. An effervescence tablet-assisted microextraction based on the solidification of deep eutectic solvents for the determination of strobilurin fungicides in water, juice, wine, and vinegar samples by HPLC. *Food Chem.* **2020**, *317*, 126424. [CrossRef]

37. Momotko, M.; Łuczak, J.; Przyjazny, A.; Boczkaj, G. First deep eutectic solvent-based (DES) stationary phase for gas chromatography and future perspectives for DES application in separation techniques. *J. Chromatogr. A* **2020**, *1635*, 461701. [CrossRef]

38. Makoś, P.; Boczkaj, G. Deep eutectic solvents based highly efficient extractive desulfurization of fuels–Eco-friendly approach. *J. Mol. Liq.* **2019**, *296*, 111916. [CrossRef]

39. Harifi-Mood, A.R.; Mohammadpour, F.; Boczkaj, G. Solvent dependency of carbon dioxide Henry's constant in aqueous solutions of choline chloride-ethylene glycol based deep eutectic solvent. *J. Mol. Liq.* **2020**, *319*, 114173. [CrossRef]

40. Web of Science. Available online: https://www.webofknowledge.com (accessed on 12 June 2021).

41. Martins, M.A.R.; Pinho, S.P.; Coutinho, J.A.P. Insights into the nature of eutectic and deep eutectic mixtures. *J. Solution Chem.* **2019**, *48*, 962–982. [CrossRef]

42. Zhang, Q.; De Oliveira Vigier, K.; Royer, S.; Jérôme, F. Deep eutectic solvents: Syntheses, properties and applications. *Chem. Soc. Rev.* **2012**, *41*, 7108–7146. [CrossRef] [PubMed]

43. Vilková, M.; Płotka-Wasylka, J.; Andruch, V. The role of water in deep eutectic solvent-base extraction. *J. Mol. Liq.* **2020**, *304*, 112747. [CrossRef]

44. Choi, Y.H.; Verpoorte, R. Green solvents for the extraction of bioactive compounds from natural products using ionic liquids and deep eutectic solvents. *Curr. Opin. Food Sci.* **2019**, *26*, 87–93. [CrossRef]

45. Dai, Y.; Varypataki, E.M.; Golovina, E.A.; Jiskoot, W.; Witkamp, G.J.; Choi, Y.H.; Verpoorte, R. Natural deep eutectic solvents in plants and plant cells: In Vitro evidence for their possible functions. In *Advances in Botanical Research*; Elsevier Ltd.: Amsterdam, The Netherlands, 2021; Volume 97, pp. 159–184, ISBN 9780128216910.

46. Dwamena, A.K. Recent advances in hydrophobic deep eutectic solvents for extraction. *Separations* **2019**, *6*, 9. [CrossRef]

47. Choi, Y.H.; van Spronsen, J.; Dai, Y.; Verberne, M.; Hollmann, F.; Arends, I.W.C.E.; Witkamp, G.J.; Verpoorte, R. Are natural deep eutectic solvents the missing link in understanding cellular metabolism and physiology? *Plant Physiol.* **2011**, *156*, 1701–1705. [CrossRef] [PubMed]

48. Smith, E.L.; Abbott, A.P.; Ryder, K.S. Deep eutectic solvents (DESs) and their applications. *Chem. Rev.* **2014**, *114*, 11060–11082. [CrossRef]

49. El Achkar, T.; Greige-Gerges, H.; Fourmentin, S. Understanding the basics and properties of deep eutectic solvents. In *Deep Eutectic Solvents for Medicine, Gas Solubilization and Extraction of Natural Substances*; Fourmentin, S., Costa Gomes, M., Lichtfouse, E., Eds.; Springer International Publishing: New York, NY, USA, 2021; pp. 1–40, ISBN 978-3-030-53068-6.

50. Abbott, A.P.; Barron, J.C.; Ryder, K.S.; Wilson, D. Eutectic-based ionic liquids with metal-containing anions and cations. *Chem. A Eur. J.* **2007**, *13*, 6495–6501. [CrossRef]

51. Florindo, C.; Branco, L.C.; Marrucho, I.M. In the quest for green—Solvent design: From hydrophilic to hydrophobic (Deep) eutectic solvents. *ChemSusChem* **2019**, *12*, 1549–1559. [CrossRef]

52. Meneses, L.; Santos, F.; Gameiro, A.R.; Paiva, A.; Duarte, A.R.C. Preparation of binary and ternary deep eutectic systems. *J. Vis. Exp.* **2019**, *152*, e60326. [CrossRef] [PubMed]

53. Mbous, Y.P.; Hayyan, M.; Hayyan, A.; Wong, W.F.; Hashim, M.A.; Looi, C.Y. Applications of deep eutectic solvents in biotechnology and bioengineering—Promises and challenges. *Biotechnol. Adv.* **2017**, *35*, 105–134. [CrossRef] [PubMed]
54. Abbott, A.P.; Capper, G.; Gray, S. Design of improved deep eutectic solvents using hole theory. *ChemPhysChem* **2006**, *7*, 803–806. [CrossRef]
55. Abbott, A.P.; Capper, G.; Davies, D.L.; Rasheed, R.K.; Tambyrajah, V. Novel solvent properties of choline chloride/urea mixtures. *Chem. Commun.* **2003**, 70–71. [CrossRef]
56. Migliorati, V.; Sessa, F.; D'Angelo, P. Deep eutectic solvents: A structural point of view on the role of the cation. *Chem. Phys. Lett. X* **2019**, *2*, 100001. [CrossRef]
57. Alsaud, N.; Shahbaz, K.; Farid, M. Evaluation of deep eutectic solvents in the extraction of β-caryophyllene from New Zealand Manuka leaves (*Leptospermum scoparium*). *Chem. Eng. Res. Des.* **2021**, *166*, 97–108. [CrossRef]
58. Sun, B.; Zheng, Y.L.; Yang, S.K.; Zhang, J.R.; Cheng, X.Y.; Ghiladi, R.; Ma, Z.; Wang, J.; Deng, W.W. One-pot method based on deep eutectic solvent for extraction and conversion of polydatin to resveratrol from Polygonum cuspidatum. *Food Chem.* **2021**, *343*, 128498. [CrossRef]
59. Torres-Vega, J.; Gómez-Alonso, S.; Pérez-Navarro, J.; Alarcón-Enos, J.; Pastene-Navarrete, E. Polyphenolic compounds extracted and purified from *Buddleja Globosa* hope (Buddlejaceae) leaves using natural deep eutectic solvents and centrifugal partition chromatography. *Molecules* **2021**, *26*, 2192. [CrossRef]
60. Wojeicchowski, J.P.; Marques, C.; Igarashi-Mafra, L.; Coutinho, J.A.P.; Mafra, M.R. Extraction of phenolic compounds from rosemary using choline chloride–based deep eutectic solvents. *Sep. Purif. Technol.* **2021**, *258*, 117975. [CrossRef]
61. Đorđević, B.S.; Todorović, Z.B.; Troter, D.Z.; Stanojević, L.P.; Stojanović, G.S.; Đalović, I.G.; Mitrović, P.M.; Veljković, V.B. Extraction of phenolic compounds from black mustard (*Brassica nigra* L.) seed by deep eutectic solvents. *J. Food Meas. Charact.* **2021**, *15*, 1931–1938. [CrossRef]
62. Nia, N.N.; Hadjmohammadi, M.R. Amino acids- based hydrophobic natural deep eutectic solvents as a green acceptor phase in two-phase hollow fiber-liquid microextraction for the determination of caffeic acid in coffee, green tea, and tomato samples. *Microchem. J.* **2021**, *164*, 106021. [CrossRef]
63. Cui, Z.; Enjome Djocki, A.V.; Yao, J.; Wu, Q.; Zhang, D.; Nan, S.; Gao, J.; Li, C. COSMO-SAC-supported evaluation of natural deep eutectic solvents for the extraction of tea polyphenols and process optimization. *J. Mol. Liq.* **2021**, *328*, 115406. [CrossRef]
64. Huang, A.; Deng, W.; Li, X.; Zheng, Q.; Wang, X.; Xiao, Y. Long-chain alkanol–alkyl carboxylic acid-based low-viscosity hydrophobic deep eutectic solvents for one-pot extraction of anthraquinones from rhei radix et rhizoma. *J. Pharm. Anal.* **2021**. [CrossRef]
65. Doldolova, K.; Bener, M.; Lalikoğlu, M.; Aşçı, Y.S.; Arat, R.; Apak, R. Optimization and modeling of microwave-assisted extraction of curcumin and antioxidant compounds from turmeric by using natural deep eutectic solvents. *Food Chem.* **2021**, *353*, 129337. [CrossRef] [PubMed]
66. Pan, C.; Zhao, L.; Zhao, D. Microwave-assisted green extraction of antioxidant components from *Osmanthus fragrans* (Lour) flower using natural deep eutectic solvents. *J. Appl. Res. Med. Aromat. Plants* **2021**, *20*, 100285. [CrossRef]
67. Panić, M.; Andlar, M.; Tišma, M.; Rezić, T.; Šibalić, D.; Cvjetko Bubalo, M.; Radojčić Redovniković, I. Natural deep eutectic solvent as a unique solvent for valorisation of orange peel waste by the integrated biorefinery approach. *Waste Manag.* **2021**, *120*, 340–350. [CrossRef]
68. López-Linares, J.C.; Campillo, V.; Coca, M.; Lucas, S.; García-Cubero, M.T. Microwave-assisted deep eutectic solvent extraction of phenolic compounds from brewer's spent grain. *J. Chem. Technol. Biotechnol.* **2021**, *96*, 481–490. [CrossRef]
69. Fernandes, C.; Melro, E.; Magalhães, S.; Alves, L.; Craveiro, R.; Filipe, A.; Valente, A.J.M.; Martins, G.; Antunes, F.E.; Romano, A.; et al. New deep eutectic solvent assisted extraction of highly pure lignin from maritime pine sawdust (*Pinus pinaster* Ait.). *Int. J. Biol. Macromol.* **2021**, *177*, 294–305. [CrossRef]
70. Su, Y.; Huang, C.; Lai, C.; Yong, Q. Green solvent pretreatment for enhanced production of sugars and antioxidative lignin from poplar. *Bioresour. Technol.* **2021**, *321*, 124471. [CrossRef]
71. Ong, V.Z.; Wu, T.Y.; Chu, K.K.L.; Sun, W.Y.; Shak, K.P.Y. A combined pretreatment with ultrasound-assisted alkaline solution and aqueous deep eutectic solvent for enhancing delignification and enzymatic hydrolysis from oil palm fronds. *Ind. Crops Prod.* **2021**, *160*, 112974. [CrossRef]
72. Fu, X.; Wang, D.; Belwal, T.; Xie, J.; Xu, Y.; Li, L.; Zou, L.; Zhang, L.; Luo, Z. Natural deep eutectic solvent enhanced pulse-ultrasonication assisted extraction as a multi-stability protective and efficient green strategy to extract anthocyanin from blueberry pomace. *Lwt* **2021**, *144*, 111220. [CrossRef]
73. Chen, L.; Yang, Y.Y.; Zhou, R.R.; Fang, L.Z.; Zhao, D.; Cai, P.; Yu, R.; Zhang, S.H.; Huang, J.H. The extraction of phenolic acids and polysaccharides from: Lilium lancifolium Thunb. using a deep eutectic solvent. *Anal. Methods* **2021**, *13*, 1226–1231. [CrossRef]
74. Mansinhos, I.; Gonçalves, S.; Rodríguez-Solana, R.; Ordóñez-Díaz, J.L.; Moreno-Rojas, J.M.; Romano, A. Ultrasonic-assisted extraction and natural deep eutectic solvents combination: A green strategy to improve the recovery of phenolic compounds from *Lavandula pedunculata* subsp. *lusitanica* (chaytor) franco. *Antioxidants* **2021**, *10*, 582. [CrossRef]
75. Jakovljević, M.; Jokić, S.; Molnar, M.; Jerković, I. Application of deep eutectic solvents for the extraction of carnosic acid and carnosol from sage (*Salvia officinalis* L.) with response surface methodology optimization. *Plants* **2021**, *10*, 80. [CrossRef]
76. Patil, S.S.; Pathak, A.; Rathod, V.K. Optimization and kinetic study of ultrasound assisted deep eutectic solvent based extraction: A greener route for extraction of curcuminoids from *Curcuma longa*. *Ultrason. Sonochem.* **2021**, *70*, 105267. [CrossRef] [PubMed]

77. Ünlü, A.E. Green and non—Conventional extraction of bioactive compounds from olive leaves: Screening of novel natural deep eutectic solvents and investigation of process parameters. *Waste Biomass Valorization* **2021**, 1–18. [CrossRef]

78. Ivanović, M.; Albreht, A.; Krajnc, P.; Vovk, I.; Razboršek, M.I. Sustainable ultrasound-assisted extraction of valuable phenolics from inflorescences of *Helichrysum arenarium* L. using natural deep eutectic solvents. *Ind. Crops Prod.* **2021**, *160*. [CrossRef]

79. Rodríguez-Juan, E.; Rodríguez-Romero, C.; Fernández-Bolaños, J.; Florido, M.C.; Garcia-Borrego, A. Phenolic compounds from virgin olive oil obtained by natural deep eutectic solvent (NADES): Effect of the extraction and recovery conditions. *J. Food Sci. Technol.* **2021**, *58*, 552–561. [CrossRef]

80. Shishov, A.; Volodina, N.; Gagarionova, S.; Shilovskikh, V.; Bulatov, A. A rotating disk sorptive extraction based on hydrophilic deep eutectic solvent formation. *Anal. Chim. Acta* **2021**, *1141*, 163–172. [CrossRef] [PubMed]

81. Dai, J.; Mumper, R.J. Plant phenolics: Extraction, analysis and their antioxidant and anticancer properties. *Molecules* **2010**, *15*, 7313–7352. [CrossRef]

82. Khezeli, T.; Daneshfar, A.; Sahraei, R. A green ultrasonic-assisted liquid-liquid microextraction based on deep eutectic solvent for the HPLC-UV determination of ferulic, caffeic and cinnamic acid from olive, almond, sesame and cinnamon oil. *Talanta* **2016**, *150*, 577–585. [CrossRef] [PubMed]

83. Farhadi, K.; Esmaeilzadeh, F.; Hatami, M.; Forough, M.; Molaie, R. Determination of phenolic compounds content and antioxidant activity in skin, pulp, seed, cane and leaf of five native grape cultivars in west azerbaijan province, Iran. *Food Chem.* **2015**, *199*, 847–855. [CrossRef] [PubMed]

84. Chong, Y.; Yan, A.; Yang, X.; Cai, Y.; Chen, J. An optimum fermentation model established by genetic algorithm for biotransformation from crude polydatin to resveratrol. *Appl. Biochem. Biotechnol.* **2012**, *166*, 446–457. [CrossRef]

85. Mensah, A.Y.; Sampson, J.; Houghton, P.J.; Hylands, P.J.; Westbrook, J.; Dunn, M.; Hughes, M.A.; Cherry, G.W. Effects of buddleja globosa leaf and its constituents relevant to wound healing. *J. Ethnopharmacol.* **2001**, *77*, 219–226. [CrossRef]

86. Estomba, D.; Ladio, A.; Lozada, M. Medicinal wild plant knowledge and gathering patterns in a mapuche community from north-western patagonia. *J. Ethnopharmacol.* **2006**, *103*, 109–119. [CrossRef]

87. García, A.; Rodríguez-Juan, E.; Rodríguez-Gutiérrez, G.; Rios, J.J.; Fernández-Bolaños, J. Extraction of phenolic compounds from virgin olive oil by deep eutectic solvents (DESs). *Food Chem.* **2016**, *197*, 554–561. [CrossRef]

88. Jangir, A.K.; Mandviwala, H.; Patel, P.; Sharma, S.; Kuperkar, K. Acumen into the effect of alcohols on choline chloride: L-lactic acid-based natural deep eutectic solvent (NADES): A spectral investigation unified with theoretical and thermophysical characterization. *J. Mol. Liq.* **2020**, *317*, 113923. [CrossRef]

89. Dai, Y.; Witkamp, G.J.; Verpoorte, R.; Choi, Y.H. Tailoring properties of natural deep eutectic solvents with water to facilitate their applications. *Food Chem.* **2015**, *187*, 14–19. [CrossRef]

90. Xue, J.; Wang, R.-Q.; Chen, X.; Hu, S.; Bai, X.-H. Three-phase hollow-fiber liquid-phase microextraction based on deep eutectic solvent as acceptor phase for extraction and preconcentration of main active compounds in a traditional Chinese medicinal formula. *J. Sep. Sci.* **2019**, *42*, 2239–2246. [CrossRef]

91. Khataei, M.M.; Yamini, Y.; Nazaripour, A.; Karimi, M. Novel generation of deep eutectic solvent as an acceptor phase in three-phase hollow fiber liquid phase microextraction for extraction and preconcentration of steroidal hormones from biological fluids. *Talanta* **2018**, *178*, 473–480. [CrossRef]

92. Afshar Mogaddam, M.R.; Farajzadeh, M.A.; Mohebbi, A.; Nemati, M. Hollow fiber–liquid phase microextraction method based on a new deep eutectic solvent for extraction and derivatization of some phenolic compounds in beverage samples packed in plastics. *Talanta* **2020**, *216*, 120986. [CrossRef]

93. Esrafili, A.; Baharfar, M.; Tajik, M.; Yamini, Y.; Ghambarian, M. Two-phase hollow fiber liquid-phase microextraction. *TrAC-Trends Anal. Chem.* **2018**, *108*, 314–322. [CrossRef]

94. Zargar, B.A.; Masoodi, M.H.; Ahmed, B.; Ganie, S.A. Phytoconstituents and therapeutic uses of rheum emodi wall. ex Meissn. *Food Chem.* **2011**, *128*, 585–589. [CrossRef]

95. Tanvir, E.M.; Hossen, M.S.; Hossain, M.F.; Afroz, R.; Gan, S.H.; Khalil, M.I.; Karim, N. Antioxidant properties of popular turmeric (*Curcuma longa*) varieties from bangladesh. *J. Food Qual.* **2017**, *2017*. [CrossRef]

96. Satari, B.; Karimi, K. Citrus processing wastes: Environmental impacts, recent advances, and future perspectives in total valorization. *Resour. Conserv. Recycl.* **2018**, *129*, 153–167. [CrossRef]

97. Kleinert, M.; Barth, T. Phenols from lignin. *Chem. Eng. Technol.* **2008**, *31*, 736–745. [CrossRef]

98. Xu, J.; Li, C.; Dai, L.; Xu, C.; Zhong, Y.; Yu, F.; Si, C. Biomass fractionation and lignin fractionation towards lignin valorization. *ChemSusChem* **2020**, *13*, 4284–4295. [CrossRef]

99. Tribot, A.; Amer, G.; Abdou Alio, M.; de Baynast, H.; Delattre, C.; Pons, A.; Mathias, J.D.; Callois, J.M.; Vial, C.; Michaud, P.; et al. Wood-lignin: Supply, extraction processes and use as bio-based material. *Eur. Polym. J.* **2019**, *112*, 228–240. [CrossRef]

100. Xu, J.F.; Han, Y.T.; WenBo, Z.; Zhao, Z.M. Multivariable analysis of the effects of factors in the pretreatment and enzymolysis processes on saccharification efficiency. *Ind. Crops Prod.* **2019**, *142*, 111824. [CrossRef]

101. Lim, H.Y.; Yusup, S.; Loy, A.C.M.; Samsuri, S.; Ho, S.S.K.; Manaf, A.S.A.; Lam, S.S.; Chin, B.L.F.; Acda, M.N.; Unrean, P.; et al. Review on conversion of lignin waste into value-added resources in tropical countries. *Waste Biomass Valorization* **2020**, 1–18. [CrossRef]

102. Lai, C.; Jia, Y.; Wang, J.; Wang, R.; Zhang, Q.; Chen, L.; Shi, H.; Huang, C.; Li, X.; Yong, Q. Co-production of xylooligosaccharides and fermentable sugars from poplar through acetic acid pretreatment followed by poly (ethylene glycol) ether assisted alkali treatment. *Bioresour. Technol.* **2019**, *288*, 121569. [CrossRef] [PubMed]

103. Li, P.; Zhang, Q.; Zhang, X.; Zhang, X.; Pan, X.; Xu, F. Subcellular dissolution of xylan and lignin for enhancing enzymatic hydrolysis of microwave assisted deep eutectic solvent pretreated Pinus bungeana Zucc. *Bioresour. Technol.* **2019**, *288*, 121475. [CrossRef] [PubMed]

104. Gągol, M.; Przyjazny, A.; Boczkaj, G. Wastewater treatment by means of advanced oxidation processes based on cavitation—A review. *Chem. Eng. J.* **2018**, *338*, 599–627. [CrossRef]

105. Martín, J.; Navas, M.J.; Jiménez-Moreno, A.M.; Asuero, A.G. Anthocyanin pigments: Importance, sample preparation and extraction. In *Phenolic Compounds-Natural Sources, Importance and Applications*; Soto-Hernandez, M., Palma-Tenango, M., Garcia-Mateos, M.R., Eds.; IntechOpen: London, UK, 2017.

106. Giusti, M.M.; Wrolstad, R.E. Characterization and measurement of anthocyanins by UV-visible spectroscopy. *Curr. Protoc. Food Anal. Chem.* **2001**, F1.2.1–F1.2.13. [CrossRef]

107. Panić, M.; Gunjević, V.; Cravotto, G.; Radojčić Redovniković, I. Enabling technologies for the extraction of grape-pomace anthocyanins using natural deep eutectic solvents in up-to-half-litre batches extraction of grape-pomace anthocyanins using NADES. *Food Chem.* **2019**, *300*, 1–8. [CrossRef]

108. Lopes, C.L.; Pereira, E.; Soković, M.; Carvalho, A.M.; Barata, A.M.; Lopes, V.; Rocha, F.; Calhelha, R.C.; Barros, L.; Ferreira, I.C.F.R. Phenolic composition and bioactivity of *Lavandula pedunculata* (Mill.) cav. samples from different geographical origin. *Molecules* **2018**, *23*, 1037. [CrossRef]

109. Nunes, R.; Pasko, P.; Tyszka-Czochara, M.; Szewczyk, A.; Szlosarczyk, M.; Carvalho, I.S. Antibacterial, antioxidant and anti-proliferative properties and zinc content of five south portugal herbs. *Pharm. Biol.* **2017**, *55*, 114–123. [CrossRef]

110. Shirsath, S.R.; Sable, S.S.; Gaikwad, S.G.; Sonawane, S.H.; Saini, D.R.; Gogate, P.R. Intensification of extraction of curcumin from Curcuma amada using ultrasound assisted approach: Effect of different operating parameters. *Ultrason. Sonochem.* **2017**, *38*, 437–445. [CrossRef]

111. Romero, M.; Toral, M.; Gómez-Guzmán, M.; Jiménez, R.; Galindo, P.; Sánchez, M.; Olivares, M.; Gálvez, J.; Duarte, J. Antihypertensive effects of oleuropein-enriched olive leaf extract in spontaneously hypertensive rats. *Food Funct.* **2016**, *7*, 584–593. [CrossRef]

112. Castro-Muñoz, R.; Boczkaj, G.; Gontarek, E.; Cassano, A.; Fíla, V. Membrane technologies assisting plant-based and agro-food by-products processing: A comprehensive review. *Trends Food Sci. Technol.* **2020**, *95*, 219–232. [CrossRef]

113. Skarpalezos, D.; Detsi, A. Deep eutectic solvents as extraction media for valuable flavonoids from natural sources. *Appl. Sci.* **2019**, *9*, 4169. [CrossRef]

114. Ali Redha, A. Review on extraction of phenolic compounds from natural sources using green deep eutectic solvents. *J. Agric. Food Chem.* **2021**, *69*, 878–912. [CrossRef] [PubMed]

Article

Phenolic Profiles of Ten Australian Faba Bean Varieties

Joel B. Johnson [1,*], Daniel J. Skylas [2], Janice S. Mani [1], Jinle Xiang [3], Kerry B. Walsh [1] and Mani Naiker [1]

[1] School of Health, Medical & Applied Sciences, CQUniversity Australia, Bruce Hwy,
North Rockhampton, QLD 4701, Australia; janice.mani@gmail.com (J.S.M.); k.walsh@cqu.edu.au (K.B.W.);
m.naiker@cqu.edu.au (M.N.)

[2] Australian Export Grains Innovation Centre, North Ryde, NSW 2113, Australia; daniel.skylas@aegic.org.au

[3] Faculty of Food & Bioengineering, Henan University of Science & Technology, Luoyang 471023, China;
xjl5013@haust.edu.cn

[*] Correspondence: joel.johnson@cqumail.com

Abstract: Although Australia is the largest exporter of faba bean globally, there is limited information available on the levels of bioactive compounds found in current commercial faba bean varieties grown in this country. This study profiled the phenolic acid and flavonoid composition of 10 Australian faba bean varieties, grown at two different locations. Phenolic profiling by HPLC-DAD revealed the most abundant flavonoid to be catechin, followed by rutin. For the phenolic acids, syringic acid was found in high concentrations (72.4–122.5 mg/kg), while protocatechuic, vanillic, *p*-hydroxybenzoic, chlorogenic, *p*-coumaric, and trans-ferulic acid were all found in low concentrations. The content of most individual phenolics varied significantly with the variety, while some effect of the growing location was also observed. This information could be used by food processors and plant breeders to maximise the potential health benefits of Australian-grown faba bean.

Keywords: phenolic acids; flavonoids; *Vicia faba*; functional food

Citation: Johnson, J.B.; Skylas, D.J.; Mani, J.S.; Xiang, J.; Walsh, K.B.; Naiker, M. Phenolic Profiles of Ten Australian Faba Bean Varieties. *Molecules* **2021**, *26*, 4642. https://doi.org/10.3390/molecules26154642

Academic Editors: Mirella Nardini and Jesus Simal-Gandara

Received: 21 June 2021
Accepted: 27 July 2021
Published: 30 July 2021

Publisher's Note: MDPI stays neutral with regard to jurisdictional claims in published maps and institutional affiliations.

1. Introduction

Faba bean (*Vicia faba* L.) is reported to be the third most important legume crop [1], with over 5.4 million tonnes being harvested globally in 2019 [1]. After China and Ethiopia, Australia is the third-largest faba bean producer in the world and the largest exporter, providing at least one-third of the internationally traded crop volume [2]. The crop is primarily consumed in China, Southeast Asian countries, and countries in the Middle East.

In recent years, there has been an increasing interest in faba bean due to its nutritional content [1,3] and health-benefitting properties [4]. It is a valuable source of protein, containing twice the protein content of cereal grains in addition to a number of essential amino acids [1]. Furthermore, reported health benefits include improving cardiovascular health [5], providing anti-obesity effects [5,6], anti-cancer activity [5], anti-inflammatory activity [7] and inhibiting xanthine oxidase [8]. This has led to an interest in using faba bean or its isolates in functional food applications [3]. One of the major groups of health-benefiting compounds found in faba bean are phenolics [7].

Due to genotypic variations, which can influence the phenolic and flavonoid biosynthetic pathways [9], the phenolic content may vary significantly between different varieties of grains and pulses [10]. Hence, there has been recent interest in identifying faba bean varieties with high levels of phenolic content. For example, Valente et al. [11] and Valente et al. [12] profiled the phenolic content and antioxidant of seven European faba bean varieties, finding that the levels of total and individual phenolic acids and flavonoids differed significantly between varieties. Similarly, Baginsky et al. [13] found clear differences in the phenolic composition of 10 faba bean varieties grown in Chile, although it should be noted that this study was performed on immature seed material. Another earlier study highlighted the range in total phenolic contents and antioxidant activity among 13 Tunisian faba bean cultivars [14]. However, despite Australia's international importance as a faba bean producer, there are few comparative studies reporting the phenolic contents of faba bean varieties commonly grown in this country.

Nasar-Abbas et al. [15] reported on the phenolics found in one Australian faba bean variety, while Siah et al. [16] and Siah et al. [5] investigated two and three varieties, respectively. Siah et al. [17] studied the phenolic content and antioxidant activity in five Australian faba bean varieties. Recently, Johnson et al. [18] reported on the antioxidant activity, total anthocyanin content, and total phenolic content of 10 Australian faba bean varieties, although individual phenolic acids or flavonoids were not investigated. This work aims to aid in filling this knowledge gap, presenting phenolic profiles on the 10 faba bean varieties studied by Johnson et al. [18]. It is hoped that this will provide further insight into the nutritional and health-benefitting properties of common Australian faba bean cultivars, as well as providing valuable information on the extent of their genotypic variation present in terms of phenolic acid and flavonoid biosynthesis pathways.

2. Results and Discussion

2.1. Total Phenolic Contents

The total phenolic content (TPC), as measured by the Folin–Ciocalteu assay, ranged from 258 to 570 mg GAE/100 g (DW) in the different faba bean varieties (Figure 1). The PBA Rana variety contained a significantly higher total phenolic content compared to all remaining varieties. A two-way ANOVA revealed that the site had no significant impact on the total phenolic content ($p < 0.05$), with samples grown at Charlick having a mean TPC of 322 ± 96 mg GAE/100 g ($n = 30$), compared to 324 ± 107 mg GAE/100 g for samples grown at Freeling ($n = 30$). In addition, there was no significant interaction between growing site × variety ($p > 0.05$).

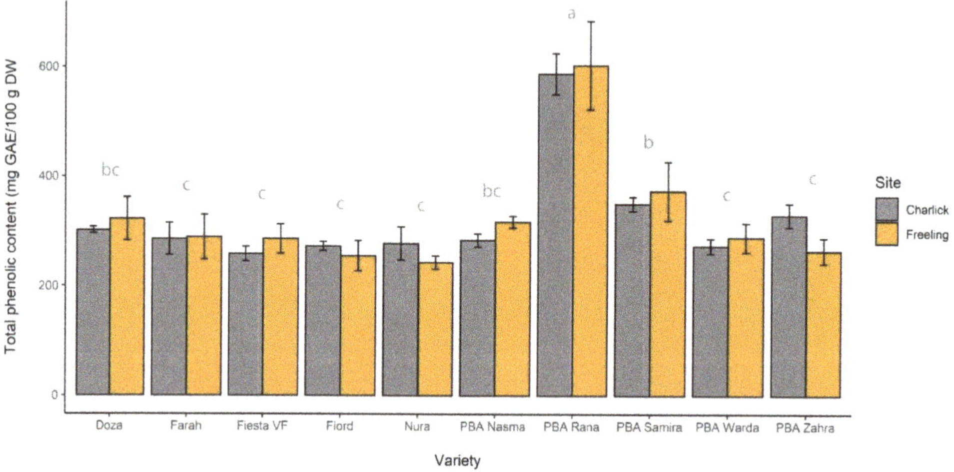

Figure 1. Total phenolic content of the 10 faba bean varieties at each of the growing sites ($n = 3$ field replicates for each bar). The letters (a–c) above each variety show the statistical significance of an ANOVA by variety averaged across both growing locations. Varieties with the same letter were not statistically different from one another at $\alpha = 0.05$.

2.2. Phenolic acid Profiling by HPLC

A total of 10 phenolic compounds were identified in the faba bean extracts (Figure 2), comprising of four hydroxybenzoic acids, three hydroxycinnamic acids and three flavonoid-related compounds (Table 1). The following compounds were all determined as being either absent or below the limit of detection: gallic acid, gentisic acid, isovanillic acid, caffeic acid, sinapic acid, cinnamic acid and quercetin-3-glucoside.

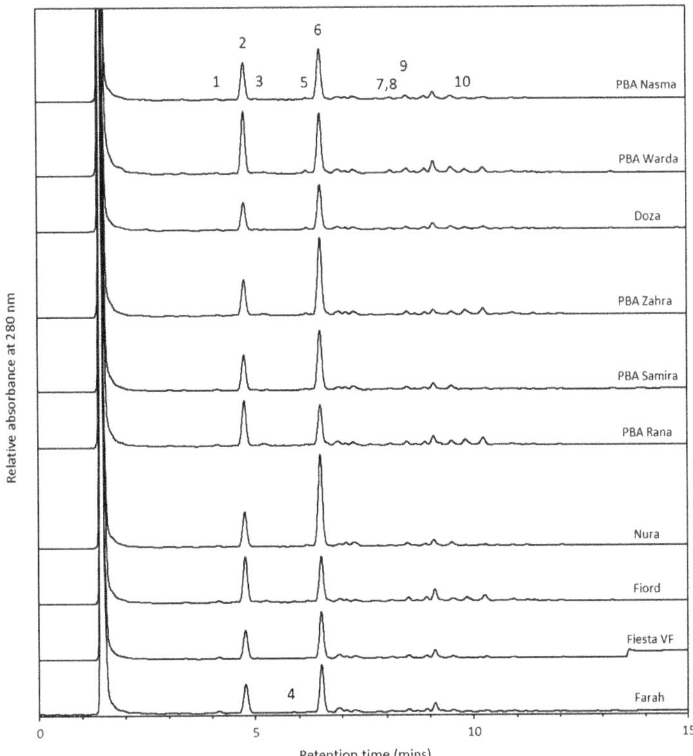

Figure 2. HPLC chromatograms of the phenolic compounds in the 10 faba bean varieties. The compounds indicated are (1) protocatechuic acid, (2) catechin, (3) chlorogenic acid, (4) *p*-hydroxybenzoic acid, (5) vanillic acid, (6) syringic acid, (7) *p*-coumaric acid, (8) vitexin, (9), *trans*-ferulic acid, (10) rutin.

The hydroxybenzoic acids found here (protocatechuic acid, *p*-hydroxybenzoic acid, vanillic acid and syringic acid) have all been previously reported from faba bean, as have two of the hydroxycinnamic acids (*p*-coumaric acid and *trans*-ferulic acid) [19–21]. In addition, several of these phenolic acids have been found in faba bean pods [11]. The concentrations of free *p*-coumaric and ferulic acids found here were similar to that reported by Liu et al. [21] in Canadian faba bean, although only one variety was included in that study. However, the concentration of syringic acid was much higher compared to previous studies [21,22]. Although the reason for this difference is unclear, it is worth noting that levels of this compound varied significantly between different genotypes (Table 1) and that the soil microbiota composition can also have a significant impact [23]. Similarly, although chlorogenic acid does not appear to have been previously found in faba bean seed, it is produced in the roots of this plant [24]. The low concentrations and high levels of environmental variability may account for its absence in previous work.

The levels of catechin reported here (ranging from 191–297 mg/kg for different varieties) were at the lower range of concentrations reported by Baginsky et al. [13] in the immature seed material of 10 Chilean faba bean varieties, with catechin contents varying between 85–978 mg/kg.

Table 1. Mean phenolic acid and flavonoid contents in the 10 faba bean varieties. Values given in µg/g (mean ± SD from 6 replicates, comprising within-field triplicates from two field locations). *p*-value indicates the significance between varieties, with results obtained from a two-way ANOVA between site × variety. Note that entries in the same row containing the same superscript letter (a–d) were not significantly different from one another at α = 0.05.

Compound	Doza	Farah	Fiesta VF	Fiord	Nura	PBA Nasma	PBA Rana	PBA Samira	PBA Warda	PBA Zahra	*p* Value
Protocatechuic acid	1.88 ± 0.83 [b]	1.45 ± 0.61 [b]	1.44 ± 0.50 [b]	1.66 ± 0.77 [b]	1.29 ± 0.22 [b]	1.65 ± 0.60 [b]	2.93 ± 1.07 [a]	2.09 ± 0.77 [ab]	1.83 ± 0.56 [b]	1.81 ± 0.36 [b]	***
p-hydroxybenzoic acid	0.57 ± 0.06 [bcd]	0.52 ± 0.07 [cd]	0.52 ± 0.11 [cd]	0.64 ± 0.18 [bcd]	0.44 ± 0.08 [d]	0.62 ± 0.13 [bcd]	1.11 ± 0.21 [a]	0.73 ± 0.15 [bc]	0.61 ± 0.10 [bcd]	0.79 ± 0.11 [b]	***
Vanillic acid	2.46 ± 0.59 [ab]	1.87 ± 0.37 [b]	1.96 ± 0.36 [b]	1.88 ± 0.43 [b]	2.40 ± 0.24 [ab]	1.96 ± 0.29 [b]	2.24 ± 0.68 [ab]	2.81 ± 0.52 [a]	2.71 ± 0.40 [a]	2.76 ± 0.85 [a]	***
Syringic acid	77.6 ± 11.2 [c]	72.4 ± 7.9 [c]	80.5 ± 13.1 [c]	77.6 ± 11.8 [c]	149.8 ± 14.8 [a]	77.8 ± 9.5 [c]	72.5 ± 7.3 [c]	109.6 ± 21.6 [b]	80.9 ± 14.1 [c]	122.5 ± 14.4 [b]	***
Sum of hydroxybenzoic acids	82.5 ± 12.6 [c]	76.2 ± 8.5 [c]	84.4 ± 13.7 [c]	81.8 ± 12.7 [c]	153.9 ± 14.9 [a]	82.0 ± 10.3 [c]	78.8 ± 8.8 [c]	115.2 ± 22.8 [b]	86.0 ± 14.8 [c]	127.9 ± 15.4 [b]	***
Chlorogenic acid	0.89 ± 0.96	0.85 ± 0.52	1.27 ± 1.30	2.88 ± 2.56	0.78 ± 0.41	0.89 ± 0.44	3.02 ± 3.31	1.14 ± 0.73	1.70 ± 2.48	1.98 ± 3.41	NS
p-coumaric acid	1.21 ± 0.21 [bc]	1.64 ± 0.25 [ab]	1.86 ± 0.40 [a]	1.69 ± 0.28 [ab]	1.26 ± 0.16 [abc]	1.52 ± 0.37 [abc]	1.70 ± 0.27 [ab]	1.52 ± 0.38 [abc]	1.62 ± 0.54 [ab]	0.95 ± 0.17 [c]	***
trans-ferulic acid	1.27 ± 0.22 [b]	0.96 ± 0.21 [b]	1.11 ± 0.27 [b]	1.42 ± 0.42 [b]	1.36 ± 0.32 [b]	1.22 ± 0.35 [b]	2.99 ± 0.65 [a]	1.34 ± 0.25 [b]	1.26 ± 0.30 [b]	1.82 ± 0.18 [b]	***
Sum of hydroxycinnamic acids	3.37 ± 1.09 [b]	3.45 ± 0.76 [b]	4.24 ± 1.81 [ab]	5.99 ± 2.97 [ab]	3.40 ± 0.68 [b]	3.63 ± 0.85 [ab]	7.71 ± 2.83 [a]	4.00 ± 1.17 [ab]	4.58 ± 3.22 [ab]	4.11 ± 3.60 [ab]	*
Catechin	216 ± 64 [ab]	191 ± 37 [b]	215 ± 52 [ab]	245 ± 52 [ab]	232 ± 27 [ab]	207 ± 37 [b]	297 ± 53 [a]	240 ± 55 [ab]	258 ± 63 [ab]	220 ± 33 [ab]	*
Vitexin	0.88 ± 0.24 [b]	1.70 ± 1.82 [ab]	0.97 ± 0.39 [b]	1.52 ± 1.82 [ab]	0.58 ± 0.41 [b]	0.80 ± 0.28 [b]	3.50 ± 1.43 [a]	0.75 ± 0.43 [b]	1.21 ± 0.82 [b]	1.43 ± 1.52 [ab]	**
Rutin	5.55 ± 5.02	7.34 ± 5.11	7.66 ± 6.36	13.91 ± 11.81	4.04 ± 3.00	4.50 ± 2.30	15.87 ± 14.22	7.67 ± 4.09	10.48 ± 10.60	9.43 ± 16.29	NS
Sum of flavonoids	223 ± 61 [ab]	200 ± 37 [b]	223 ± 57 [ab]	261 ± 61 [ab]	237 ± 28 [ab]	212 ± 37 [b]	316 ± 45 [a]	248 ± 56 [ab]	269 ± 71 [ab]	231 ± 35 [ab]	*

NS = not significant ($p > 0.05$), * $p < 0.05$, ** $p < 0.01$, *** $p < 0.001$.

Although vitexin is more commonly known to occur in mungbean [25], it has been previously reported from faba bean using UHPLC-ESI-QTOF-MS-based metabolic profiling [26,27], although it was not quantified. Similarly, although rutin (quercetin-3-rutinoside) does not appear to have been previously reported from faba bean seed, numerous other types of quercetin glycosides have been found in this matrix [8,11]. In addition, rutin has been reported from the flower tissue of several faba bean genotypes, indicating that the synthetic pathways for the production of this compound do occur in faba bean [9].

A two-way ANOVA revealed that the content of all constituents, apart from chlorogenic acid and rutin, varied significantly with variety (Table 1). For most of these compounds, the highest concentrations were found in PBA Rana, which also contained the highest total phenolic content (Figure 1). However, Nura showed the highest levels of syringic acid and total hydroxybenzoic acids.

Similarly, the two-way ANOVA demonstrated that in the case of the 2017 growing season, the site had a significant impact on the content of protocatechuic acid, vanillic acid, chlorogenic acid, vitexin and rutin, as well as on the total amounts of hydroxybenzoic acids and hydroxycinnamic acids (Table 2). For both hydroxybenzoic acids (protocatechuic acid and vanillic acid), samples grown at the Freeling site showed higher contents; while for chlorogenic acid and the flavonoids catechin and rutin, the Charlick samples showed higher concentrations.

Table 2. Impact of the growing site on phenolic acid and flavonoid contents. Values given in μg/g (mean ± SD from 3 replicates for each location). The *p*-value column indicates the significance between sites, with results obtained from a two-way ANOVA between site × variety.

Compound	Charlick (*n* = 30)	Freeling (*n* = 30)	Site *p* Value	Variety × Site Interaction
Protocatechuic acid	1.43 ± 0.36	2.17 ± 0.87	***	NS
p-hydroxybenzoic acid	0.67 ± 0.19	0.64 ± 0.25	NS	NS
Vanillic acid	2.11 ± 0.41	2.50 ± 0.67	***	**
Syringic acid	89.3 ± 28.2	94.9 ± 27.9	NS	*
Sum of hydroxybenzoic acids	93.5 ± 28.3	100.2 ± 28.4	*	*
Chlorogenic acid	2.22 ± 2.57	0.86 ± 0.68	**	NS
p-coumaric acid	1.45 ± 0.42	1.55 ± 0.38	NS	NS
trans-ferulic acid	1.45 ± 0.47	1.38 ± 0.77	NS	**
Sum of hydroxycinnamic acids	5.11 ± 3.00	3.78 ± 1.39	*	NS
Catechin	224 ± 45	240 ± 60	NS	NS
Vitexin	1.67 ± 1.56	0.98 ± 0.86	*	NS
Rutin	12.21 ± 11.59	5.07 ± 3.54	**	NS
Sum of flavonoids	238 ± 52	246 ± 62	NS	NS

* $p < 0.05$, ** $p < 0.01$, *** $p < 0.001$

For the 2017 growing season, no significant effects of growing location were found for *p*-hydroxybenzoic acid, syringic acid, *p*-coumaric acid, *trans*-ferulic acid, catechin or the total amount of flavonoids. Significant interactions were found between the variety and growing site for several parameters, namely vanillic acid, syringic acid, *trans*-ferulic acid and the sum of hydroxybenzoic acids. It should be noted that the present study investigated only one growing season; hence the results found here may not be generalizable across a wider range of seasons and locations.

There appears to be limited previous literature investigating the impact of growing site and variety × growing site interaction on phenolic acid content in faba bean; however, Mpofu et al. [28] found a significant impact of growing location on six phenolic acids in wheat. In contrast, Oomah et al. [29] found very little impact of growing location on the total phenolic content of 13 faba bean genotypes grown at two locations in Canada.

2.3. Principal Component Analysis and Correlation Analysis

Overall, PBA Rana appeared to have the highest levels of most phenolic acids and flavonoids, possessing a distinct chemical profile compared to the other varieties. This observation was supported by the results of the principal component analysis performed on the normalised phenolic data, which revealed that most samples of PBA Rana were clustered toward the lower right of the scores plot, separated from the majority of other genotypes (Figure 3). Examination of the loadings plot revealed that this corresponded with higher concentrations of catechin and protocatechuic acid, and lower concentrations of syringic acid. In addition, this concurred with previous work highlighting the unique bioactive profile of this genotype [18].

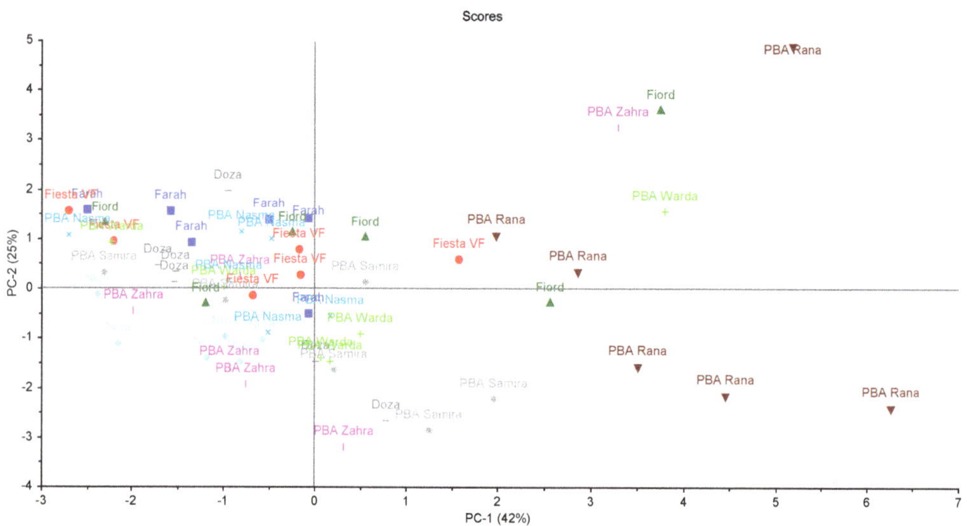

Figure 3. Scores plot showing the results of the principal component analysis performed on the normalised phenolic data. Each faba bean variety is indicated by a different symbol color and shape.

In contrast, the variety Nura was clustered toward the lower left of the scores plot, with syringic acid also weighted on this region of the two PCS. The remaining varieties were more or less clustered around the centre of the scores plot, indicating a relatively similar phenolic composition between them.

Finally, correlation analysis was performed between the 10 phenolic compounds to ascertain if the concentrations of any specific compounds were closely linked to the concentrations of another compound. This may occur due to similar synthesis pathways between the compounds [30] or result from regulatory genes controlling multiple synthesis pathways. The correlation results demonstrated moderate to strong correlations between several compounds, most notably between rutin and chlorogenic acid (r_{60} = 0.979, $p < 0.001$), and between ferulic and p-hydroxybenzoic acid (r_{60} = 0.812, $p < 0.001$) (Figure 4). Rutin is a quercetin glycoside, while chlorogenic acid is the ester of caffeic acid and quinic acid, hence these compounds are not closely structurally related. However, both can be synthesised through the phenylpropanoid pathway [31], suggesting that a regulatory gene may be responsible for the correlation between these compounds. Similarly, ferulic acid is a hydroxycinnamic acid, while p-hydroxybenzoic acid is a simple hydroxybenzoic acid; however, both can be produced through the shikimate biosynthesis pathway [32].

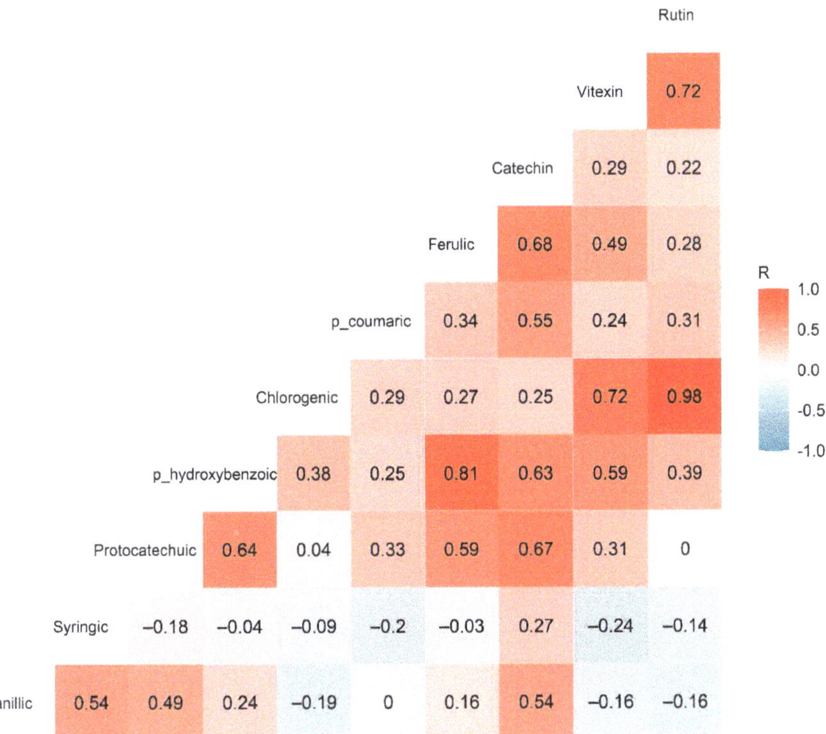

Figure 4. Correlogram showing the correlations between the various phenolic acids and flavonoids quantified in the faba bean samples (*n* = 60 replicates). The numbers inside each square show the Pearson R correlation values.

3. Materials and Methods

3.1. Seed Material

Faba bean seed material was sourced from growing field trials in South Australia, as previously described [33]. This comprised material from 10 varieties, each of which was grown at 2 different sites (Charlick and Freeling) during the 2017 season. Three within-field replicates (each comprising approximately 50 g) were subsampled from the mechanically harvested composite sample for each treatment [33], for a total of 60 samples. Whole seeds were coarsely ground (Van Gelder grinder with a 3 mm screen) before being finely ground to a fine flour (Falling Number grinder with 0.8 mm screen) [33].

3.2. Extraction of Phenolic Compounds

Polar phenolic compounds were extracted from the faba bean flour using 90% methanol, following previously reported methods [18]. The total phenolic content was determined using the Folin–Ciocalteu method, as previously described [18].

3.3. Phenolic Profiling by HPLC

Each methanol extract (20 mL) was concentrated using a rotary evaporator with the water bath temperature limited to 27 °C, before being reconstituted in 1 mL of methanol and syringe filtered (Livingstone 0.45 μm PTFE). The phenolics were separated on an Agilent 1100 HPLC system (Waldbronn, Germany) using previously described methods [27]. Briefly, a reversed-phase C_{18} column (Agilent Eclipse XDB-C18; 150 × 4.6 mm; 5 μm pore size) and guard cartridge (Gemini C_{18} 4 × 2 mm) were used, with an injection volume of 5 μL and column temperature of 27 ± 0.8 °C. The mobile phase comprised 0.01 M phosphoric acid (A) and methanol (B) at a flow rate of 1 mL/min, with the gradient beginning at 20% B and ramping linearly to reach 100% B at 20 min. The total run time was 25 min, with a post-run equilibration time of 7 min.

Compounds were identified based on a comparison of their retention time and UV spectra with authentic standards (Sigma Aldrich Australia). The purity of the peaks was confirmed by examination of the UV spectra at different points throughout the peak. Quality-of-analysis parameters are shown in Table 3.

Table 3. Quality-of-analysis parameters for the phenolic standards. All calibrations were performed at concentrations between 1–100 mg L^{-1}.

No.	Compound	Retention Time (min)	Wavelength (nm)	Slope	LOD (mg L^{-1})	LOQ (mg L^{-1})	Calibration R^2
			Hydroxybenzoic acids				
1	Protocatechuic acid	3.94	250 nm	13.5	0.1	0.4	1
4	*p*-hydroxybenzoic acid	5.78	250 nm	25.9	0.1	0.2	1
5	Vanillic acid	6.26	250 nm	12.3	0.1	0.4	1
6	Syringic acid	6.59	280 nm	15.1	0.1	0.3	1
			Hydroxycinnamic acids				
3	Chlorogenic acid	5.26	320 nm	14.1	0.1	0.4	1
7	*p*-coumaric acid	8.12	320 nm	32.3	0.05	0.2	1
9	*trans*-ferulic acid	8.44	320 nm	26.9	0.1	0.2	1
			Flavonoids				
2	Catechin	4.55	280 nm	4.0	0.4	1.3	1
8	Vitexin	8.17	320 nm	7.3	0.2	0.7	0.9999
10	Rutin	9.82	250 nm	6.2	0.2	0.8	1

3.4. Data Analysis

Statistical analysis was performed in R Studio running R 4.0.2 [34]. A two-way ANOVA was performed to assess the impact of variety and growing site on the content of various constituents. This was considered appropriate as the majority of data were approximately normally distributed and due to the large sample size ($n = 60$), the Central limit theorem could be applied to the dataset. Principal component analysis was performed in the Unscrambler X 10.5 software (Camo ASA, Oslo, Norway). Where applicable, results were presented as mean ± 1 standard deviation.

4. Conclusions

This study profiled the phenolic acid and flavonoid composition in 10 commercial varieties of Australian faba bean for the first time. The most abundant compounds were catechin and syringic acid, with rutin, vitexin, protocatechuic, vanillic, *p*-hydroxybenzoic, chlorogenic, *p*-coumaric, and *trans*-ferulic acid all found in low concentrations. The content of most individual phenolics varied significantly with the variety while growing location had a significant effect for around half of these compounds. Genotype × location interactions were only observed for vanillic, syringic, and *trans*-ferulic acids. Significant correlations were observed between a number of constituents, including between rutin and chlorogenic acid, and between ferulic and *p*-hydroxybenzoic acid. Notably, PBA Rana showed a distinct phenolic profile compared to the remaining nine varieties, supporting the findings of earlier research. In addition to providing baseline information on the typical phenolic contents of Australian-grown faba bean varieties, this study may inform plant breeders and growers in optimising the potential health benefits of the Australian faba bean crop.

Author Contributions: Conceptualization, J.B.J. and M.N.; methodology, J.B.J. and M.N; software, J.B.J. and K.B.W.; validation, J.B.J.; formal analysis, J.B.J.; investigation, J.B.J.; resources, J.B.J., D.J.S. and M.N.; data curation, J.B.J.; writing—original draft preparation, J.B.J.; writing—review and editing, J.B.J., D.J.S., J.S.M., J.X., K.B.W. and M.N.; visualization, J.B.J.; supervision, M.N.; project administration, M.N.; funding acquisition, M.N. All authors have read and agreed to the published version of the manuscript.

Funding: This research was funded by CQUniversity, in the form of a New Staff Grant (RSH/5343) awarded to one of the authors (M.N.). One of the authors (J.B.J.) also acknowledges support from the Australian Government in the form of a Research Training Program Scholarship.

Institutional Review Board Statement: Not applicable.

Informed Consent Statement: Not applicable.

Data Availability Statement: The data presented in this study are available on request from the corresponding author.

Acknowledgments: The authors acknowledge the assistance of Tania Collins in the laboratory work. We also thank the Australian Export Grains Innovation Centre (AEGIC) for supplying the seed material used in this study.

Conflicts of Interest: The authors declare no conflict of interest. The funders had no role in the design of the study; in the collection, analyses, or interpretation of data; in the writing of the manuscript, or in the decision to publish the results.

References

1. Rahate, K.A.; Madhumita, M.; Prabhakar, P.K. Nutritional composition, anti-nutritional factors, pretreatments-cum-processing impact and food formulation potential of faba bean (*Vicia faba* L.): A comprehensive review. *LWT-Food Sci. Technol.* **2021**, *138*, 110796. [CrossRef]
2. AEGIC. Australian Pulses: Quality, Versatility, Nutrition. Available online: https://www.aegic.org.au (accessed on 6 February 2019).
3. Sharan, S.; Zanghelini, G.; Zotzel, J.; Bonerz, D.; Aschoff, J.; Saint-Eve, A.; Maillard, M.-N. Fava bean (*Vicia faba* L.) for food applications: From seed to ingredient processing and its effect on functional properties, antinutritional factors, flavor, and color. *Compr. Rev. Food Sci. Food Saf.* **2021**, *20*, 401–428. [CrossRef]
4. Turco, I.; Ferretti, G.; Bacchetti, T. Review of the health benefits of Faba bean (*Vicia faba* L.) polyphenols. *J. Food Nutr. Res.* **2016**, *55*, 283–293.
5. Siah, S.D.; Konczak, I.; Agboola, S.; Wood, J.A.; Blanchard, C.L. In vitro investigations of the potential health benefits of Australian-grown faba beans (*Vicia faba* L.): Chemopreventative capacity and inhibitory effects on the angiotensin-converting enzyme, α-glucosidase and lipase. *Br. J. Nutr.* **2012**, *108*, S123–S134. [CrossRef] [PubMed]
6. Jakubczyk, A.; Karaś, M.; Złotek, U.; Szymanowska, U.; Baraniak, B.; Bochnak, J. Peptides obtained from fermented faba bean seeds (*Vicia faba*) as potential inhibitors of an enzyme involved in the pathogenesis of metabolic syndrome. *LWT-Food Sci. Technol.* **2019**, *105*, 306–313. [CrossRef]
7. Boudjou, S.; Oomah, B.D.; Zaidi, F.; Hosseinian, F. Phenolics content and antioxidant and anti-inflammatory activities of legume fractions. *Food Chem.* **2013**, *138*, 1543–1550. [CrossRef] [PubMed]
8. Spanou, C.; Veskoukis, A.S.; Kerasioti, T.; Kontou, M.; Angelis, A.; Aligiannis, N.; Skaltsounis, A.-L.; Kouretas, D. Flavonoid Glycosides Isolated from Unique Legume Plant Extracts as Novel Inhibitors of Xanthine Oxidase. *PLoS ONE* **2012**, *7*, e32214. [CrossRef]
9. Zanotto, S.; Khazaei, H.; Elessawy, F.M.; Vandenberg, A.; Purves, R.W. Do Faba Bean Genotypes Carrying Different Zero-Tannin Genes (zt1 and zt2) Differ in Phenolic Profiles? *J. Agric. Food Chem.* **2020**, *68*, 7530–7540. [CrossRef] [PubMed]
10. Xiang, J.; Zhang, M.; Apea-Bah, F.B.; Beta, T. Hydroxycinnamic acid amide (HCAA) derivatives, flavonoid C-glycosides, phenolic acids and antioxidant properties of foxtail millet. *Food Chem.* **2019**, *295*, 214–223. [CrossRef] [PubMed]
11. Valente, I.M.; Maia, M.R.G.; Malushi, N.; Oliveira, H.M.; Papa, L.; Rodrigues, J.A.; Fonseca, A.J.M.; Cabrita, A.R.J. Profiling of phenolic compounds and antioxidant properties of European varieties and cultivars of *Vicia faba* L. pods. *Phytochemistry* **2018**, *152*, 223–229. [CrossRef]
12. Valente, I.M.; Cabrita, A.R.J.; Malushi, N.; Oliveira, H.M.; Papa, L.; Rodrigues, J.A.; Fonseca, A.J.M.; Maia, M.R.G. Unravelling the phytonutrients and antioxidant properties of European *Vicia faba* L. seeds. *Food Res. Int.* **2019**, *116*, 888–896. [CrossRef] [PubMed]
13. Baginsky, C.; Peña-Neira, Á.; Cáceres, A.; Hernández, T.; Estrella, I.; Morales, H.; Pertuzé, R. Phenolic compound composition in immature seeds of fava bean (*Vicia faba* L.) varieties cultivated in Chile. *J. Food Compos. Anal.* **2013**, *31*, 1–6. [CrossRef]
14. Chaieb, N.; González, J.L.; López-Mesas, M.; Bouslama, M.; Valiente, M. Polyphenols content and antioxidant capacity of thirteen faba bean (*Vicia faba* L.) genotypes cultivated in Tunisia. *Food Res. Int.* **2011**, *44*, 970–977. [CrossRef]
15. Nasar-Abbas, S.; Siddique, K.; Plummer, J.; White, P.; Harris, D.; Dods, K.; D'antuono, M. Faba bean (*Vicia faba* L.) seeds darken rapidly and phenolic content falls when stored at higher temperature, moisture and light intensity. *LWT-Food Sci. Technol.* **2009**, *42*, 1703–1711. [CrossRef]
16. Siah, S.; Konczak, I.; Wood, J.A.; Agboola, S.; Blanchard, C.L. Effects of roasting on phenolic composition and in vitro antioxidant capacity of Australian grown faba beans (*Vicia faba* L.). *Plant Foods Hum. Nutr.* **2014**, *69*, 85–91. [CrossRef] [PubMed]
17. Siah, S.; Wood, J.A.; Agboola, S.; Konczak, I.; Blanchard, C.L. Effects of soaking, boiling and autoclaving on the phenolic contents and antioxidant activities of faba beans (*Vicia faba* L.) differing in seed coat colours. *Food Chem.* **2014**, *142*, 461–468. [CrossRef]
18. Johnson, J.; Collins, T.; Skylas, D.; Quail, K.; Blanchard, C.; Naiker, M. Profiling the varietal antioxidative content and macrochemical composition in Australian faba beans (*Vicia faba* L.). *Legume Sci.* **2020**, *2*, e28. [CrossRef]

19. Ryszard, A.; Fereidoon, S. Antioxidant activity of faba bean extract and fractions thereof. *J. Food Bioact.* **2018**, *2*, 112–118. [CrossRef]
20. Choudhary, D.K.; Chaturvedi, N.; Singh, A.; Mishra, A. Characterization, inhibitory activity and mechanism of polyphenols from faba bean (gallic-acid and catechin) on α-glucosidase: Insights from molecular docking and simulation study. *Prep. Biochem. Biotechnol.* **2020**, *50*, 123–132. [CrossRef]
21. Liu, Y.; Ragaee, S.; Marcone, M.F.; Abdel-Aal, E.-S.M. Composition of Phenolic Acids and Antioxidant Properties of Selected Pulses Cooked with Different Heating Conditions. *Foods* **2020**, *9*, 908. [CrossRef] [PubMed]
22. Sosulski, F.W.; Dabrowski, K.J. Composition of free and hydrolyzable phenolic acids in the flours and hulls of ten legume species. *J. Agric. Food Chem.* **1984**, *32*, 131–133. [CrossRef]
23. Dong, Y.; Dong, K.; Yang, Z.X.; Zheng, Y.; Tang, L. Microbial and physiological mechanisms for alleviating fusarium wilt of faba bean in intercropping system. *Ying Yong Sheng Tai Xue Bao* **2016**, *27*, 1984–1992. [CrossRef]
24. El-Akkad, S.S.; Hassan, E.A.; Ali, M.E. Phenolic acid changes during Orobanche parasitism on faba bean and some other hosts. *Egypt. J. Biol.* **2002**, *4*, 37–44.
25. Yao, Y.; Cheng, X.-Z.; Ren, G.-X. Contents of D-chiro-Inositol, Vitexin, and Isovitexin in Various Varieties of Mung Bean and Its Products. *Agric. Sci. China* **2011**, *10*, 1710–1715. [CrossRef]
26. Abu-Reidah, I.M.; del Mar Contreras, M.; Arráez-Román, D.; Fernández-Gutiérrez, A.; Segura-Carretero, A. UHPLC-ESI-QTOF-MS-based metabolic profiling of *Vicia faba* L. (Fabaceae) seeds as a key strategy for characterization in foodomics. *Electrophoresis* **2014**, *35*, 1571–1581. [CrossRef] [PubMed]
27. Johnson, J.B.; Mani, J.S.; Skylas, D.; Walsh, K.B.; Bhattarai, S.P.; Naiker, M. Phenolic profiles and nutritional quality of four new mungbean lines grown in northern Australia. *Legume Sci.* **2020**. early online. [CrossRef]
28. Mpofu, A.; Sapirstein, H.D.; Beta, T. Genotype and Environmental Variation in Phenolic Content, Phenolic Acid Composition, and Antioxidant Activity of Hard Spring Wheat. *J. Agric. Food Chem.* **2006**, *54*, 1265–1270. [CrossRef]
29. Oomah, B.D.; Luc, G.; Leprelle, C.; Drover, J.C.G.; Harrison, J.E.; Olson, M. Phenolics, Phytic Acid, and Phytase in Canadian-Grown Low-Tannin Faba Bean (*Vicia faba* L.) Genotypes. *J. Agric. Food Chem.* **2011**, *59*, 3763–3771. [CrossRef]
30. Santos-Sánchez, N.F.; Salas-Coronado, R.; Hernández-Carlos, B.; Villanueva-Cañongo, C. Shikimic acid pathway in biosynthesis of phenolic compounds. In *Plant Physiological Aspects of Phenolic Compounds*; Soto-Hernández, M., García-Mateos, R., Palma-Tenango, M., Eds.; IntechOpen: London, UK, 2019. [CrossRef]
31. Fraser, C.M.; Chapple, C. The Phenylpropanoid Pathway in Arabidopsis. *Arab. Book* **2011**, *2011*. [CrossRef]
32. Marchiosi, R.; dos Santos, W.D.; Constantin, R.P.; de Lima, R.B.; Soares, A.R.; Finger-Teixeira, A.; Mota, T.R.; de Oliveira, D.M.; Foletto-Felipe, M.d.P.; Abrahão, J.; et al. Biosynthesis and metabolic actions of simple phenolic acids in plants. *Phytochem. Rev.* **2020**, *19*, 865–906. [CrossRef]
33. Skylas, D.J.; Paull, J.G.; Hughes, D.G.; Gogel, B.; Long, H.; Williams, B.; Mundree, S.; Blanchard, C.L.; Quail, K.J. Nutritional and anti-nutritional seed-quality traits of faba bean (*Vicia faba*) grown in South Australia. *Crop Pasture Sci.* **2019**, *70*, 463–472. [CrossRef]
34. R Core Team. *R: A Language and Environment for Statistical Computing*, version 4.0.2; R Foundation for Statistical Computing: Vienna, Austria, 2020.

 *molecules*

Review

Flavonoids as Promising Antiviral Agents against SARS-CoV-2 Infection: A Mechanistic Review

Mohammad Amin Khazeei Tabari [1,2], **Amin Iranpanah** [3,4,5], **Roodabeh Bahramsoltani** [6,7,8] **and Roja Rahimi** [6,8,*]

1   Student Research Committee, Mazandaran University of Medical Sciences, Sari, Iran; aminkhazeeitabari@gmail.com
2   USERN Office, Mazandaran University of Medical Sciences, Sari, Iran
3   Pharmaceutical Sciences Research Center, Health Institute, Kermanshah University of Medical Sciences, Kermanshah, Iran; amin.iranpanah75@gmail.com
4   Student Research Committee, Kermanshah University of Medical Sciences, Kermanshah, Iran
5   Kermanshah USERN Office, Universal Scientific Education and Research Network (USERN), Kermanshah, Iran
6   Department of Traditional Pharmacy, School of Persian Medicine, Tehran University of Medical Sciences, Tehran P.O. Box 1417653761, Iran; Roodabeh.b.s.88@hotmail.co.uk
7   Research Center for Clinical Virology, Tehran University of Medical Sciences, Tehran, Iran
8   PhytoPharmacology Interest Group (PPIG), Universal Scientific Education and Research Network (USERN), Tehran, Iran
*   Correspondence: rahimi_r@tums.ac.ir; Tel.: +98-21-8899-0831

Abstract: A newly diagnosed coronavirus in 2019 (COVID-19) has affected all human activities since its discovery. Flavonoids commonly found in the human diet have attracted a lot of attention due to their remarkable biological activities. This paper provides a comprehensive review of the benefits of flavonoids in COVID-19 disease. Previously-reported effects of flavonoids on five RNA viruses with similar clinical manifestations and/or pharmacological treatments, including influenza, human immunodeficiency virus (HIV), severe acute respiratory syndrome (SARS), Middle East respiratory syndrome (MERS), and Ebola, were considered. Flavonoids act via direct antiviral properties, where they inhibit different stages of the virus infective cycle and indirect effects when they modulate host responses to viral infection and subsequent complications. Flavonoids have shown antiviral activity via inhibition of viral protease, RNA polymerase, and mRNA, virus replication, and infectivity. The compounds were also effective for the regulation of interferons, pro-inflammatory cytokines, and sub-cellular inflammatory pathways such as nuclear factor-κB and Jun N-terminal kinases. Baicalin, quercetin and its derivatives, hesperidin, and catechins are the most studied flavonoids in this regard. In conclusion, dietary flavonoids are promising treatment options against COVID-19 infection; however, future investigations are recommended to assess the antiviral properties of these compounds on this disease.

Keywords: inflammation; lung; oxidative damage; antiviral; polyphenol

Citation: Khazeei Tabari, M.A.; Iranpanah, A.; Bahramsoltani, R.; Rahimi, R. Flavonoids as Promising Antiviral Agents against SARS-CoV-2 Infection: A Mechanistic Review. *Molecules* **2021**, *26*, 3900. https://doi.org/10.3390/molecules26133900

Academic Editor: Mirella Nardini

Received: 2 June 2021
Accepted: 23 June 2021
Published: 25 June 2021

1. Introduction

By the end of 2019, an unusual pneumonia was reported from China which was further diagnosed as a novel coronavirus (CoV) causing severe acute respiratory syndrome (SARS) and was called COVID-19 [1,2]. Later, the virus (SARS-CoV-2) spread to other countries and was declared a pandemic by WHO on 11 March 2020 [3]. The virus is transmitted mostly by respiratory droplets, and the severity ranges from mild to severe lethal symptoms. The asymptomatic cases in the incubation period are thought to be an important source of contagion [4]. In most cases, mild symptoms take 1–2 weeks to resolve, whereas severe cases can lead to death [5].

SARS-CoV-2 affects the respiratory system, causing fever and dry cough [4]; however, the virus can cause organ failure, mainly in the heart and kidneys, as well as causing cytokine storms, which further increase mortality. The viral life cycle of SARS-CoV-2 includes attachment, penetration, biosynthesis, maturation, and release. After the attachment, viral RNA enters the cell nucleus for replication, and viral mRNA starts generating viral structural proteins, including the spike (S), membrane (M), envelope (E), and nucleocapsid (N) proteins [6]. The angiotensin-converting enzyme 2 (ACE2) receptor, being highly expressed in the lungs, has also been shown to act as a co-receptor for SARS-CoV [7].

SARS, MERS, and SARS-CoV-2 are all RNA βCoVs. SARS-CoV-2 genome is 88% identical to bat-derived severe acute respiratory syndromes (SARS)-like CoVs, 79% similar to SARS-CoV, and 50% similar to MERS-CoV [8,9]. SARS-CoV-2 proteins are 90%–100% homologous to SARS-CoV; though, orf10 and orf8 are different between SARS-CoV-2 and SARS-CoV. Orf8 is an accessory protein for βCoVs. It forms a six strand alpha helix protein that enhances the virus's ability to spread. SARS-CoV-2 orf1a/b, spike, envelope, membrane, and nucleoprotein are also closely related to those of SARS-CoV [10].

Since there is no specific antiviral agent against SARS-CoV-2, currently available antiviral drugs are considered for the treatment of COVID-19. Remdesivir is a new antiviral drug specifically introduced for the Ebola virus in 2015. It has an inhibitory effect on viral RNA polymerase and has recently been used in some trials for COVID-19 treatment [11]. Favipiravir has an inhibitory effect on influenza and the Ebola virus with the same mechanism and is also assessed in SARS-CoV-2 [12–14]. Lopinavir is a viral protease inhibitor and was firstly developed for HIV treatment. In vitro studies showed an inhibitory effect of lopinavir in CoV-infected cells [15,16]; however, systematic reviews failed to show any beneficial effect against SARS-CoV-2 [17].

Medicinal plants have been a reliable source of natural drugs, including antiviral agents, since ancient times. Traditional medicine and ethnopharmacological studies of different countries all over the world have always opened new ways for drug discovery [18–22]. Oseltamivir which is a conventional antiviral agent, is a derivative of shikimic acid, a secondary metabolite of star anise (*Illicium verum* Hook.f.). In the case of SARS-CoV-2, in silico studies have revealed the possible antiviral properties of herbal ingredients [23–25]. Flavonoids are a large class of phytochemicals commonly found in several foods and vegetables in the human diet with numerous valuable pharmacological activities, including antiviral properties. It is demonstrated that these compounds can inhibit viral pathogenesis targeting essential stages of the viral life cycle [26]. Quercetin, catechins, kaempferol, and baicalein are examples of the most important flavonoids exhibiting antiviral properties [18,27,28]. This study aims to discuss the available antiviral evidence of flavonoids as a possible treatment against SARS-CoV-2 considering the previously-reported effects of these compounds on five RNA viruses with similar clinical manifestations and/or pharmacological treatments, including influenza, human immunodeficiency virus (HIV), severe acute respiratory syndrome (SARS), Middle East respiratory syndrome (MERS), and Ebola.

2. Results

Antiviral activity of flavonoids can be categorized into direct antiviral effects where the virus is directly affected by the flavonoid, and indirect effects where the flavonoid improves host defense mechanisms against viral infection. Here, the underlying antiviral mechanisms of flavonoids are discussed with reference to the most important flavonoids demonstrating these mechanisms.

2.1. Direct Antiviral Mechanisms

2.1.1. Inhibition of Viral Protease

Viral proteases are used for cleaving the viral polyprotein precursors at certain places to release functional proteins. Specific viral proteases have also been shown to cleave host cell proteins, including translation initiation factors (eIF4 and eIF3d) in HIV, to prevent host protein translation [29,30]. Coronaviruses generate three types of viral proteases, including 3-chymotrypsin-like cysteine (3CLpro), papain-like protease (PLpro), and main protease (M pro) [31]. 3CLpro is important for the SARS-CoV life cycle, PLpro plays a role in SARS-CoV-2 replication, and Mpro is responsible for the maturation of functional proteins in SARS-CoV-2 [32,33]. As a result, these molecules are suitable drug targets in antiviral research and drug discovery [34].

Kaempferol which is an abundant flavonoid in several foods decreased CPE in Vero E6 cells infected with clinical isolates of SARS-CoV-2 with around 88% of inhibition at 125 µM concentration. A coupled in silico investigation suggests inhibition of SARS-CoV-2 3CLpro enzyme to be the main mechanism of action [35]. The addition of kaempferol to 3CLpro and PLpro of SARS-CoV and MERS-CoV expressed in *E. coli* caused antiviral effects via inhibition of these enzymes [36]. Park and coworkers (2017) showed that the presence of the hydroxyl group in kaempferol causes a more potent antiviral activity through inhibition of 3CLpro and PLpro [37]. Epigallocatechin gallate (EGCG) is a flavonoid found in tea (*Camellia sinensis* L.) with antifungal, antibacterial, and antiviral properties [38,39]. Studies indicated that EGCG inhibits reverse transcriptase (RT) activity, protease activity, p24, viral entry, and viral production in THP-1 and H9 cells infected with HIV-1, and liposome modification of EGCG amplified its inhibitory effects. Cell-free studies also showed significant downregulation in protease kinetics after treatment with EGCG. The galloyl group in EGCG is considered to be responsible for its antibacterial and antiviral activities [40]. Isoliquiritigenin, a chalcone, can be used as a therapeutic agent in bacterial and viral infections [41]. This compound has shown an inhibitory effect on SARS-CoV, and MERS-CoV 3CLpro and PLpro expressed in *E. coli*. The presence of a prenyl functional group on the resorcinol ring allows the formation of hydrophobic interactions with proteases [37]. Theaflavins are polyphenols found in various kinds of tea [42]. Experiments on SARS-CoV recombinant protease showed a significant reduction in 3CLpro activity after treatment with theaflavin-3,3'-digallate with an IC50 value of 9.5 µM. The gallate group attached to the 3' position in some theaflavins might be essential for interaction with the 3CL proactive site [43]. Prenylisoflavonoids extracted from *Erythrina senegalensis* DC. were used to evaluate anti-protease activity against recombinant HIV-1 protease. The results showed that the compounds could inhibit HIV-1 recombinant protease in vitro with 0.5 to 30 µM IC50 values. Hydroxy and pernyl groups might be responsible for the inhibition of HIV protease [44]. Quercetin and quercetin-β-galactoside downregulated PLpro, 3CLpro, deubiquitination, and DeISGylation activity in SARS-CoV, MERS-CoV, and HIV-1. The position of hydroxyl groups might be effective in anti-protease activity. Quercetin with five hydroxyl groups at the 3,5,7,3' and 4' positions exhibited a strong inhibitory effect on viral proteases; while the presence of glycosyl group at position 3,7,4' reduced the inhibitory effect [37,45].

2.1.2. Inhibition of Viral RNA Polymerase and Viral mRNA

RNA-dependent RNA polymerase (RdRp) is an important enzyme catalyzing the replication of RNA from an RNA sequence [46]. This enzyme is encoded in all RNA viruses, as well as some eukaryotes [47]. Viruses are obligate intracellular parasites, i.e., they cannot independently survive out of cells. They must use cellular translational equipment to translate mRNAs for protein production, which is required for replication. Thus, any interference with mRNA translation would inhibit viral replication, spread, and evolution [48].

A recent investigation by Zandi et al. revealed the in vitro antiviral effect of baicalin and baicalein against SARS-CoV-2 infection in Vero CCL-81 cell line through inhibition of RdRp, with a higher potency by baicalein. Further in silico evaluations showed these two compounds to have a higher affinity to RdRp in comparison to remdesivir. The attachment site of baicalin and baicalein also seems to be different from that of remdesivir; thus, these flavonoids can be used as an adjuvant treatment along with remdesivir [49]. The effects of quercetin-7-O-glucoside (Q7G) were assessed in an in vitro study on MDCK cells infected with influenza viruses A and B in comparison to the standard antiviral agent oseltamivir. Oseltamivir was used as a control drug and showed moderate antiviral activity with IC50 values 25.4 to 42.2 µg/mL; while Q7G inhibited influenza A and B virus with IC50 value

3.10 µg/mL to 8.19 µg/mL. Oseltamivir also showed a weaker activity against influenza B than influenza A; whereas Q7G demonstrated strong activity against all influenza viruses. Additionally, quantitative PCR assays reported a higher decrease in viral RNA synthesis after Q7G treatment compared with oseltamivir, indicating the inhibitory effect of Q7G on viral RNA polymerase. Molecular docking analysis revealed this interaction to be due to the attachment of Q7G to the PB2 subunit of viral RNA polymerase [50].

Oroxylin A (OA) is a flavonoid found in *Oroxylum indicum* (L.) Kurz. It has been shown that OA can inhibit several influenza A strains in MDCK cells in a dose-dependent manner. Oral treatment of mice infected with influenza virus H1N1 with OA also decreased virus-induced death, bodyweight loss, and lung injury, with a survival rate of 60.0% at 100 mg/kg daily dose. Antiviral effects of OA were reported to be due to downregulation of H1N1 matrix 1 (M1) mRNA transcription and protein synthesis (Jin et al. 2018). The M1 protein is a protein within the viral envelope that binds to the viral RNA and can mediate encapsidation of RNA nucleoprotein cores into the membrane envelope [51]. Although OA could inhibit protein synthesis, it could not block viral entry to host cell or nucleoprotein (NP) entrance to host cell nucleus [52]. Baicalin and biochanin A could inhibit influenza H5N1 infection in A549 cells with the IC50 values of 18.79 and 8.92 µM, respectively. This effect was mediated by suppressing nuclear viral ribonucleoprotein (RNP) export [53]. Other studies have also shown that baicalin can downregulate influenza M1 protein expression [54,55].

Host cdc2-like kinase 1 (CLK1) has a key role in the splicing of the H1N1 influenza virus M2 gene and is an important anti-influenza target. M2 is a proton channel in the viral envelope of the influenza A virus [56,57]. It was demonstrated that gallocatechin-7-gallate isolated from *Pithecellobium clypearia* is an inhibitor of host cdc2-like kinase 1 (CLK1), an anti-influenza target due to its role in viral M2 mRNA alternative splicing. Investigations on the effect of gallocatechin-7-gallate at the daily dose of 30 mg/kg on ICR mice infected with H1N1 virus showed a significantly higher survival up to 8 days. It also inhibited virus-induced acute lung injury and weight loss. Additionally, assessments on H1N1-infected A549 cells demonstrated a significant downregulation of viral NP and M2 mRNAs. Moreover, the phosphorylation of splicing factors SF2/ASF and SC35, key factors for virus M2 gene alternative splicing, was significantly decreased after treatment with gallocatechin-7-gallate [58]. Cirsimaritin (CST), a flavonoid from *Artemisia scoparia* Waldst. and Kitam was assessed regarding its in vitro antiviral effects on MDCK and THP-1 cells infected with three influenza virus strains which showed IC50 values ranging from 5.8 to 11.1 µg/mL, compared with 3.4 to 8.9 µg/mL for ribavirin. Data demonstrated that CST could effectively reduce influenza M2 and protein expression in a dose-dependent manner so that the potency of CST at 20 µg/mL was higher than 10 µM of the standard antiviral ribavirin [59]. Luteolin is another flavonoid with an inhibitory effect on M2 mRNA expression. In MDCK cells infected with different influenza strains, 15 µM of luteolin was more effective than 10 µM of oseltamivir in both H1N1 and H2N3 infected cells. Luteolin also downregulated influenza virus coat protein I (COPI) expression, mediating virus entry and endocytic pathway, in infected cells [60]. Santin is a flavonoid extracted from *Artemisia rupestris* L., which was also suggested to have anti-influenza virus effects through suppression of M2 mRNA expression in a dose-dependent manner [61].

It was indicated that quercetin could be a probable therapeutic agent against influenza infection at the early stages of infection so that it can be used for influenza virus prophylaxis. Investigations on the effects of quercetin on MDCK and A549 cells infected with influenza virus A strains revealed that it could inhibit viral NP mRNA in a dose-dependent manner, with the highest activity at 50 µM concentration [62].

Research has shown that tricin (4′,5,7-trihydroxy-3′,5′-dimethoxyflavone) exhibits antiviral activities against influenza A and B strains. RT-PCR tests indicated that tricin could suppress M protein mRNA synthesis in MDCK cells infected with influenza virus; with no significant effects on neuraminidase and hemagglutinin biosynthesis. The 50% effective concentration of tricin, which could inhibit viral mRNA synthesis, was 3.4–10 µM for influenza A virus strains and 4.9 µM for influenza B virus. In mice infected with influenza virus, tricin with a dose of 20 µg/kg ameliorated body weight loss and survival time [63].

2.1.3. Viral Entry, Replication, and Infectivity

RNA viruses encode proteins utilizing the host cellular machinery for their life cycle. Understanding these host cell necessities not only informs us of the molecular pathways used by the virus, but also presents additional targets for drug development [64].

An in vitro study assessed the effect of quercetin and isorhamnetin on SARS-CoV-2 entry to ACE2h cells. ACE2 expressed on lung cells is a co-receptor of viral spike protein and, thus, is a main target of antiviral agents against SARS-CoV-2. It was observed that these two flavonoids have a high binding affinity to ACE2 and subsequently decrease viral entry via the inhibition of spike protein attachment to this receptor [65]. Another study assessed the effect of baicalein on SARS-CoV-2 infection in Vero E6 cells and hACE2 transgenic mice. A significant reduction was observed in in vitro and in vivo viral replication, as well as body weight loss and lung injury of animals [66]. Dihydroxy-6′-methoxy-3′,5′-dimethylchalcone and myricetin-3′,5′-dimethyl ether 3-*O*-β-D-galactopyranoside are flavonoids derived from *Cleistocalyx operculatus* (Roxb.) Merr. and L.M. Perry. Cytopathic effect (CPE) reduction assay showed that these flavonoids inhibit viral replication of influenza virus H1N1 in MDCK cells. Structure-activity relationship (SAR) studies indicated that OH groups at C-7 and C-4, a double bond between C-2 and C-3, and especially a carbonyl group at the C-4 position, are critical functional groups that significantly improve the antiviral properties of flavonoids [67]. 3-deoxysappanchalcone (3DSC) isolated from *Caesalpinia sappan* L. could inhibit influenza virus replication in high concentrations via inhibition of viral NP expression in MDCK cells infected with the H1N1 virus. At an equal concentration (30 µM), both ribavirin and 3DSC showed significant inhibition of NP expression, though ribavirin had a stronger effect [68]. Studies have demonstrated that biochanin A and baicalein inhibited caspase-3 activation, an enzyme involved in viral replication [53,69]. These compounds could also inhibit the nuclear export of viral RNP complexes, which is critical in viral replication [53]. Biochanin A showed an inhibitory effect against $\frac{1}{2}$ mitogen-activated p38 and NF-κB, which were shown to be involved in viral replication. NF κB and p38 are activated due to oxidative stress and are known to affect influenza A virus replication and pathology [53,70]. Investigations on cell cultures of MDCK cells and A549 cells infected with influenza virus showed that baicalein could inhibit viral replication at 20–80 µg/mL concentrations. Interestingly, baicalin showed similar antiviral activity to ribavirin and oseltamivir at concentrations of 40 µg/mL and 60 µg/mL, respectively. Baicalin also inhibited viral replication in the lungs of mice in vivo [71]. CST was shown to downregulate NF-κB protein and NF-κB phosphorylation in the nucleus [59]. It is already known that NF-κB has an important role in inflammation, oxidative stress, and host immunity suppression [72]. The downregulation of NF-κB also inhibits replication in various types of viruses, including influenza virus [73].

In Vero E6 cells infected with SARS-CoV-2, naringenin could inhibit CPE in a time- and concentration-dependent manner. This effect was mediated through inhibition of endo-lysosomal Two-Pore Channels (TPCs), a pathway involves in infectivity of SARS-CoV-2, Ebola, and MERS via facilitating viral entry [74]. EGCG has shown a dose-dependent inhibitory effect (25, 50 µM) on HIV replication in T-cells; however, inhibition of viral replication was not directly affected by RT inhibition. Fassina et al. analyzed the p24 enzyme, which is involved in packaging viral particles. The results showed a downregulation of p24 concentration and RT activity in HIV-infected T lymphoblasts. Based on the following results, it was noted that EGCG inhibited viral replication through the downregulation of viral infection. There is yet no certainty about the exact effect of EGCG on viral infection [75]. gp120 signaling is commonly associated with increased HIV-1 replication in previously infected cells [75]. Studies show the inhibitory effect of genistein on gp120 and, subsequently, HIV-1 viral replication. There was no change in viral replication after administration of genistein with a concentration of 1–2.5 μg/mL, but in a range of 5–10 μg/mL, genistein could suppress viral replication [76]. Herbacitrin is a flavonoid derived from *Drosera peltata* Thunb. and was previously known as an antiviral agent. It was shown that herbacitrin inhibits both RT and integrase in HIV-1 infected MT-4 and MT-2 cell cultures, resulting in viral replication blockade at different stages. At a concentration of 21.5 μg/M, herbacitrin could suppress RT activity, while it could inhibit integrase at a lower concentration, 2.15 μM [77]. Scutellarin purified from *Erigeron breviscapus* is a flavonoid with anti-HIV-1 activity. This flavonoid inhibited HIV-1 RT activity and cell fusion as major participants of viral replication [78]. Hesperidin and linarin are flavonoids with rutinose at the A ring and methoxy (-OCH3) substitution at the B ring. Isoquercetin has been shown to inhibit influenza A and B virus replication in infected MDCK cells. The combination of isoquercetin with amantadine also showed a synergistic effect on viral replication in MDCK cells infected with influenza A virus only in low doses (0.5 μM for isoquercetin and 1 μM for amantadine). Virus titer values after administration of isoquercetin and amantadine were about 7.5; while increasing isoquercetin and amantadine concentrations lowered the synergistic effect on virus titers to the value of 5 [79]. Q3R derived from *Houttuynia cordata* exerts anti-influenza virus effects. The effects of Q3R on MDCK cells infected with influenza virus A were compared to oseltamivir. Pulmonary lesions and edema were inhibited by Q3R more than oseltamivir. Q3R also had a higher efficacy compared to oseltamivir. The inhibitory effect of Q3R on influenza virus replication was indirect and through interaction with viral particles. Oseltamivir demonstrated moderate antiviral activity, about 58% against influenza A virus, and weak antiviral activity less than 49% with doses under 10 μg/mL; while Q3R showed 86% viral inhibition at 100 μg/mL and 66% inhibition in 10 μg/mL concentrations [80]. Quercetin 3-β-O-D-Glucoside (Q3G) was shown to prevent Ebola virus replication in vitro. Prophylactic administration of Q3G 30 min before the infection showed significant prevention of the Ebola virus. Q3G could also inhibit viral entry at the early stages. So Q3G could be an effective flavonoid for Ebola virus prophylaxis [81].

2.2. Indirect Antiviral Effects

2.2.1. Effect on Interferons

Interferons (IFNs) comprise a group of proteins produced by several immune cells in response to many pathogens like viruses, parasites, bacteria, and tumor cells. There are three major classes of IFNs, including type I or acid-stable interferons (IFN-α subtypes, IFN-β, IFN-κ, IFN-ε, IFN-ω, and IFN-τ), type II (IFN-γ), and type III IFNs that known as IFN-λ [82–84]. They show a wide range of biological activities like activation of the innate immune response, increasing the expression of major histocompatibility complex (MHC) molecules, suppressing angiogenesis. Their most important role is to interfere with viral infections [84,85].

In the early phases of viral infection, IFNs activate the innate immune system. Recent studies have reported a decrease in the type I and type II IFN induction and signaling in COVID-19 patients [86]. These types of IFN have demonstrated antiviral effects by decreasing neutrophils immigration to the inflammation site, increasing antigen presentation, suppression of mononuclear macrophage-mediated pro-inflammation, and activating the acquired immunity for the progression of antigen-specific B and T cell responses [86–90]. Thus, IFNs usage at the early phase of the disease could decrease symptoms of the COVID-19 by reducing viral replication. Researchers also reported that IFN-γ levels could increase in COVID-19 patients with ARDS. The rapid rise in IFNs levels could invite pro-inflammatory cytokines into the alveolar tissue and resulting in pulmonary inflammation and lung injury [90,91]. Therefore, it seems that either upregulation or dysregulation of IFNs and other pro-inflammatory cytokines responses or both could exert a significant role in the progression and pathological features of SARS-CoV-2.

Li et al. investigated the anti-influenza effects of baicalin, a glycosyloxyflavone that is the 7-O-glucuronide of baicalein, in the in vitro and in vivo model of influenza A virus infection. TNF receptor-associated factor 6 (TRAF6) is an effective mediator in the IFN production signaling pathway. Overexpression of TRAF6 leads to increased production of type I IFN [54]. MicroRNAs (miRNAs) are small molecules that control gene regulation post-transcriptionally [92]. miR-146a has been shown to have a regulatory role in inflammation [93]. miR-146a could enhance the replication of H1N1 and H3N2 through the downregulation of TRAF6. Baicalin (20 µg/mL) indicated a significant reduction in the miR-146a expression, viral NP, M1 protein levels, viral titer, and also increased mice survival rate [54]. In another study, Nayak and colleagues represented anti-influenza virus (H1N1-pdm09) activity of baicalin through regulating viral protein NS1, resulting in up-regulation of interferon regulatory factor 3 (IRF-3), IFN-γ, and IFN-β. This IFN up-regulation decreased viral replication that could reduce viral transcripts and pro-inflammatory cytokines expression, including IL-8 and TNF-α [55].

Ding et al. designed a study to investigate the effects of hesperidin, a flavanone glycoside, in the influenza A virus (H1N1)-induced lung injury in male rats. The results showed that hesperidin attenuated lung injury via decreasing pro-inflammatory cytokine production, including IFN-α, TNF-α, and IL-6, through suppressing MAPK signaling pathways. Hesperidin also decreased IFN-α in the H1N1 infected pulmonary microvascular endothelial cells [94]. In another study by Kim et al. isoquercetin effectively attenuated lung injury induced by the H1N1 virus in mice via reducing IFN-γ, iNOS, RANTES, virus titers, viral bronchitis, and bronchiolitis [79].

Oroxylin A (OA) from *Oroxylum indicum* (L.) Kurz prevented the lung injury induced by influenza A H1N1 virus in mice via up-regulation of IFN-β and IFN-γ [52]. Wogonin, another flavonoid isolated from *Scutellaria baicalensis* Georgi, exhibited a significant anti-influenza activity by regulation of AMPK pathways. Wogonin also increased the regulation of IFN-β, IFN-λ1, and IFN downstream molecules, including myxovirus resistance gene A (MxA) and 2-5' oligoadenylate synthetase (OAS), in MDCK and A549 infected cells [95].

2.2.2. Effect on Pro-Inflammatory Cytokines (TNF, IL, and MCP)

CoVs contain some open reading frames which encode a few accessory proteins. These accessory proteins have been shown to modulate inflammatory pathways such as IFN signaling and pro-inflammatory cytokines [96]. It has been elucidated that the prognosis of COVID-19 could be worsened by the secretion of pro-inflammatory cytokines, including interleukins, IFNγ, and TNF-α [97]. Blanco-Mello et al. indicated that inappropriate immune response might help virus replication and complications due to severe types of COVID-19 [98]. Ruan et al. also showed that an elevation in inflammatory cytokines such as IL-6 is associated with ARDS, respiratory failure, and adverse clinical outcomes [99]. Respiratory failure caused by lung damage is a result of the overproduction of pro-inflammatory cytokines after the infiltration of immune cells into the lung [100]. Cytokine storm is a systemic inflammatory response associated with a broad range of factors like infections and certain drugs. Several studies showed a significant connection between the cytokine storm, severe inflammation, and multiple organ failure in COVID-19 patients [101–103]. SARS-CoV-2 virus recognition with innate and adaptive immune systems could result in the activation and production of inflammatory cytokines. According to recent studies, plasma levels of pro-inflammatory cytokines are enhanced in COVID-19 patients. These inflammatory cytokines like TNF-α, IL 6, IL 2, IL-1β, IL 7, IL 10, and IL-18, as well as monocyte chemoattractant protein-1 (MCP-1), have pivotal roles in pathological progression and severity of COVID-19 through an increase in viral load, pneumonia, lung damage, neurological disorders, and mortality [97,101].

These events could lead to multi-organ failure and lung injury as the main complication of SARS-CoV-2; therefore, modulation of pro-inflammatory cytokines can be considered as a reasonable treatment goal in COVID-19. In addition, significant anti-inflammatory effects of flavonoids have been demonstrated in many studies; thus, they may be promising compounds in combating inflammation-related complications of COVID-19 [97].

Yang et al. proved the protective effect 3-deoxysappanchalcone (30 µM) on in vitro influenza H1N1 virus-induced inflammation and apoptosis by decreasing IL-1β and IL-6 levels [68]. Baicalein, a flavone, and biochanin A, an O-methylated isoflavone, reduced pro-inflammatory cytokine expression in A549 cells and primary human monocyte-derived macrophages (MDM) infected with influenza H5N1 virus strains, that could prevent inflammatory pathway activation and tissue damages [53]. In influenza A-infected A549 and MDCK cells, baicalin, the glycosylated form of baicalein (baicalein-7-glucuronide), could increase IFN levels, resulting in a reduction of pro-inflammatory cytokines production. Thus, IL-8 and TNF-α were significantly lower in baicalin-treated cells compared with the untreated control cells [55].

An in vitro study has shown that 2.5, 5, and 10 µg/mL concentrations of CST, a dimethoxyflavone, has a significant effect on the attenuation of NF-κB signal transduction pathway in THP-1 cells infected with influenza A (H1N1) virus. Following NF-κB inhibition, the production of pro-inflammatory cytokines including IL-1β, IL-8, IL-10, and TNF-α, as well as the inflammation-related protein COX-2, were suppressed by CST in a dose-dependent manner [59]. In an in vitro study by Yonekawa et al. on the antiviral properties of hesperidin and linarin, these flavonoid glycosides inhibited R5-HIV-1-NL(AD8) viral replication in CD4+ NKT cells by increasing the production of anti-inflammatory cytokines including IL-2, IL-5, and IL-13. It was observed that the stimulatory effect of these two flavonoids are critically dependent on the sugar moiety as the aglycones (hesperetin and acacetin) failed to show such activity. Furthermore, methoxy (-OCH3) substitution at the B ring is essential for the stimulatory activity of hesperidin and linarin on CD4+ NKT cells. They could also induce RANTES, MIP-1α, and MIP-1β secretion from Vδ1+ expressing T cell receptors which subsequently suppressed viral replication in CD4+ NKT cells [104]. Kang et al. reported anti-influenza effects of purified flavonoids from *Pithecellobium clypearia* Benth on the in vitro model of influenza A virus infection. These Purified flavonoids suppressed the production of IL-6 and MCP-1 in H1N1-infected human A549 lung cells [105]. Mehrbod et al. investigated the anti-inflammatory effect of quercetin-3-O-α-L-rhamnopyranoside (Q3R), a glycosylated flavone, on MDCK cells infected with influenza H1N1 virus. Q3R at 150 µg/mL concentration significantly decreased virus titer and increased IL-27 production, which could further elevate IL-10 secretion by CD4+ T cells and enhance their antiviral activity. On the other hand, Q3R suppressed TNF-α production as one of the important inflammatory mediators causing fever and triggering NF-κB pro-inflammatory pathway, further worsening the condition of patients [106]. The trimethoxyflavone santin has demonstrated anti-influenza activity in THP-1 and MDCK cells in a 60 µM concentration. Influenza A (H3N2) virus induces pro-inflammatory cytokine production in THP-1 cells that results in lung inflammation and injury [61]. Anti-inflammatory cytokines might also be altered during influenza virus infection. IL-10 is an anti-inflammatory cytokine that can be induced by influenza virus. IL-10 inhibits invariant natural killer T cells by downregulating the production of IL-12 by pulmonary monocyte-derived dendritic cells [107]. The levels of IL-6, IL-8, IL-10, IL-1β, and TNF-α were significantly decreased in the santin-treated group through downregulation of MAPKs and NF-κB signaling pathways [61].

In addition to the above, gallocatechin-7-gallate, genistein and theaflavins are other flavonoids with modulating effects on the production of pro-inflammatory cytokines [58,76,108]; thus, these molecules seem to have a desirable anti-inflammatory effect, helpful in controlling viral infection-related inflammation.

2.2.3. Effect on Sub-Cellular Inflammatory Pathways (NF-kB, PI3K/Akt, and MAPK/JNK)

When a virus enters a host cell, the host cell recognizes its replication via pattern recognition receptors (PRRs) [109]. Virus RNA structure is involved in oligomerization of PRRs and activation of downstream transcription factors, in particular, interferon regulatory factors (IRF) and NF-κB. Activation of NF-κB and IRFs leads to engagement of cellular antiviral defense by the induction of type I and III interferons and chemokine secretion [110].

Chiou et al. investigated the effects of 8-prenylkaempferol (8-PK) in A549 cells infected with the influenza A (H1N1) virus. Results showed that interfering with the PI3K-Akt pathway is the main mechanism of 8-PK lead to protective effects against the influenza A virus. 8-PK decreased NF-κB and IRF-3 nuclear translocation through attenuation of Akt phosphorylation and PI3K activity. Finally, reduced production of regulated activation, normal T cell expressed and secreted (RANTES) through H1N1-infected A549 cells [111]. Zhu et al. represented that influenza A (H3N2) virus-induced autophagy in the A549 and Ana-1 infected cells via suppressing the mTOR signaling pathway. Baicalin could increase mTOR phosphorylation and rescued H3N2 virus effects in a dose-dependent manner [112]. In another study, baicalin was found to exert anti-influenza virus (H1N1-pdm09) activity by downregulation of the PI3k/Akt pathway caused through modulating viral protein NS1 expression [55]. Besides, biochanin A, an O-methylated isoflavone, indicated protective effects on H5N1 influenza A virus-infected cells via decreasing AKT, ERK1/2, JNK, and p38 phosphorylation. It could also modulate cellular signaling pathways, decrease IL-6, IL-8, CXCL10 (IP-10), TNF-α, and improved IκB levels [53].

CST represented inhibitory effects on the in vitro model of influenza A virus infection through inhibition of the NF-κB/p65 signal pathway, resulting in the downregulation of pro-inflammatory cytokines. CST also decreased phospho-p38 MAPK and phospho-JNK levels [59]. In another study by Ding et al., administration of hesperidin at the daily doses of 200 and 500 mg/kg for five days could inhibit pulmonary inflammation in influenza A virus (H1N1)-induced lung injury in rats. This effect was mediated via attenuating pro-inflammatory cytokine production, including IL-6 and TNF-α. Hesperidin also decreased IL-6 and TNF-α expression in H1N1 infected pulmonary microvascular endothelial cells through inhibition of MAPK signaling pathways [94]. Further, studies suggested ERK signaling pathway as a main modulator of the MAPK signaling pathway. Isorhamnetin (50 μM), a monomethoxyflavone, decreased ERK phosphorylation in MDCK cells after influenza A (H1N1) virus infection [113]. Jeong and colleagues investigated the cytotoxic effects of oroxylin A and tectorigenin in the CHME5 cells and primary human macrophages infected with HIV-1-D3. These flavonoids exert their effects via reducing the phosphorylation of PI3K, Akt, m-TOR, PDK1, GSK-3β, and Bad in the lipopolysaccharide/cycloheximide treated cells santin suppressed influenza A virus replication in the MDCK and THP-1 infected cells [114]. At the concentration of 60 μM, santin attenuated phosphorylation of p38 MAPK, ERK, JNK/SAPK, and NF-κB [61].

3. Discussion

Flavonoids as a class of safe and abundant phytoconstituents have attracted a lot of attention regarding their beneficial effects in COVID-19, and several attempts have been made to assess the structure-activity relationship of these compounds against SARS-CoV-2 proteins [115,116]. This paper reviewed the potential antiviral mechanisms of flavonoids based on the in vitro and in vivo studies on different viruses that follow the same pathogenic mechanisms as SARS-CoV-2, including HIV, influenza virus, ebola virus, SARS, and MERS. Available data on all virus and host targets were included in this study. Figures 1 and 2 provide an overview of the direct and indirect mechanisms of flavonoids.

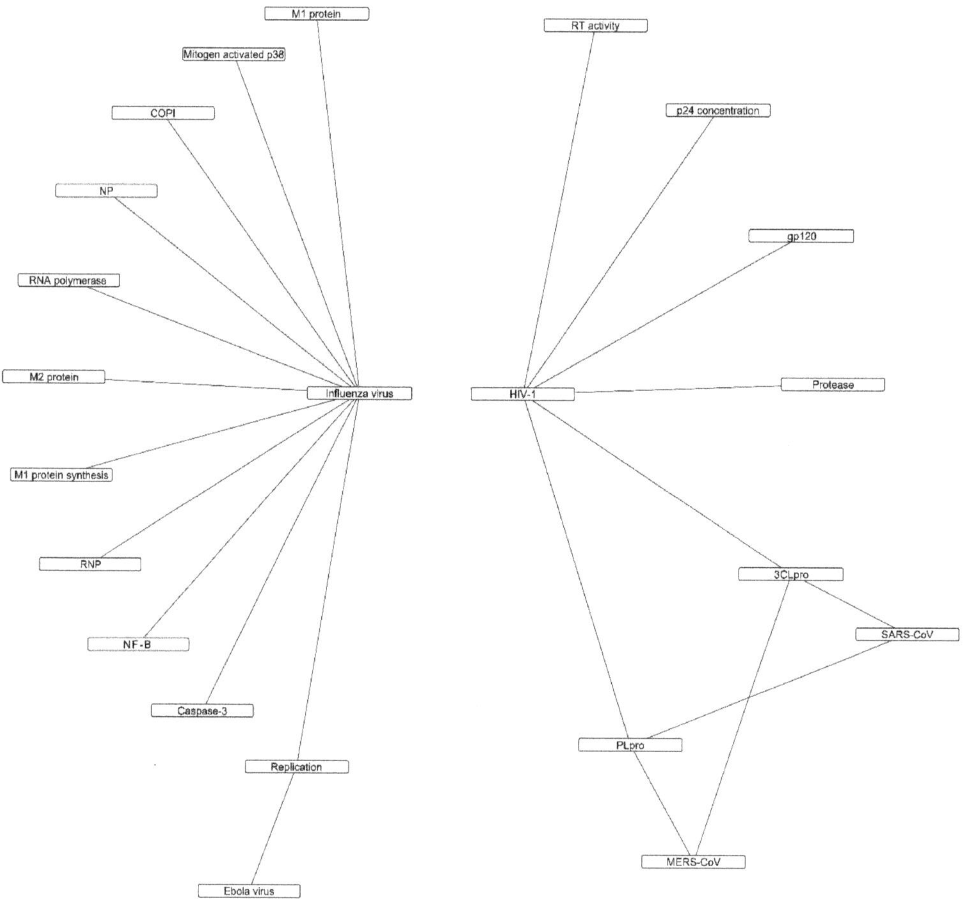

Figure 1. Direct antiviral mechanisms of flavonoids against viral infections with similar pathogenesis to SARS-CoV-2.

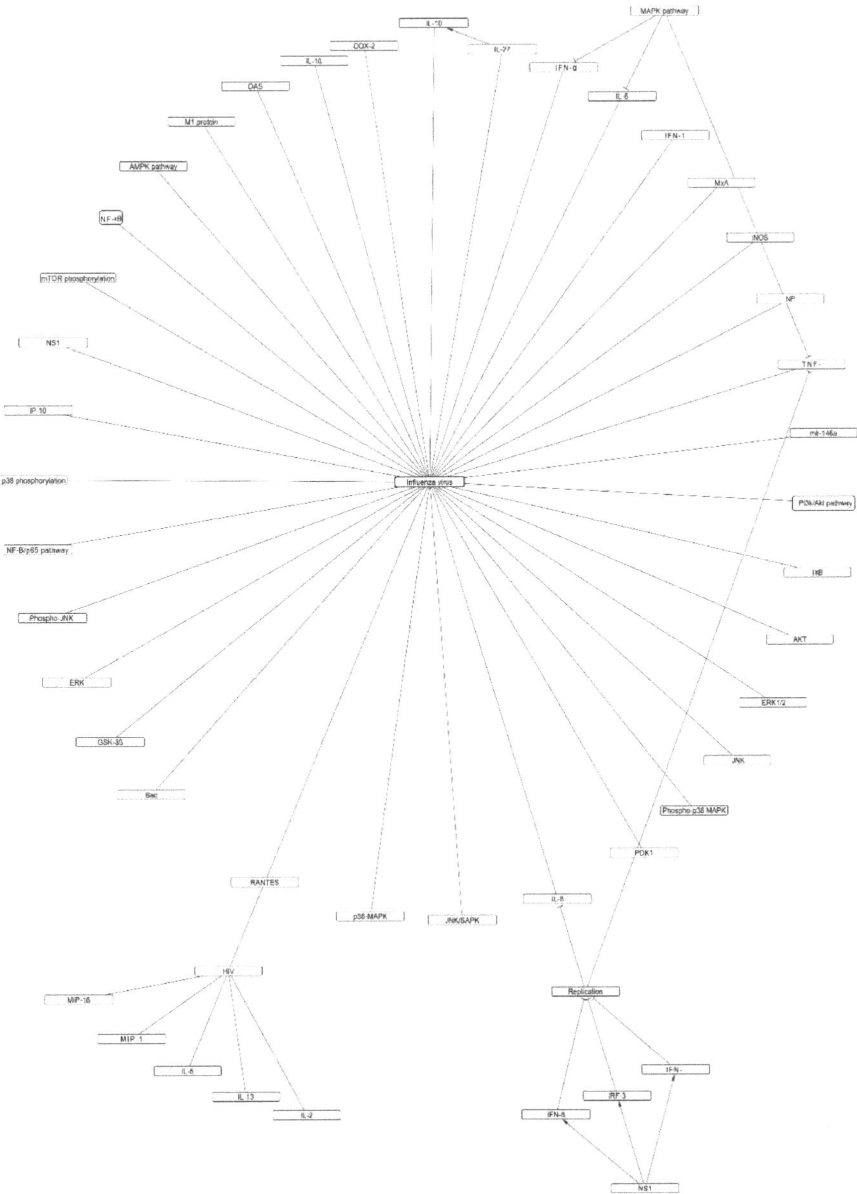

Figure 2. Indirect antiviral mechanisms of flavonoids against viral infections with similar pathogenesis to SARS-CoV-2.

Amongst direct antiviral mechanisms, inhibition of viral proteases are the most frequently reported property of flavonoids. Due to the high similarity of SARS-CoV-2 proteases to those of SARS, flavonoids with inhibitory effects on these enzymes, such as isoliquiritigenin, kaempferol, and its derivatives, quercetin and its derivatives, theaflavins, flavonoids derived from *Angelica keiskei* (Miq.) Koidz. and *Broussonetia papyrifera* (L.) L'Hér. ex Vent. can be considered as candidates for future antiviral assessments against SARS-CoV-2 (Table 1). On the other hand, modulation of inflammatory host responses to the viral infections by the flavonoids seems to be the most important mechanism by which the complications of viral infection are managed. Baicalin and baicalein, biochanin A, cirsimaritin, gallocatechin-7-gallate, and hesperidin are flavonoids with modulating effects on both TNF-α and ILs and thus, can regulate severe conditions due to malfunction of host immune system such as cytokine storm.

According to the current literature, theaflavins, quercetin, luteolin, myricetin, kaempferol, catechins, hesperidin, and baicalin were the most promising flavonoids against the afore-mentioned viruses. Regarding the herbal sources of flavonoids, the most studied plants were *Camellia sinensis* (L.) Kuntze (tea) and *Scutellaria baicalensis* Georgi (skullcap). Green tea is a rich source of catechins, whereas black tea mostly contains theaflavins. Flavonoids from both types of tea have shown direct antiviral properties. Since tea is a popular drink in the human diet, it can be suggested as a safe dietary intervention for COVID-19 patients with mild to moderate symptoms. Due to its acceptable safety profile, tea can also be introduced as a suitable candidate for investigation in future clinical trials. Skullcap is a medicinal plant mostly used in Chinese medicine and is the natural source of baicalin, baicalein, oroxylin A, and wogonin. These flavonoids have demonstrated significant effects on the immune response of infected cells and animals via modulation of IFNs, endoge-nous antioxidant defense mechanisms, and inflammatory responses, as well as direct antiviral properties.

Some of the flavonoids reviewed in this study, such as cirsimaritin were shown to have antiviral activity higher than standard chemically synthesized drugs like ribavirin [59]. It should be mentioned that the results of in vitro antiviral studies do not necessarily guaran-tee the same potency and efficacy in clinical settings; though, they can be considered as a screening method to select the most effective compounds amongst numerous candidates for further in vivo and mechanistic evaluations. As previously mentioned, oseltamivir which is an anti-influenza agent has been designed and synthesized using shikimic acid, a plant-derived compound; thus, the introduced flavonoids in this review can be used as molecular backbones for the design and development of novel semisynthetic medicines with better bioavailability and clinical efficacy.

Despite hundreds of flavonoids evaluated against SARS-CoV-2 through virtual screen-ings, the experimental evidence on the in vitro or in vivo antiviral effect of these com-pounds against this exact type of virus is limited. Amongst the included flavonoids in our review, only four compounds, including baicalin, baicalein, quercetin, and isorhamnetin, were experimentally assessed in SARS-CoV-2-infected cells or animals.

Previous in silico studies and molecular analysis of different CoVs showed the po-tential antiviral effects of phytochemicals at different stages of viral biogenesis, including binding to ACE2, surface gangliosides, RdRp, viral spike protein, and viral protease in host cells and paved the way for more clinical and experimental studies [9,117–123]. Nev-ertheless, it should be considered that an acceptable antiviral activity in virtual screenings does not necessarily guarantee in vivo antiviral activity, and that is why an overview of flavonoids with antiviral properties in experimental studies is a further step toward the selection of natural antiviral agents. On the other hand, several of the mechanisms sug-gested for antiviral flavonoids in virtual screenings are not yet experimentally evaluated. In vitro and in vivo evidence discussed in this review, together with the results of virtual screenings, provides a better overview of the proper compounds for further investigations.

Additionally, there are some recently-published review articles that have focused on the effect of flavonoids on one specific target (e.g., ACE-2) or clinical manifestation (cytokine storms or lung injury) of SARS-CoV-2 infection [124–127]. Such points of view can put a focus on the development of natural medicines against one specific viral target; however, we preferred a more general approach in our study. We considered no limitation for antiviral/symptoms relieving mechanisms of flavonoids, and all experimental evidence of flavonoids on the above-mentioned viruses were included.

In conclusion, flavonoids can be considered as promising plant-derived compounds to manage SARS-CoV-2 infection via direct antiviral properties or management of host immune response to viral infection. Future experimental mechanistic and clinical studies are needed to further clarify the role of these compounds in primary and secondary prevention of SARS-CoV-2 infection.

4. Materials and Methods

Electronic databases, including PubMed, Scopus, and Web of Science, were searched from inception until April 2021 with the following formula: (COVID-19 OR SARS OR MERS OR corona OR HIV OR ebola OR influenza (title/abstract)) AND (plant OR extract OR herb OR phytochemical OR flavonoid (all fields)). As a supplementary search, the names of popular flavonoids including catechin, quercetin, rutin, hesperidin, hesperetin, naringenin, naringin, baicalin, bailaein, and epigallocatechin gallate (EGCG) were also individually searched in order to collect all related papers. After excluding duplicates, primary retrieved results were screened by two independent investigators based on the title and abstract. Selected papers were then checked based on their full text. Inclusion criteria were any in vitro or in vivo study in which the antiviral effect and mechanism of a flavonoid were evaluated. Studies on phytochemicals other than flavonoids, antiviral assessments of flavonoids without clarifying the mechanisms, and studies with non-English full-texts were excluded from our review. In silico studies were excluded unless coupled with an in vitro/in vivo experiment. We also did not discuss antiviral mechanisms such as inhibition of hemagglutinin and neuraminidase of influenza virus since these proteins are not mutual with SARS-CoV-2 and cannot be extrapolated to this virus. Those studies included in the final article are summarized in Table 1.

Table 1. Flavonoids with antiviral properties against SARS-CoV-2 and viral infections with similar pathogenesis.

Phytochemical Name	Plant Source	Model	Dose/Concentration	Mechanisms/Outcomes	Reference
(−)-epigallocatechin 3-*O*-gallate	*Limonium morisianum* Arrigoni	Anti-HIV-1 RT and IN	-	↓ HIV-1 RT-associated RNase H activity: IC50 = 0.21 µM ↓ IN catalytic function and IN-LEDGF-dependent activity: IC50 = 1.92 µM	[128]
2′,4′-dihydroxy-6′-methoxy-3′,5′-dimethylchalcone	*Cleistocalyx operculatus* (Roxb.) Merr. and L.M.Perry	HEK293 cells infected with the plasmid H1N1 or oseltamivir-resistant novel H1N1 (H274Y) MDCK cells infected with influenza H1N1 A/PR/8/34 and H9N2 A/Chicken/Korea/O1310/2001	20–40 µM	↑ cell viability, ↓ NA activity (IC50 = 5.07 to 8.84 µM), viral replication, and CPE	[67]
3-Deoxysappanchalcone	*Caesalpinia sappan* L.	MDCK, A549, and THP-1 cells infected with influenza virus A/PR/8/34 (H1N1)	30 µM	↓ viral genomic replication, DNA fragmentation, CCL5, CXCL10, IL-6, IL-1β, caspase 3/7, 8, and 9 activity and HA copy number (IC_{50} = 3.9 µM)	[68]
3-deoxysappanchalcone, Sappanchalcone and rhamnetin	*Caesalpinia sappan* L.	MDCK cells infected with influenza A/Guangdong/243/72 (H3N2), A/PR/8/34 (H1N1) and B/Jiangsu/10/2003	-	↓ NA activity: IC_{50} = 13.9–24.1 µg/mL ↓ CPE: IC_{50} = 1.06–15.4 µg/mL CC_{50} = 12.83–115.47 µg/mL SI = 6.23–16.27	[129]
5,7,4′-Trihydroxy-8-methoxyflavone	-	MDCK cells infected with influenza A/PR/8/34 (H1N1), A/Guizhou/54/89 (H3N2), and B/Ibaraki/2/85	-	↓ Sialidase activity: IC50 = 6.58–9.78 µg/mL	[130]
6-hydroxyluteolin 7-*O*-β-D-glucoside	*Salvia plebeia* R.Br.	MDCK cells infected with influenza A/PR/8/34 (H1N1)	20, 50 µM	↓ NA activity and CPE, ↑ cell viability	[131]
8-Prenylkaempferol	*Sophora flavescens* Aiton	A549 cells infected with influenza A/PR/8/34 (H1N1) virus	1–30 µM	↓ RANTES production, NF-κB, IRF-3, PI3K activity, Akt phosphorylation, and IκB degradation	[111]
Agathisflavone	*Anacardium occidentale* L.	Mice infected with wild-type and oseltamivir-resistant influenza virus		IC50 = 20 to 2.0 µM, EC50 = 1.3 µM ↓ NA activity and virus replication	[132]

Table 1. *Cont.*

Phytochemical Name	Plant Source	Model	Dose/Concentration	Mechanisms/Outcomes	Reference
Apigenin 7-*O*-β-D-(4′-caffeoyl)glucuronide	*Chrysanthemum morifolium* Ramat.	MT-4 cells infected with HIV-I$_{IIIB}$	-	↓ HIV-1 integrase activity: IC50 = 7.2 μg/mL CPE: EC50 = 41.86μg/mL, SI ≥ 3.58	[133]
Baicalein	-	A549 cells infected with influenza H5N1 virus strains (A/Thailand/1(Kan-1)/04 and A/Vietnam/1203/04) Primary human monocyte-derived macrophages (MDM) infected with influenza A/Thailand/1(Kan-1)/04	40–100 μM	↓ viral nucleoprotein: IC50 = 18.79 μM CC50 = 109.41 μM SI = 5.82 ↓ virus titer, caspase-3 activation, NA activity, ↓ IL-6 and IL-8 ↓ viral replication, IL6, CXCL10, and TNF-α	[53]
Baicalin	-	Human lung epithelial A549 cells infected with influenza A/Jingfang/01/1986 (H1N1) and A/Lufang/09/1993 (H3N2) Balb/C mice inoculated intranasally with the influenza A H1N1 virus	20 μg/mL	↓ viral NP, M1 protein levels, viral titer, miR-146a expression, virus replication and viral copy number (EC50 = 17.04, 19.31 μg/mL), ↑ TRAF6 level, IFN-α, and IFN-β ↓miR-146a expression and virus copy number ↑ survival rate, IFN-α and IFN-β	[54]
Baicalin	*Scutellaria baicalensis* Georgi	A549 and Ana-1 cells infected with influenza virus A3/Beijing/30/95 (H3N2)	12.5–50 μg/mL	↑ mTOR phosphorylation, ↓ autophagy, Atg5–Atg12 complex and LC3-II expression, EC$_{50}$ = 15–15.6 μg/mL	[112]

Table 1. *Cont.*

Phytochemical Name	Plant Source	Model	Dose/Concentration	Mechanisms/Outcomes	Reference
Baicalin	-	A549 and MDCK cells infected with influenza virus A/H1N1/Eastern India/66/pdm09 (H1N1-pdm09)	0.5–320 μM	TD50 = 220 μM IC50 = 0.5 and 18 μM ↓ NP transcription, RIG-1, PKR, NS1 expression, viral replication, TNF-α, IL-8, p-85b–NS1 binding, p-Akt, M1 protein, ↑ IRF-3, IFN-γ, and IFN-β	[55]
		BALB/c mice infected intranasally with H1N1-pdm09	(10–120 mg/kg/day) twice daily for 3 days	↓ viral titer: MIC50 ≈ 80 mg/kg/day ↓ p-Akt and M1 protein expression	
Baicalin	*Scutellaria baicalensis* Georgi	MDCK and A549 cells infected with influenza A/FM1/1/47 (H1N1) and A/Beijing/32/92 (H3N2)	20–80 μg/mL (in MDCK cells) 5–40 μg/mL (in A549 cells)	↑ cell viability, ↓ virus replication, and CPE: EC50 = 40.3 and 104.9 μg/mL SI: 2.1–8.6 ↓ NA activity: IC50 = 52.3 and 85.8μg/mL ↓ death rate, weight loss, ↑ mean day to death, survival rates, and improved the lung parameters	[71]
		ICR mice infected with influenza A/FM1/1/47 (H1N1) virus	50–200 mg/kg/d for 5 days, i.v injection		
Baicalin	*Scutellaria baicalensis* Georgi	Hos/CD4/CCR5 or Hos/CD4/CXCR4 cells infected with recombinant vaccinia virus vTF7-3	0.04 to 400 μM	↓ X4 and R5 HIV-1 Env-mediated fusion, CAT activity, CD4/CXCR4, CD4/CCR5, and HIV-1 entry	[134]
Baicalein, Baicalin	-	Vero CCL-81 cell line infected with SARS-CoV-2	20 μM	↓ viral RdRp and viral replication CC50 = 86–100 μM EC50 = 1–9 μM	[49]
Baicalein	-	Vero E6 cells infected with SARS-CoV-2	0.1 μM	↓ body weight loss, the replication of the virus, relieved the lesions of lung tissue, inflammatory cell infiltration, IL-1β and TNF-α ↑ respiratory function	[66]
		hACE2 transgenic mice infected with SARS-CoV-2	200 mg/kg		

Table 1. *Cont.*

Phytochemical Name	Plant Source	Model	Dose/Concentration	Mechanisms/Outcomes	Reference
Biochanin A	-	A549 cells infected with influenza H5N1 virus strains (A/Thailand/1(Kan-1)/04 and A/Vietnam/1203/04) Primary human monocyte-derived macrophages (MDM) infected with influenza A/Thailand/1(Kan-1)/04	40 μM	↓ viral nucleoprotein: IC50 = 8.92 μM CC50 = 49.91 μM SI = 5.60 ↓ virus titer, caspase-3 activation, NFκB p65 accumulation, IL-6, IL-8, CXCL10 production, phosphorylation of AKT and ERK 1/2 and ↑ IκB levels ↓ IL6, CXCL10, and TNF-α	[53]
Catechin	-	A549 cells infected with influenza A H1N1	5–50 μM	↓ NA and HA activity, viral load, and virus-induced autophagy	[135]
Catechins	*Camellia sinensis* (L.) Kuntze	MDCK cell infected with influenza A/Chile/1/83 (H1N1), A/Sydney/5/97 (H3N2), and B/Yamagata/16/88	30–1200 μM	↓ plaque formation: EC50 = 22.2–318 μM ↓ NA activity, HA activity, and viral RNA synthesis	[136]
Catechins (EGCG, ECG and C5G)	*Camellia sinensis* (L.) Kuntze	MDCK cells infected with influenza A/Victoria/503/2013, A/SouthAustralia/21/2013 and A/Perth/25/2013	50–100 μM	↓ NA activity IC50 = 100.3–173 μM CC50 = 274–551.3 μM ↓ plaque number EC50 = 28.4–34.3 μM	[137]
Catechins with a galloyl moiety	-	HIV-1 integrase assay kit	0.1–100 μM	↓ HIV-1 integrase activity IC$_{50}$ = 0.56–3.02 μM	[138]
Cirsimaritin	-	MDCK and THP-1 cells infected with influenza A/Fort Monmouth/1/1947(H1N1), A/tianjinjinnan/15/2009(H1N1) and A/JiangXi/312/2006(H3N2)	2.5–20 μg/mL	↓ CPE: IC50 = 5.8, 6.3, 11.1 μg/mL, SI = 24.3, 26.4, 13.8, and TC50 = 153.3 ↓ viral replication, M2 viral protein expression, intracellular p65/NF-κB protein, p65/NF-κB phosphorylation, TNF-α, IL-1β, IL-8, IL-10, COX-2 expression, phospho-p38 MAPK, and ↓ phospho-JNK	[59]

Table 1. *Cont.*

Phytochemical Name	Plant Source	Model	Dose/Concentration	Mechanisms/Outcomes	Reference
				EC50 = 4.90–8.79 µM SI = 10.15 to > 24.49 ↓ NA activity: IC50 = 2.55–93.77 µM	
EGCG	*Camellia sinensis* (L.) Kuntze (green tea)	MDCK cells infected with influenza A/Puerto Rico/8/34 (H1N1) (PR8), A/Hong Kong/8/68 (H3N2) (HK), A/Brisbane/59/2007 (H1N1) (BB), A/Taiwan/1/1986 (H1N1) (TW), A/Korea/01/2009 (H1N1) (KR) and B/Panama/45/1990 (PNM)	10–100 µM	↓ CPE: EC50 = 8.9–17.3 µM ↓ hemifusion events, viral membrane integrity, cell penetration capacity, NP protein, viral entry, and NA activity: IC50 = 133.2 to > 500 µM	[140]
EGCG	-	PBMCs, CD4$^+$ T cells and macrophages infected with several clinical isolates of HIV-1	6–100 µM	↓ HIV-1 p24 antigen: IC50 = 4.5–12 µM ↓ HIV-1 infectivity, HIV-1–glycoprotein 120 attachment to the CD4 molecule, CC50 > 100 µM	[141]
EGCG	-	CD4$^+$ T cells	0.2–20 µM	↓ HIV-1 gp 120 binding to the CD4$^+$ T cells	[142]
EGCG	*Camelia sinensis* (L.) Kuntze	Peripheral blood CD4$^+$ T cells (by flow cytometry)	25–100 µM	↓ CD4 expression, anti-CD4 antibody binding to its antigen, gp120 binding to CD4, and HIV infection	[143]
EGCG	*Camelia sinensis* (L.) Kuntze	Peripheral blood lymphocytes infected with either LAI/IIIB or Bal HIV strains	1–50 µM	↓ virus replication, RT activity, and p24	[75]
EGCG	*Camelia sinensis* (L.) Kuntze	THP-1 and H9 cells infected with HIV-1	1–100 µM	↓ RT activity, protease activity, p24, viral entry, and viral production	[40]

Table 1. *Cont.*

Phytochemical Name	Plant Source	Model	Dose/Concentration	Mechanisms/Outcomes	Reference
Flavonoid aglycones (demethoxymatteucinol, matteucinol, matteucin, methoxymatteucin, and 3′-hydroxy-5′-methoxy-6,8-dimethylhuazhongilexone)	*Pentarhizidium orientale* (Hook.) Hayata	MDCK cells infected with influenza A/PR/8/34 (H1N1) or A/chicken/Korea/01210/2001 (H9N2)	-	↓ NA activities and CPE: IC50 = 23.1–31.3 μM, EC50 = 21.4–30.7 μM, CC50 = 77.6 μM (demethoxymatteucinol), CC50 => 100 μM (matteucin, methoxymatteucin, and 3′-hydroxy-5′-methoxy-6,8-dimethylhuazhongilexone)	[144]
Flavonoid compounds	-	SARS-CoV proteases (recombinant 3CL^{pro}) expressed in *Pichia pastoris* GS115	200 μM	↓ 3CLpro activity: IC50 = 47–381 μM	[145]
Gallocatechin-7-gallate	*Pithecellobium clypearia* Benth	A549 and MDCK cells infected with influenza A/PR/8/34 (H1N1)	3–30 μM	↓ CPE: CC50 = above 100 μM, EC50 = 1.69 μM ↓ NP and M2 expression levels, HA mRNA expression, M2/M1 levels, and phosphorylation of SF2/ASF and SC35 ↓ viral NP mRNA expression, TNF-α, IL-1β, IL-6, bodyweight loss, acute lung injury and lung virus titer, ↑ survival rate, T-lymphocyte stimulation index, B-lymphocyte stimulation index, and spleen and thymus indices	[58]
		ICR mice infected intranasally with influenza A H1N1 virus	30 mg/kg/d, i.v, for 5 days		
Genistein	-	Heterologously expression of viral protein U of HIV in Xenopus oocyte	20 μM	↓ Ba2+-sensitive current and blocked Vpu ion channels	[146]
Genistein	-	Primary human macrophages infected with HIV-1$_{Ba-L}$ Env expressed on 293 T cells	5–10 μg/mL	↓ R5 Env pseudotyped virus infection, HIV-1Ba-L Env expressing cells and macrophages cell-fusion, reporter gene expression, virus penetration, gp120-induced TNF-α secretion, virus replication	[76]
Ginkgetin	*Ginkgo biloba* L. and *Cephalotaxus harringtonia* K. Koch	MDCK cells infected with influenza A/PR/8/34 (H1N1), A/Guizhou/54/89 (H3N2), and B/Ibaraki/2/85	-	↓ sialidase activity: IC50 = 9.78 to > 100 μg/mL	[130]

Table 1. *Cont.*

Phytochemical Name	Plant Source	Model	Dose/Concentration	Mechanisms/Outcomes	Reference
Herbacitrin	*Drosera peltata* Thunb.	HIV-1 infected MT-4 and MT-2 cell culture	21.5 μM	↓ HIV-1 replication, HIV-1 RT activity, IN activity, and p24 level	[77]
Hesperidin	-	R5-type HIV-1 in CD4$^+$ NKT cells and human Vδ1$^+$ cells in PBMCs	30–100 μg/mL	↑ IL-2, IL-5, IL-13, MIP-1α, MIP-1β, RANTES, CFSE, and CD25 expression, and ↓ viral replication	[104]
Hesperidin	-	Influenza A virus (H1N1) induced lung injury in male Sprague-Dawley rats, by the intrathecal route H1N1 infected pulmonary microvascular endothelial cells	200 and 500 mg/kg/d, i.p., for 5 days 1 mg/mL	↑ pulmonary function, ↓ Local numbers of immune cells, TNF-α, IL-6, and IFN-α ↓ TNF-α, IL-6, IFN-α, phosphorylated p38 and JNK	[94]
Hexamethoxyflavone (5-Hydroxy-3,6,7,8,3′,4′-hexamethoxyflavone)	*Marcetia taxifolia* (A. St.-Hil.) DC.	MT4 cells infected with HIV-1 (HTLV-IIIB/H9)	45 μM	↓ HIV-1 RT activity: IC50 = 4.1 μM, EC50 = 0.04 μM, CC50 > 50 μM	[147]
Hispidulin	*Salvia plebeia* R. Br.	MDCK cells infected with influenza strain H1N1 A/PR/8/34 virus	20–50 μM	↓ CPE and NA activity: IC50 = 19.83 μM, EC50 = 22.62 μM, SI > 8.90 ↑ cell survival rate recovered the chromosome condensation	[148]
Homoplantaginin	*Salvia plebeia* R.Br.	MDCK cells infected with influenza A/PR/8/34 (H1N1)	20, 50 μM	↓ NA activity and CPE, ↑ cell viability	[131]
IND02	*Cinnamomum zeylanicum* Blume	MAGI cells and PBMCs infected with HIV-1 LAI and NL4-3	5–30 μM	↓ gp120 binding to HS (IC$_{50}$ = 7 μM), gp120 binding to CD4 (IC$_{50}$ = 20 μM), and envelope binding to CD4	[149]
IND02-trimer	*Cinnamomum zeylanicum* Blume	MAGI cells and PBMCs infected with HIV-1 LAI, NL4-3, Ba-L and clinical isolates (HIV-1 92UG029(A-X4), HIV-1 92HT599 (B-X4), HIV-1 96USHIPS4 (B-X4/R5) and HIV-1 98IN017 (C-X4)) HIV-1-infected CD4+ and CD8+ T cells	2–20 μM 0.46–46.3 μM	↓ gp120 binding to HS (IC50 = 7.5 μM), EC50 = 0.8–7 μM, CC50 = 96 and 23 ↓ up-modulation of Tim-3 and PD-1	[149]

Table 1. *Cont.*

Phytochemical Name	Plant Source	Model	Dose/Concentration	Mechanisms/Outcomes	Reference
Isoliquiritigenin	-	SARS-CoV proteases (3CL^{pro} and PL^{pro}) expressed in *E. coli* BL21 MERS-CoV proteases (3CL^{pro} and PL^{pro}) expressed in *E. coli* BL21	-	↓ PLpro activity: IC50 = 24.6 µM, Deubiquitination activity = 17.2, DeISGylation activity = 12.6, ↓ 3CLpro activity: IC50 = 61.9 µM ↓ PLpro activity: IC50 = 82.2 µM, ↓ 3CLpro activity: IC50 = 33.9 µM	[37]
Isoquercetin	-	MDCK or Vero cells infected with influenza A viruses from pigs (A/swine/OH/511445/2007 [H1N1], Oh7) and human (A/PR/8/34 [H1N1], PR8), and human influenza B virus (B/Lee/40) BALB/c mice infected with influenza A/PR/8/34, H1N1virus	1–5 µM 2–10 mg/kg/day, i.p.	↓ viral replication and CPE: ED50 = 1.2 µM TD50 = 45 µM SI = 38 ↓ IFN-γ, iNOS, RANTES, virus titers, viral bronchitis, and bronchiolitis	[79]
Isorhamnetin	-	MDCK cells infected with influenza virus A/PR/08/34 (H1N1)Embryonated chicken eggs infected with influenza virus A/PR/08/34 (H1N1) C57BL/6 mice infected with influenza A/PR/8/34 (H1N1)	50 µM 1 mg/kg/day for 5 days (intranasal route)	↑ cell viability EC50 = 23 µM CC50 > 280 µM SI > 12 ↓ autophagy, ROS generation, ERK phosphorylation, cytoplasmic lysosome acidification, NA and HA expression, and NA activity ↓ virus titer, adsorption onto RBCs and RBCs hemolysis ↓ lung virus titer and body weight loss, ↑ survival rate	[113]
Quercetin Isorhamnetin	*Elaeagnus rhamnoides* (L.) A.Nelson (synonym: *Hippophae rhamnoides* L.)	ACE2h cells infected with SARS-CoV-2	50 µM	↓ Viral entry ↓ Viral Binding affinity to ACE2	[65]

Table 1. *Cont.*

Phytochemical Name	Plant Source	Model	Dose/Concentration	Mechanisms/Outcomes	Reference
Kaempferol	-	SARS-CoV proteases (3CL[pro] and PL[pro]) expressed in *E. coli* BL21 MERS-CoV proteases (3CL[pro] and PL[pro]) expressed in *E. coli* BL21	-	↓ PLpro activity: IC50 = 16.3 µM, Deubiquitination activity = 61.7, DeISGylation activity = 71.7, ↓ 3CLpro activity: IC50 = 116.3 µM ↓ PLpro activity: IC50 = 206.6 µM, ↓ 3CLpro activity: IC50 = 35.3 µM	[37]
Kaempferol derivatives	-	Heterologously expression of 3a protein of SARS-CoV in Xenopus oocyte	10–20 µM	↓ Ba^{2+}-sensitive current and 3a-mediated current, blocked 3a-protein channel IC_{50} = 2.3 µM (for juglanin)	[36]
Linarin	-	R5-type HIV-1 in $CD4^+$ NKT cells and human $V\delta1^+$ cells in PBMCs	10–100 µg/mL	↑ IL-2, IL-5, IL-13, MIP-1α, MIP-1β, RANTES, CFSE, and CD25 expression, ↓ viral replication	[104]
Luteolin	-	MDCK, Calu-3, and Vero cells infected with influenza A/Jiangxi/312/2006 (H3N2) and A/Fort Monmouth/1/1947 (H1N1)	3.75–240 µM	↓ CPE: IC50 = 6.89, 7.15 µM, CC50 = 148–240 µM ↓ M2 viral protein expression, virus absorption, and internalization	[60]
Luteolin	*Salvia plebeia* R. Br.	MDCK cells infected with influenza strain H1N1 A/PR/8/34 virus	50 µM	↓ NA activity: IC_{50} = 17.96 µM, EC_{50} = cytotoxic	[148]
Luteolin	-	PBMCs, TZM-bl reporter, and Jurkat cellsinfected with wild–type HIV (NLENY1) or VSV-HIV-1	5–10 µM	↓ clade-B- and -C -Tat-driven LTR transactivation, reactivation of latent HIV-1 infection, HIV-1 gene expression, LTR activity, PBMC cell aggregation/syncytia, viral entry	[150]

Table 1. *Cont.*

Phytochemical Name	Plant Source	Model	Dose/Concentration	Mechanisms/Outcomes	Reference
Luteolin and luteolin 7-methyl ether	*Coleus parvifolius* Benth.	MT-4 cells infected with HIV-1 (HTLV III$_B$)	-	↓ HIV-1 integrase activity: IC50 = 11–70 µM ↓ viral replication	[151]
Myricetin	-	TZM-bl cell infected with HIV-1 BaL (R5 tropic), H9 and PBMC cells infected with HIV-1 MN (X4 tropic), and the dual tropic (X4R5) HIV-1 89.6, Anti-HIV-1 RT	0.01–100 µM	↓ p24 antigen: IC50 = 1.76–22.91 µM, CC50 = 804.94–1214.72 µM, ↓ HIV-1 RT: IC50 = 203.65 µM	[152]
Myricetin	-	TZM-bl cell infected with HIV-1 BaL (R5 tropic), H9 and PBMC cells infected with HIV-1 MN (X4 tropic), and the dual tropic (X4R5) HIV-1 89.6, Anti-HIV-1 RT	0.01–100 µM	↑ cell viability, ↓ p24 antigen: IC50 = 1.76–22.91 µM, CC50 = 804.94–1214.72 µM, ↓ HIV-1 RT: IC50 = 203.65 µM	[152]
Myricetin-3-*O*-(6″-*O*-galloyl)-β-D-galactopyranoside	*Limonium morisianum* Arrigoni	Anti-HIV-1 RT and IN	-	↓ HIV-1 RT-associated RNase H activity: IC50 = 10.9 µM ↓ IN catalytic function and IN-LEDGF-dependent activity: IC50 = 6.47 µM	[128]
Myricetin derivatives	*Marcetia taxifolia* (A. St.-Hil.) DC.	MT4 cells infected with HIV-1 (HTLV-IIIB/H9)	-	↓ HIV-1 RT activity: IC$_{50}$ = 7.6–13.8 µM, EC$_{50}$ = 45–230 µM, SI > 1.3–7	[153]
Myricetin-3′,5′-dimethylether 3-*O*-β-D-galactopyranoside	*Cleistocalyx operculatus* (Roxb.) Merr. and L.M.Perry	HEK293 cells infected with the plasmid H1N1 or oseltamivir-resistant novel H1N1 (H274Y) MDCK cells infected with influenza H1N1 A/PR/8/34 and H9N2 A/Chicken/Korea/O1310/2001	40 µM	↑ cell viability, ↓ NA activity (IC50 = 6.50 to 9.34 µM), viral replication and CPE	[67]
Naringenin	-	Vero E6 cells infected with HCoVOC43, HCoV229E, and SARS-CoV-2	62.5, 250 µM	↓ TPC2, CPE activity	[74]
Nepetin	*Salvia plebeia* R. Br.	MDCK cells infected with influenza strain H1N1 A/PR/8/34 virus	20–50 µM	↓ CPE and NA activity: IC50 = 11.18 µM, EC50 = 17.45 µM, SI = ~11.47, ↑ cell survival rate	[148]

Table 1. *Cont.*

Phytochemical Name	Plant Source	Model	Dose/Concentration	Mechanisms/Outcomes	Reference
Nepitrin	*Salvia plebeia* R.Br.	MDCK cells infected with influenza A/PR/8/34 (H1N1)	20, 50 μM	↓ NA activity and CPE, ↑ cell viability	[131]
Oroxylin A	-	MDCK and A549 cells infected with influenza A/FM/1/47 (H1N1), A/Beijing/32/92 (H3N2), and oseltamivir-resistant A/FM/1/47-H275Y (H1N1-H275Y) viruses H1N1-H275Y and A/Anhui/1/2013-R294 K (H7N9-R294 K)	40–50 μM	↓ CPE: IC50 = 270.9, 245.0, 241.4 μM, EC50 = 44.6, 36.1, 109.4 μM	[52]
			100 μM	↓ viral mRNA and M1 protein expression ↓ NA activity, IC50 = 241.4 and 203.6 μM	
		ICR mice infected intranasally with the A/FM/1/47 (H1N1)	100 mg/kg/d, p.o.	↑ IFN-β, IFN-γ and survival rate ↓ body weight loss, lung injury, lung indexes and lung scores	
Oroxylin A	*Scutellaria baicalensis* Georgi	CHME5 cells and primary human macrophages infected with HIV-1-D3	5–20 μM	↓ phosphorylation of PI3K, PDK1, Akt, activation of GSK3β, m-TOR, and Bad	[114]
Pentamethoxyflavone(5,3′-dihydroxy-3,6,7,8,4′-pentamethoxyflavone)	*Marcetia taxifolia* (A. St.-Hil.) DC.	MT4 cells infected with HIV-1 (HTLV-IIIB/H9)	45 μM	↓ HIV-1 RT activity: IC_{50} = 0.4 μM, EC_{50} = 0.05 μM, CC_{50} > 50 μM	[154]
Pongamones A–E	*Pongamia pinnata* (L.) Pierre	In vitro inhibitory activity against HIV-1 RT	-	↓ RT activity IC50 > 10 μg/mL	[155]
Prenylisoflavonoids	*Erythrina senegalensis* DC.	In vitro inhibitory activity against recombinant HIV-1 protease	-	↓ HIV-1 protease activity IC50 = 0.5–30.1 μM	[44]
Purified chalcones	*Angelica keiskei* (Miq.) Koidz.	SARS-CoV proteases (3CL[pro] and PL[pro]) expressed in *E. coli* BL21	-	↓ 3CLpro activity: Cell-free cleavage: IC50 = 11.4–129.8 μM, Cell-based cleavage: IC50 = 5.8–50.8 μM, SI = 0.4–9.2 ↓ PLpro activity: IC50 = 1.2–46.4 μM, Deubiquitination activity = 2.6–44.1, DeISGylation activity = 1.1–11.3	[156]

Table 1. *Cont.*

Phytochemical Name	Plant Source	Model	Dose/Concentration	Mechanisms/Outcomes	Reference
Purified flavanone glucosides	*Thevetia peruviana* SCHUM.	HIV-1 IN protein expressed in *E. coli*,		↓ HIV-1 RDDP activity: IC50 = 20–43 μM	[157]
		RDDP and DDDP inhibitory activity assay		↓ HIV-1 DDDP activity: IC50 = 42 and 69 μM ↓ HIV-1 IN activity: IC50 = 5–45 μM	
Purified flavones	*Kaempferia parviflora* Wall. ex Baker	In vitro inhibitory activity against HIV-1 protease	-	↓ HIV-1 protease IC50 = 19.04–160.07 μM	[158]
Purified flavonoids	*Pithecellobium clypearia* Benth	A549 cells infected with influenza A/PR/8/34 (H1N1), A/Sydney/5/97 (H3N2) and B/Jiangsu/10/2003	3–30 μg/mL	↓ NA activity: IC50 = 29.77–39.15 μg/mL ↓ IL-6 and MCP-1	[105]
Purified flavonoids	-	NAs from influenza A/PR/8/34 (H1N1), A/Jinan/15/90 (H3N2), andB/Jiangshu/10/2003	-	↓ NA activity: IC50 = 22–87.6 μM ↓ CPE: IC50 = 4.74–24.70 μM SI = 1.82–9.64	[159]
		MDCK cells infected with influenza A/Jinan/15/90 (H3N2)			
Purified flavonoids	*Elsholtzia rugulosa* Hemsl.	NAs from influenza viruses A/PR/8/34(H1N1), A/Jinan/15/90(H3N2) and B/Jiangsu/10/2003	-	↓ NA activity: IC50 = 7.81–28.49 μM ↓ viral replication and CPE: IC50 = 1.43 to > 500 μM, SI = 1.73–7.48	[160]
		MDCK cells infected with influenza A/Jinan/15/90 (H3N2)			
Purified flavonoids (Ochnaflavone 7″-O-methyl ether and 2″,3″dihydroochnaflavone 7″ methyl ether)	*Ochna integerrima* (Lour.) Merr.	1A2 cell line infected with [ΔTat/rev]MC99 virus	200 μg/mL	↓ RT activity: IC50 = 2.0 and 2.4 μg/mL ↓ HIV-1 activities: EC50 = 2 and 0.9 μg/mL IC50 = 6.3 and 2.9 μg/mL SI = 3.1 and 3.2	[161]
Purified flavonol glycosides	*Zanthoxylum piperitum* (L.)	MDCK cells infected with influenza A/NWS/33 (H1N1)	7.8–1000 μg/mL	↓ NA activity: IC50 = 211–434 μg/mL ↓ PFU	[162]

Table 1. *Cont.*

Phytochemical Name	Plant Source	Model	Dose/Concentration	Mechanisms/Outcomes	Reference
Purified flavonols	*Rhodiola rosea* L.	MDCK cells infected with influenza A/PR/8/34 (H1N1) and A/Chicken/Korea/MS96/96 (H9N2) Recombinant influenza A virus H1N1(rvH1N1)	-	↓ CPE: EC50 = 6.25–145.4 µM SI = 1.6 to > 48 ↓ NA activity: IC50 = 2.2–56.9 µg/mL	[163]
Purified flavonoids	*Broussonetia papyrifera* (L.) L'Hér. ex Vent.	SARS-CoV proteases (3CL^{Pro} and PL^{pro}) expressed in *E. coli* BL21 MERS-CoV proteases (3CL^{Pro} and PL^{pro}) expressed in *E. coli* BL21	-	↓ PLpro activity: IC50 = 3.7–66.2 µM, Deubiquitination activity = 7.6–74.8, DeISGylation activity = 8.5–70.8, ↓ 3CLpro activity: IC50 = 30.2–233.3 µM ↓ PLpro activity: IC50 = 39.5–171.6 µM, ↓ 3CLpro activity: IC50 = 27.9–193.7 µM	[37]
Quercetin	-	SARS-CoV proteases (3CLpro and PLpro) expressed in *E. coli* BL21 MERS-CoV proteases (3CLpro and PLpro) expressed in *E. coli* BL21	-	↓ PLpro activity: IC50 = 8.6 µM, Deubiquitination activity = 20.7, DeISGylation activity = 34.4, ↓ 3CLpro activity: IC50 = 52.7 µM ↓ 3CLpro activity: IC50 = 34.8 µM	[37]
Kaempferol		Vero E6 cells infected with SARS-CoV	125, 62.5, and 31.25 µM	↓ Virus-induced cell death, 3CLprotease	[35]
Quercetin	-	MDCK and A549 cells infected with influenza A/Puerto Rico/8/34 (H1N1), A/FM-1/47/1 (H1N1) and A/Aichi/2/68 (H3N2)	12.5–100 µg/mL (50 µg/mL)	↓ CPE: IC50 = 2.738–7.756 µg/mL, IC90 = 8.24–24.58 µg/mL, ↓ HA mRNA transcription, viral NP protein synthesis, viral HA expression and virus infection rate, target the membrane fusion process during virus entry	[62]

Table 1. *Cont.*

Phytochemical Name	Plant Source	Model	Dose/Concentration	Mechanisms/Outcomes	Reference
Quercetin	-	Inhibitory activity against recombinant HIV-1 protease	-	↓ HIV-1 protease activity: IC50 = 58.8 μM	[45]
Quercetin 3-O-(6″-feruloyl)-β-D-galactopyranoside	*Polygonum viscosum* Buch.-Ham. ex D. Don	In vitro anti-HIV-1 activity	-	↓ RT activity IC50 = 25.61 μg/mL	[164]
Quercetin 3-rhamnoside	*Houttuynia cordata* Thunb.	MDCK cells infected with influenza A/WS/33	10–100 μg/mL	↓ CPE, viral mRNA synthesis, virus replication and virus infection	[80]
Quercetin 3-β-O-D-glucoside	-	Vero E6 epithelial cells infected with EBOV-Kikwit-GFP, EBOV-Makona	10 μM	↓ virus replication (EC$_{50}$ = 5.3 μM, EC$_{90}$ = 9.3 μM), viral titers and entry of Ebola viruses	[81]
		and SUDV or VSV-EBOV, and VSV-SUDV	50 mg/kg every other day, i.p.	↓ virus replication and body weight loss,	
		BALB/c or C57BL/6 mice infected with mouse-adapted Ebola virus		↑ survival rate	
Quercetin-3-O-α-L-rhamnopyranoside	*Rapanea melanophloeos* (L.) Mez	MDCK cells infected with A/Puerto Rico/8/1934 (H1N1)	150 μg/mL	↓ CPE: CC50 = 200 μg/mL, EC50 = 25 μg/mL, EC90 = 100 μg/mL ↓ NP and M2 genes copy numbers, viral titer, HA titer and TNF-α, ↑ IL-27 protein level and cell viability	[106]
Quercetin-7-O-glucoside	*Dianthus superbus var. longicalycinus* (Maxim.) F.N.Williams	MDCK cells infected with influenza A/Vic/3/75 (H3N2), A/PR/8/34 (H1N1), B/Maryland/1/59 and B/Lee/40 viruses	10 μg/mL	↓ CPE: IC50 = 3.10 μg/mL to 8.19 μg/mL, CC50 > 100 μg/mL, SI = 12.21 to 32.25, ↓ ROS, autophagy, viral RNA synthesis and viral RNA polymerase	[50]

Table 1. *Cont.*

Phytochemical Name	Plant Source	Model	Dose/Concentration	Mechanisms/Outcomes	Reference
Quercetin-β-galactoside	-	SARS-CoV proteases (3CL^{pro} and PL^{pro}) expressed in *E. coli* BL21	-	↓ PLpro activity: IC50 = 51.9 μM, Deubiquitination activity = 136.9, DeISGylation activity = 67.7, ↓ 3CLpro activity: IC50 = 128.8 μM	[37]
		MERS-CoV proteases (3CL^{pro} and PL^{pro}) expressed in *E. coli* BL21		↓ PLpro activity: IC50 = 129.4 μM, ↓ 3CLpro activity: IC50 = 68.0 μM	
Santin	*Artemisia rupestris* L.	MDCK and THP-1 cells infected with influenza strain A/Fort Monmouth/1/1947 (H1N1) and A/Wuhan/359/1995 (H3N2)	60 μM	↓ CPE: IC50 = 27.68, 37.20 μM, SI = 14.45, 10.75, TC50 > 400 μM ↓ M2 viral protein expression, phosphorylation of p38 MAPK, JNK/SAPK, ERK, NF-κB, TNF-α, IL-1β, IL-6, IL-8, and IL-10 production	[61]
Scutellarin	*Erigeron breviscapus* (Vant.) Hand.-Mazz	C8166 cells infected with HIV-1$_{IIIB}$ and HIV-1$_{IIIB}$/H9, MT-2 cells infected with HIV-1$_{74V}$, PBMC cells infected with HIV-1$_{KM018}$ Purified recombinant HIV-1 RT	54–541 μM	↓ HIV-1 replication: EC50 = 15–253 μM CC50 = 336 -> 1082 μM ↓ RT activity, HIV-1 particle attachment and fusion	[165]
Tectorigenin	*Pueraria thunbergiana* (Siebold and Zucc.) Benth.	CHME5 cells and primary human macrophages infected with HIV-1-D3	5–20 μM	↓ phosphorylation of PI3K, PDK1, Akt, activation of GSK3β, m-TOR, and Bad	[114]
Theaflavins	*Camellia sinensis* (L.) Kuntze (black tea)	MDCK andA549 cells infected with influenza A/PR/8/34(A/H1N1), A/Sydney/5/97(A/H3N2) and B/Jiangsu/10/2003	0.1 to 30 μg/mL	↓ NA: IC50 = 10.67–49.6 μM ↓ CPE: CC50 = 76.7–177.1 μM ↓ HA activity, IL-6 and vRNP nuclear localization,	[108]

Table 1. *Cont.*

Phytochemical Name	Plant Source	Model	Dose/Concentration	Mechanisms/Outcomes	Reference
Theaflavins	-	SARS-CoV proteases (3CL^{pro}) expressed in *E. coli*	-	↓ 3CLpro activity IC50 = 3–9.5 μM	[43]
Tricin	-	MDCK cells infected with influenza A/Solomon islands/3/2006 (H1N1), A/Hiroshima/52/2005 (H3N2), A/California/07/2009 (H1N1pdm), A/Narita/1/2009 (H1N1pdm) and B/Malaysia/2506/2004	3.3–30 μM	↓ HA and matrix protein, mRNA expression, virus titer (EC50 = 3.4–10.2 μM)	[63]
		DBA/2 Cr mice infected intranasally with influenza A/PR/8/34 virus	20–100 μg/kg, p.o.	↓ Body weight loss, ↑ Survival rate	
Wogonin	*Scutellaria baicalensis* Georgi	MDCK and A549 cells infected with human influenza virus A/Puerto-Rico/8/34 (H1N1) PR8, seasonal H1N1, H3N2 and B (yamagata lineage)	10 μg/mL	↓ NA and NS1 levels, viral replication, Akt phosphorylation, ↑ IFN-β, IFN-λ1, MxA, OAS, AMPK phosphorylation, phospho-IRF-3 expression, cleaved PARP, and caspase-3 expression and apoptosis ↓ Plaque formation: IC50 = 10 μg/mL	[95]

Abbreviations: HA: hemagglutinin; NA: neuraminidase; MDCK: Madin-Darby canine kidney; HIV: human immunodeficiency virus; CD: cluster of differentiation; NKT: natural killer T cells; CPE: cytopathic effect; IC50: inhibitory concentration 50%; CC50: cytotoxic concentration 50%; THP-1: Human acute monocytic leukemia; SI: selectivity index; TC50: 50% toxicity concentration; MAPK: Mitogen-activated protein Kinase; JNK: c-Jun N-terminal kinase; SAPK: stress-activated protein kinase; ERK: extracellular signal-regulated kinase; NF-κB: Nuclear factor kappa B; IL: interleukin; TNF-α: tumor necrosis factor; PBMCs: Peripheral Blood Mononuclear Cells; RT: reverse transcriptase; RNase H: Ribonuclease H; IN: integrase; NP: nucleoprotein; mir-146a: microRNA-146a; EC50: effective concentration 50%; TRAF6: TNF receptor-associated factor 6; IFN: Interferon; COX-2: Cyclooxygenase-2; NS1: nonstruc-tural protein 1; AMPK: 5′ adenosine monophosphate-activated protein kinase; IRF-3: Interferon regulatory factor 3; PARP: Poly (ADP-ribose) polymerase; SARS-CoV: Severe acute respiratory syndrome coronavirus; PLpro: papain-like protease; 3CLpro: 3-chymotripsin-like protease; SPR: Surface plasmon resonance; IC90: inhibitory concentration 90%; EBOV: Ebola virus; ROS: Reactive oxygen species; HS: heparan sulphate; HA: hemagglutinin; NA: neuraminidase; HIV: human immunodeficiency virus; CD: cluster of differentiation; NKT: natural killer T cells; CPE: cytopathic effect; IC50: inhibitory concentration 50%; CC50: cytotoxic concentration 50%; EGCG: Epigallocatechin gallate; ECG: Epicatechin gallate; C5G: Catechin-5-gallate; PI3K: phosphoinositide 3-kinase; PDLKII: pyruvate dehydrogenase lipoamid kinase isozyme 1; GSK3 β:glycogen synthase kinase-3β; LC3B:light chain 3-B; ROS: reactive oxygen species; MAP1: microtubule associated protein1; SI: selective index; G1,2: group 1,2; orf3a: open-reading-frame 3a; RT: reverse transcriptase; PBMC: peripheral blood mononuclear cells; IFN:interferon; NS1: non-structural protein 1; TF: theaflavin; TF-3-G:theaflavin-3-gallate; TF-3′-G:theaflavin-3′-G; TF-3,3′-DG: theaflavin-3,3′-DG; 3-DSC: 3-deoxysappanchalcone; GCG: gallocatechin gallate; Q3R: quercetin 3-rhamnoside; HMB: 2-hydroxy-3-methyl-3-butenyl alkyl; RANTES: regulated activation; normal T cell expressed and secreted; CG: catechin gallate; SEVI: semen-derived enhancer of virus infection; Env: envelope protein; gp: glycoprotein; TF2B: 3-isotheaflavin-3-gallate; TF3: theaflavin-3,3′-digallate; TPC: endo-lysosomal Two-Pore Channels.

Author Contributions: Conceptualization, R.R. and M.A.K.T.; methodology, M.A.K.T. and R.B.; data curation, A.I. and M.A.K.T.; writing—original draft preparation, A.I. and M.A.K.T.; writing—review and editing, R.B. and R.R.; supervision, R.R.; project administration, R.B. and R.R.; All authors have read and agreed to the published version of the manuscript.

Funding: This research received no external funding.

Institutional Review Board Statement: Not applicable.

Informed Consent Statement: Not applicable.

Data Availability Statement: Not applicable.

Conflicts of Interest: The authors declare no conflict of interest.

References

1. Huang, Y.F.; Bai, C.; He, F.; Xie, Y.; Zhou, H. Review on the potential action mechanisms of Chinese medicines in treating Coronavirus Disease 2019 (COVID-19). *Pharmacol. Res.* **2020**, *158*, 104939. [CrossRef]
2. Samieefar, N.; Yari Boroujeni, R.; Jamee, M.; Lotfi, M.; Golabchi, M.R.; Afshar, A.; Miri, H.; Khazeei Tabari, M.A.; Darzi, P.; Abdullatif Khafaie, M.; et al. Country Quarantine During COVID-19: Critical or Not? *Disaster Med. Public Health Prep.* **2020**, 1–2. [CrossRef] [PubMed]
3. Park, S.E. Epidemiology, virology, and clinical features of severe acute respiratory syndrome -coronavirus-2 (SARS-CoV-2; Coronavirus Disease-19). *Clin. Exp. Pediatr.* **2020**, *63*, 119–124. [CrossRef] [PubMed]
4. Singhal, T. A review of coronavirus disease-2019 (COVID-19). *Indian J. Pediatr.* **2020**, *87*, 281–286. [CrossRef]
5. Novel, C.P.E.R.E. The epidemiological characteristics of an outbreak of 2019 novel coronavirus diseases (COVID-19) in China. *Zhonghua Liu Xing Bing Xue Za Zhi* **2020**, *41*, 145. [CrossRef]
6. Bosch, B.J.; Van der Zee, R.; De Haan, C.A.; Rottier, P.J.M. The coronavirus spike protein is a class I virus fusion protein: Structural and functional characterization of the fusion core complex. *J. Virol.* **2003**, *77*, 8801–8811. [CrossRef]
7. Li, H.Y.; Li, F.; Sun, H.Z.; Qian, Z.M. Membrane-inserted conformation of transmembrane domain 4 of divalent-metal transporter. *Biochem. J.* **2003**, *372*, 757–766. [CrossRef] [PubMed]
8. Lu, R.; Zhao, X.; Li, J.; Niu, P.; Yang, B.; Wu, H.; Wang, W.; Song, H.; Huang, B.; Zhu, N.; et al. Genomic characterisation and epidemiology of 2019 novel coronavirus: Implications for virus origins and receptor binding. *Lancet* **2020**, *395*, 565–574. [CrossRef]
9. Khazeei Tabari, M.A.; Khoshhal, H.; Tafazoli, A.; Khandan, M.; Bagheri, A. Applying computer simulations in battling with COVID-19, using pre-analyzed molecular and chemical data to face the pandemic. *Inf. Med. Unlocked* **2020**, *21*, 100458. [CrossRef]
10. Chan, J.F.; Kok, K.H.; Zhu, Z.; Chu, H.; To, K.K.; Yuan, S.; Yuen, K.Y. Genomic characterization of the 2019 novel human-pathogenic coronavirus isolated from a patient with atypical pneumonia after visiting Wuhan. *Emerg. Microbes Infect.* **2020**, *9*, 221–236. [CrossRef]
11. Grein, J.; Ohmagari, N.; Shin, D.; Diaz, G.; Asperges, E.; Castagna, A.; Feldt, T.; Green, G.; Green, M.L.; Lescure, F.X.; et al. Compassionate Use of Remdesivir for Patients with Severe Covid-19. *N. Engl. J. Med.* **2020**. [CrossRef] [PubMed]
12. Sissoko, D.; Laouenan, C.; Folkesson, E.; M'lebing, A.B.; Beavogui, A.H.; Baize, S.; Camara, A.M.; Maes, P.; Shepherd, S.; Danel, C. Experimental treatment with favipiravir for Ebola virus disease (the JIKI Trial): A historically controlled, single-arm proof-of-concept trial in Guinea. *PLoS. Med.* **2016**, *13*. [CrossRef] [PubMed]
13. Furuta, Y.; Komeno, T.; Nakamura, T. Favipiravir (T-705), a broad spectrum inhibitor of viral RNA polymerase. *Proc. Jpn. Acad. Ser. B Phys. Biol. Sci.* **2017**, *93*, 449–463. [CrossRef] [PubMed]
14. Du, Y.X.; Chen, X.P. Favipiravir: Pharmacokinetics and Concerns about Clinical Trials for 2019-nCoV Infection. *Clin. Pharmacol. Ther.* **2020**, *108*, 242–247. [CrossRef]
15. Choy, K.T.; Wong, A.Y.; Kaewpreedee, P.; Sia, S.F.; Chen, D.; Hui, K.P.Y.; Chu, D.K.W.; Chan, M.C.W.; Cheung, P.P.; Huang, X.; et al. Remdesivir, lopinavir, emetine, and homoharringtonine inhibit SARS-CoV-2 replication in vitro. *Antivir. Res.* **2020**, *178*, 104786. [CrossRef] [PubMed]
16. Cao, B.; Wang, Y.; Wen, D.; Liu, W.; Wang, J.; Fan, G.; Ruan, L.; Song, B.; Cai, Y.; Wei, M.; et al. A Trial of Lopinavir-Ritonavir in Adults Hospitalized with Severe Covid-19. *N. Engl. J. Med.* **2020**, *382*, 1787–1799. [CrossRef]
17. Yousefifard, M.; Zali, A.; Mohamed Ali, K.; Madani Neishaboori, A.; Zarghi, A.; Hosseini, M.; Safari, S. Antiviral therapy in management of COVID-19: A systematic review on current evidence. *Arch. Acad. Emerg. Med.* **2020**, *8*, e45.
18. Bahramsoltani, R.; Sodagari, H.R.; Farzaei, M.H.; Abdolghaffari, A.H.; Gooshe, M.; Rezaei, N. The preventive and therapeutic potential of natural polyphenols on influenza. *Expert. Rev. Anti Infect. Ther.* **2016**, *14*, 57–80. [CrossRef]
19. Bahramsoltani, R.; Rahimi, R. An Evaluation of Traditional Persian Medicine for the Management of SARS-CoV-2. *Front. Pharmacol.* **2020**, *11*, 571434. [CrossRef]
20. Siahpoosh, M.B. How Can Persian Medicine (Traditional Iranian Medicine) Be Effective to Control COVID-19? *Tradit. Integr. Med.* **2020**, *5*, 46–48.
21. Siahpoosh, M.B. Clinically Improvement in a Case of COVID-19 with Gastrointestinal Manifestations through Using Natural Therapy based on Persian Medicine: A Case Report. *Tradit. Integr. Med.* **2020**, *5*, 114–117.

22. Zargaran, A.; Karimi, M.; Rezaeizadeh, H. COVID 19: Natural Products and Traditional Medicines; Opportunity or Threat? *Tradit. Integr. Med.* **2021**, *6*, 1–2.
23. Zhang, Z.L.; Zhong, H.; Liu, Y.X.; Le, K.J.; Cui, M.; Yu, Y.T.; Gu, Z.C.; Gao, Y.; Lin, H.W. Current therapeutic options for coronavirus disease 2019 (COVID-19)-lessons learned from severe acute respiratory syndrome (SARS) and Middle East Respiratory Syndrome (MERS) therapy: A systematic review protocol. *Ann. Transl. Med.* **2020**, *8*, 1527. [CrossRef]
24. Gurung, A.B.; Ali, M.A.; Lee, J.; Farah, M.A.; Al-Anazi, K.M. Unravelling lead antiviral phytochemicals for the inhibition of SARS-CoV-2 M(pro) enzyme through in silico approach. *Life. Sci.* **2020**, *255*, 117831. [CrossRef] [PubMed]
25. Kumar, A.; Choudhir, G.; Shukla, S.K.; Sharma, M.; Tyagi, P.; Bhushan, A.; Rathore, M. Identification of phytochemical inhibitors against main protease of COVID-19 using molecular modeling approaches. *J. Biomol. Struct. Dyn.* **2020**, 1–21. [CrossRef]
26. Sadati, S.M.; Gheibi, N.; Ranjbar, S.; Hashemzadeh, M.S. Docking study of flavonoid derivatives as potent inhibitors of influenza H1N1 virus neuraminidase. *Biomed. Rep.* **2019**, *10*, 33–38. [CrossRef] [PubMed]
27. Kumar, S.; Pandey, A.K. Chemistry and biological activities of flavonoids: An overview. *Sci. World J.* **2013**, *2013*, 162750. [CrossRef] [PubMed]
28. Ahmadian, R.; Rahimi, R.; Bahramsoltani, R. Kaempferol: An encouraging flavonoid for COVID-19. *Bol. Latinoam. Caribe Plantas Med. Aromat.* **2020**, *19*. [CrossRef]
29. Suwannarach, N.; Kumla, J.; Sujarit, K.; Pattananandecha, T.; Saenjum, C.; Lumyong, S. Natural Bioactive Compounds from Fungi as Potential Candidates for Protease Inhibitors and Immunomodulators to Apply for Coronaviruses. *Molecules* **2020**, *25*, 1800. [CrossRef]
30. Jäger, S.; Cimermancic, P.; Gulbahce, N.; Johnson, J.R.; McGovern, K.E.; Clarke, S.C.; Shales, M.; Mercenne, G.; Pache, L.; Li, K.; et al. Global landscape of HIV-human protein complexes. *Nature* **2011**, *481*, 365–370. [CrossRef] [PubMed]
31. Pillaiyar, T.; Manickam, M.; Namasivayam, V.; Hayashi, Y.; Jung, S.H. An Overview of Severe Acute Respiratory Syndrome-Coronavirus (SARS-CoV) 3CL Protease Inhibitors: Peptidomimetics and Small Molecule Chemotherapy. *J. Med. Chem.* **2016**, *59*, 6595–6628. [CrossRef] [PubMed]
32. Liang, P.H. Characterization and inhibition of SARS-coronavirus main protease. *Curr. Top. Med. Chem.* **2006**, *6*, 361–376. [CrossRef] [PubMed]
33. Rut, W.; Lv, Z.; Zmudzinski, M.; Patchett, S.; Nayak, D.; Snipas, S.J.; El Oualid, F.; Huang, T.T.; Bekes, M.; Drag, M.; et al. Activity profiling and crystal structures of inhibitor-bound SARS-CoV-2 papain-like protease: A framework for anti-COVID-19 drug design. *Sci. Adv.* **2020**, *6*, eabd4596. [CrossRef]
34. Kurt Yilmaz, N.; Swanstrom, R.; Schiffer, C.A. Improving Viral Protease Inhibitors to Counter Drug Resistance. *Trends Microbiol.* **2016**, *24*, 547–557. [CrossRef] [PubMed]
35. Khan, A.; Heng, W.; Wang, Y.; Qiu, J.; Wei, X.; Peng, S.; Saleem, S.; Khan, M.; Ali, S.S.; Wei, D.Q. In silico and in vitro evaluation of kaempferol as a potential inhibitor of the SARS-CoV-2 main protease (3CLpro). *Phytother. Res.* **2021**. [CrossRef]
36. Schwarz, S.; Sauter, D.; Wang, K.; Zhang, R.; Sun, B.; Karioti, A.; Bilia, A.R.; Efferth, T.; Schwarz, W. Kaempferol derivatives as antiviral drugs against the 3a channel protein of coronavirus. *Planta Med.* **2014**, *80*, 177. [CrossRef]
37. Park, J.Y.; Yuk, H.J.; Ryu, H.W.; Lim, S.H.; Kim, K.S.; Park, K.H.; Ryu, Y.B.; Lee, W.S. Evaluation of polyphenols from Broussonetia papyrifera as coronavirus protease inhibitors. *J. Enzyme Inhib. Med. Chem.* **2017**, *32*, 504–515. [CrossRef] [PubMed]
38. Aggarwal, V.; Tuli, H.S.; Tania, M.; Srivastava, S.; Ritzer, E.E.; Pandey, A.; Aggarwal, D.; Barwal, T.S.; Jain, A.; Kaur, G.; et al. Molecular mechanisms of action of epigallocatechin gallate in cancer: Recent trends and advancement. *Semin. Cancer Biol.* **2020**. [CrossRef]
39. Matsumoto, Y.; Kaihatsu, K.; Nishino, K.; Ogawa, M.; Kato, N.; Yamaguchi, A. Antibacterial and antifungal activities of new acylated derivatives of epigallocatechin gallate. *Front. Microbiol.* **2012**, *3*, 53. [CrossRef]
40. Yamaguchi, K.; Honda, M.; Ikigai, H.; Hara, Y.; Shimamura, T. Inhibitory effects of (-)-epigallocatechin gallate on the life cycle of human immunodeficiency virus type 1 (HIV-1). *Antivir. Res.* **2002**, *53*, 19–34. [CrossRef]
41. Ramalingam, M.; Kim, H.; Lee, Y.; Lee, Y.I. Phytochemical and Pharmacological Role of Liquiritigenin and Isoliquiritigenin From Radix Glycyrrhizae in Human Health and Disease Models. *Front. Aging Neurosci.* **2018**, *10*, 348. [CrossRef]
42. Subramanian, N.; Venkatesh, P.; Ganguli, S.; Sinkar, V.P. Role of polyphenol oxidase and peroxidase in the generation of black tea theaflavins. *J. Agric. Food Chem.* **1999**, *47*, 2571–2578. [CrossRef]
43. Chen, C.N.; Lin, C.P.C.; Huang, K.K.; Chen, W.C.; Hsieh, H.P.; Liang, P.H.; Hsu, J.T.A. Inhibition of SARS-CoV 3C-like protease activity by theaflavin-3,3'- digallate (TF3). *Evid. Based Complement. Alternat. Med.* **2005**, *2*, 209–215. [CrossRef]
44. Lee, J.; Oh, W.K.; Ahn, J.S.; Kim, Y.H.; Mbafor, J.T.; Wandji, J.; Fomum, Z.T. Prenylisoflavonoids from Erythrina senegalensis as novel HIV-1 protease inhibitors. *Planta Med.* **2009**, *75*, 268–270. [CrossRef] [PubMed]
45. Xu, H.X.; Wan, M.; Dong, H.; But, P.P.H.; Foo, L.Y. Inhibitory activity of flavonoids and tannins against HIV-1 protease. *Biol. Pharm. Bull.* **2000**, *23*, 1072–1076. [CrossRef] [PubMed]
46. Peersen, O.B. A Comprehensive Superposition of Viral Polymerase Structures. *Viruses* **2019**, *11*, 745. [CrossRef]
47. Koonin, E.V.; Gorbalenya, A.E.; Chumakov, K.M. Tentative identification of RNA-dependent RNA polymerases of dsRNA viruses and their relationship to positive strand RNA viral polymerases. *FEBS Lett.* **1989**, *252*, 42–46. [CrossRef]
48. Walsh, D.; Mathews, M.B.; Mohr, I. Tinkering with translation: Protein synthesis in virus-infected cells. *Cold Spring Harb. Perspect. Biol.* **2013**, *5*, a012351. [CrossRef]

49. Zandi, K.; Musall, K.; Oo, A.; Cao, D.; Liang, B.; Hassandarvish, P.; Lan, S.; Slack, R.L.; Kirby, K.A.; Bassit, L. Baicalein and Baicalin Inhibit SARS-CoV-2 RNA-Dependent-RNA Polymerase. *Microorganisms* **2021**, *9*, 893. [CrossRef] [PubMed]

50. Gansukh, E.; Kazibwe, Z.; Pandurangan, M.; Judy, G.; Kim, D.H. Probing the impact of quercetin-7-O-glucoside on influenza virus replication influence. *Phytomedicine* **2016**, *23*, 958–967. [CrossRef] [PubMed]

51. Sha, B.; Luo, M. Structure of a bifunctional membrane-RNA binding protein, influenza virus matrix protein M1. *Nat. Struct. Biol.* **1997**, *4*, 239–244. [CrossRef] [PubMed]

52. Jin, J.; Chen, S.; Wang, D.C.; Chen, Y.J.; Wang, Y.X.; Guo, M.; Zhou, C.L.; Dou, J. Oroxylin A suppresses influenza A virus replication correlating with neuraminidase inhibition and induction of IFNs. *Biomed. Pharmacother.* **2018**, *97*, 385–394. [CrossRef]

53. Sithisarn, P.; Michaelis, M.; Schubert-Zsilavecz, M.; Cinatl, J., Jr. Differential antiviral and anti-inflammatory mechanisms of the flavonoids biochanin A and baicalein in H5N1 influenza A virus-infected cells. *Antivir. Res.* **2013**, *97*, 41–48. [CrossRef] [PubMed]

54. Li, R.; Wang, L.X. Baicalin inhibits influenza virus A replication via activation of type I IFN signaling by reducing miR-146a. *Mol. Med. Rep.* **2019**, *20*, 5041–5049. [CrossRef] [PubMed]

55. Nayak, M.K.; Agrawal, A.S.; Bose, S.; Naskar, S.; Bhowmick, R.; Chakrabarti, S.; Sarkar, S.; Chawla-Sarkar, M. Antiviral activity of baicalin against influenza virus H1N1-pdm09 is due to modulation of NS1-mediated cellular innate immune responses. *J. Antimicrob. Chemother.* **2014**, *69*, 1298–1310. [CrossRef]

56. Zu, M.; Li, C.; Fang, J.-S.; Lian, W.-W.; Liu, A.-L.; Zheng, L.-S.; Du, G.-H. Drug Discovery of Host CLK1 Inhibitors for Influenza Treatment. *Molecules* **2015**, *20*, 19735–19747. [CrossRef]

57. Pielak, R.M.; Chou, J.J. Influenza M2 proton channels. *Biochim. Biophys. Acta* **2011**, *1808*, 522–529. [CrossRef]

58. Li, C.; Xu, L.J.; Lian, W.W.; Pang, X.C.; Jia, H.; Liu, A.L.; Du, G.H. Anti-influenza effect and action mechanisms of the chemical constituent gallocatechin-7-gallate from *Pithecellobium clypearia* Benth. *Acta Pharmacol. Sin.* **2018**, *39*, 1913–1922. [CrossRef]

59. Yan, H.Y.; Wang, H.Q.; Ma, L.L.; Ma, X.P.; Yin, J.Q.; Wu, S.; Huang, H.; Li, Y.H. Cirsimaritin inhibits influenza A virus replication by downregulating the NF-kappa B signal transduction pathway. *Virol. J.* **2018**, 15. [CrossRef]

60. Yan, H.; Ma, L.; Wang, H.; Wu, S.; Huang, H.; Gu, Z.; Jiang, J.; Li, Y. Luteolin decreases the yield of influenza A virus in vitro by interfering with the coat protein I complex expression. *J. Nat. Med.* **2019**, *73*, 487–496. [CrossRef]

61. Zhong, M.; Wang, H.Q.; Yan, H.Y.; Wu, S.; Gu, Z.Y.; Li, Y.H. Santin inhibits influenza A virus replication through regulating MAPKs and NF-kappaB pathways. *J. Asian Nat. Prod. Res.* **2019**, *21*, 1205–1214. [CrossRef] [PubMed]

62. Wu, W.J.; Li, R.C.; Li, X.L.; He, J.; Jiang, S.B.; Liu, S.W.; Yang, J. Quercetin as an Antiviral Agent Inhibits Influenza A Virus (IAV) Entry. *Viruses* **2016**, *8*, 6. [CrossRef] [PubMed]

63. Yazawa, K.; Kurokawa, M.; Obuchi, M.; Li, Y.; Yamada, R.; Sadanari, H.; Matsubara, K.; Watanabe, K.; Koketsu, M.; Tuchida, Y.; et al. Anti-influenza virus activity of tricin, 4′,5,7-trihydroxy-3′,5′-dimethoxyflavone. *Antivir. Chem. Chemother.* **2011**, *22*, 1–11. [CrossRef] [PubMed]

64. König, R.; Stertz, S.; Zhou, Y.; Inoue, A.; Hoffmann, H.H.; Bhattacharyya, S.; Alamares, J.G.; Tscherne, D.M.; Ortigoza, M.B.; Liang, Y.; et al. Human host factors required for influenza virus replication. *Nature* **2010**, *463*, 813–817. [CrossRef]

65. Zhan, Y.; Ta, W.; Tang, W.; Hua, R.; Wang, J.; Wang, C.; Lu, W. Potential antiviral activity of isorhamnetin against SARS-CoV-2 spike pseudotyped virus in vitro. *Drug. Dev. Res.* **2021**. [CrossRef]

66. Song, J.; Zhang, L.; Xu, Y.; Yang, D.; Yang, S.; Zhang, W.; Wang, J.; Tian, S.; Yang, S.; Yuan, T. The comprehensive study on the therapeutic effects of baicalein for the treatment of COVID-19 in vivo and in vitro. *Biochem. Pharmacol.* **2021**, *183*, 114302. [CrossRef]

67. Ha, T.K.Q.; Dao, T.T.; Nguyen, N.H.; Kim, J.; Kim, E.; Cho, T.O.; Oh, W.K. Antiviral phenolics from the leaves of *Cleistocalyx operculatus*. *Fitoterapia* **2016**, *110*, 135–141. [CrossRef]

68. Yang, F.; Zhou, W.L.; Liu, A.L.; Qin, H.L.; Lee, S.M.; Wang, Y.T.; Du, G.H. The protective effect of 3-deoxysappanchalcone on in vitro influenza virus-induced apoptosis and inflammation. *Planta Med.* **2012**, *78*, 968–973. [CrossRef]

69. Wurzer, W.J.; Planz, O.; Ehrhardt, C.; Giner, M.; Silberzahn, T.; Pleschka, S.; Ludwig, S. Caspase 3 activation is essential for efficient influenza virus propagation. *EMBO J.* **2003**, *22*, 2717–2728. [CrossRef]

70. Geiler, J.; Michaelis, M.; Naczk, P.; Leutz, A.; Langer, K.; Doerr, H.W.; Cinatl, J., Jr. N-acetyl-L-cysteine (NAC) inhibits virus replication and expression of pro-inflammatory molecules in A549 cells infected with highly pathogenic H5N1 influenza A virus. *Biochem. Pharmacol.* **2010**, *79*, 413–420. [CrossRef]

71. Ding, Y.; Dou, J.; Teng, Z.J.; Yu, J.; Wang, T.T.; Lu, N.; Wang, H.; Zhou, C.L. Antiviral activity of baicalin against influenza A (H1N1/H3N2) virus in cell culture and in mice and its inhibition of neuraminidase. *Arch. Virol.* **2014**, *159*, 3269–3278. [CrossRef] [PubMed]

72. Santoro, M.G.; Rossi, A.; Amici, C. NF-kappaB and virus infection: Who controls whom. *EMBO J.* **2003**, *22*, 2552–2560. [CrossRef]

73. Kumar, N.; Xin, Z.T.; Liang, Y.; Ly, H.; Liang, Y. NF-kappaB signaling differentially regulates influenza virus RNA synthesis. *J. Virol.* **2008**, *82*, 9880–9889. [CrossRef]

74. Clementi, N.; Scagnolari, C.; D'Amore, A.; Palombi, F.; Criscuolo, E.; Frasca, F.; Pierangeli, A.; Mancini, N.; Antonelli, G.; Clementi, M. Naringenin is a powerful inhibitor of SARS-CoV-2 infection in vitro. *Pharmacol. Res.* **2020**, *163*, 105255. [CrossRef] [PubMed]

75. Fassina, G.; Buffa, A.; Benelli, R.; Varnier, O.E.; Noonan, D.M.; Albini, A. Polyphenolic antioxidant (-)-epigallocatechin-3-gallate from green tea as a candidate anti-HIV agent. *AIDS* **2002**, *16*, 939–941. [CrossRef]

76. Stantchev, T.S.; Markovic, I.; Telford, W.G.; Clouse, K.A.; Broder, C.C. The tyrosine kinase inhibitor genistein blocks HIV-1 infection in primary human macrophages. *Virus Res.* **2007**, *123*, 178–189. [CrossRef] [PubMed]

77. Áy, É.; Hunyadi, A.; Mezei, M.; Minárovits, J.; Hohmann, J. Flavonol 7-O-Glucoside Herbacitrin Inhibits HIV-1 Replication through Simultaneous Integrase and Reverse Transcriptase Inhibition. *Evid. Based Complement. Alternat. Med.* **2019**, *2019*, 1064793. [CrossRef] [PubMed]

78. Zhang, H.J.; Nguyen, V.H.; Nguyen, M.C.; Soejarto, D.D.; Pezzuto, J.M.; Fong, H.H.; Tan, G.T. Sesquiterpenes and butenolides, natural anti-HIV constituents from Litsea verticillata. *Planta Med.* **2005**, *71*, 452–457. [CrossRef] [PubMed]

79. Kim, Y.J.; Narayanan, S.; Chang, K.O. Inhibition of influenza virus replication by plant-derived isoquercetin. *Antivir. Res.* **2010**, *88*, 227–235. [CrossRef]

80. Choi, H.J.; Song, J.H.; Park, K.S.; Kwon, D.H. Inhibitory effects of quercetin 3-rhamnoside on influenza A virus replication. *Eur. J. Pharm. Sci.* **2009**, *37*, 329–333. [CrossRef]

81. Qiu, X.G.; Kroeker, A.; He, S.H.; Kozak, R.; Audet, J.; Mbikay, M.; Chretienc, M. Prophylactic Efficacy of Quercetin 3-beta-O-D-Glucoside against Ebola Virus Infection. *Antimicrob. Agents Chemother.* **2016**, *60*, 5182–5188. [CrossRef]

82. Abbas, A.K.; Lichtman, A.H.; Pillai, S. *Cellular and Molecular Immunology*; Elsevier: Amsterdam, The Netherlands, 2021.

83. Bramhachari, P.V. *Dynamics of Immune Activation in Viral Diseases*; Springer: Berlin/Heidelberg, Germany, 2020.

84. Sadler, A.J.; Williams, B.R.J. Interferon-inducible antiviral effectors. *Nat. Rev. Immunol.* **2008**, *8*, 559–568. [CrossRef] [PubMed]

85. Zhou, F. Molecular mechanisms of IFN-γ to up-regulate MHC class I antigen processing and presentation. *Int. Rev. Immunol.* **2009**, *28*, 239–260. [CrossRef]

86. Abdolvahab, M.H.; Moradi-Kalbolandi, S.; Zarei, M.; Bose, D.; Majidzadeh-A, K.; Farahmand, L. Potential role of interferons in treating COVID-19 patients. *Int. Immunopharmacol.* **2021**, *90*, 107171. [CrossRef] [PubMed]

87. Parmar, S.; Platanias, L.C. Interferons: Mechanisms of action and clinical applications. *Curr. Opin. Oncol.* **2003**, *15*, 431–439. [CrossRef] [PubMed]

88. Mustafa, S.; Balkhy, H.; Gabere, M.N. Current treatment options and the role of peptides as potential therapeutic components for Middle East respiratory syndrome (MERS): A review. *J. Infect. Public Health* **2018**, *11*, 9–17. [CrossRef] [PubMed]

89. Morgenstern, B.; Michaelis, M.; Baer, P.C.; Doerr, H.W.; Cinatl, S., Jr. Ribavirin and interferon-β synergistically inhibit SARS-associated coronavirus replication in animal and human cell lines. *Biochem. Biophys. Res. Commun.* **2005**, *326*, 905–908. [CrossRef]

90. Acosta, P.L.; Byrne, A.B.; Hijano, D.R.; Talarico, L.B. Human type I interferon antiviral effects in respiratory and reemerging viral infections. *J. Immunol. Res.* **2020**, *2020*. [CrossRef]

91. Acharya, D.; Liu, G.; Gack, M.U. Dysregulation of type I interferon responses in COVID-19. *Nat. Rev. Immunol.* **2020**, *20*, 397–398. [CrossRef]

92. Mishan, M.A.; Tabari, M.A.K.; Parnian, J.; Fallahi, J.; Mahrooz, A.; Bagheri, A.J.G. Chromosomes; Cancer, Functional mechanisms of miR-192 family in cancer. *Genes Chromosomes Cancer* **2020**, *59*, 722–735. [CrossRef]

93. Mishan, M.A.; Tabari, M.A.K.; Zargari, M.; Bagheri, A. MicroRNAs in the anticancer effects of celecoxib: A systematic review. *Eur. J. Pharmacol.* **2020**, 173325. [CrossRef]

94. Ding, Z.; Sun, G.; Zhu, Z. Hesperidin attenuates influenza A virus (H1N1) induced lung injury in rats through its anti-inflammatory effect. *Antivir. Ther.* **2018**, *23*, 611–615. [CrossRef]

95. Seong, R.K.; Kim, J.A.; Shin, O.S. Wogonin, a flavonoid isolated from *Scutellaria baicalensis*, has anti-viral activities against influenza infection via modulation of AMPK pathways. *Acta Virol.* **2018**, *62*, 78–85. [CrossRef] [PubMed]

96. Liu, D.X.; Fung, T.S.; Chong, K.K.L.; Shukla, A.; Hilgenfeld, R. Accessory proteins of SARS-CoV and other coronaviruses. *Antivir. Res.* **2014**, *109*, 97–109. [CrossRef]

97. Costela-Ruiz, V.J.; Illescas-Montes, R.; Puerta-Puerta, J.M.; Ruiz, C.; Melguizo-Rodríguez, L. SARS-CoV-2 infection: The role of cytokines in COVID-19 disease. *Cytokine Growth Factor Rev.* **2020**, *54*, 62–75. [CrossRef] [PubMed]

98. Blanco-Melo, D.; Nilsson-Payant, B.E.; Liu, W.-C.; Uhl, S.; Hoagland, D.; Møller, R.; Jordan, T.X.; Oishi, K.; Panis, M.; Sachs, D.J.C. Imbalanced host response to SARS-CoV-2 drives development of COVID-19. *Cell* **2020**, *181*, 1036–1045.e9. [CrossRef] [PubMed]

99. Ruan, Q.; Yang, K.; Wang, W.; Jiang, L.; Song, J. Clinical predictors of mortality due to COVID-19 based on an analysis of data of 150 patients from Wuhan, China. *Intensive Care Med.* **2020**, *46*, 846–848. [CrossRef] [PubMed]

100. Jose, R.J.; Manuel, A. COVID-19 cytokine storm: The interplay between inflammation and coagulation. *Lancet Respir. Med.* **2020**, *8*, e46–e47. [CrossRef]

101. Roshanravan, N.; Seif, F.; Ostadrahimi, A.; Pouraghaei, M.; Ghaffari, S. Targeting cytokine storm to manage patients with COVID-19: A mini-review. *Arch. Med. Res.* **2020**, *51*, 608–612. [CrossRef] [PubMed]

102. Fara, A.; Mitrev, Z.; Rosalia, R.A.; Assas, B.M. Cytokine storm and COVID-19: A chronicle of pro-inflammatory cytokines. *Open Biol.* **2020**, *10*, 200160. [CrossRef]

103. Karki, R.; Sharma, B.R.; Tuladhar, S.; Williams, E.P.; Zalduondo, L.; Samir, P.; Zheng, M.; Sundaram, B.; Banoth, B.; Malireddi, R.K.S.; et al. COVID-19 cytokines and the hyperactive immune response: Synergism of TNF-α and IFN-γ in triggering inflammation, tissue damage, and death. *Cell* **2020**, *184*, 149–168.e17. [CrossRef]

104. Yonekawa, M.; Shimizu, M.; Kaneko, A.; Matsumura, J.; Takahashi, H. Suppression of R5-type of HIV-1 in CD4(+) NKT cells by Vdelta1(+) T cells activated by flavonoid glycosides, hesperidin and linarin. *Sci. Rep.* **2019**, *9*, 7506. [CrossRef]

105. Kang, J.; Liu, C.; Wang, H.; Li, B.; Li, C.; Chen, R.; Liu, A. Studies on the bioactive flavonoids isolated from Pithecellobium clypearia Benth. *Molecules* **2014**, *19*, 4479–4490. [CrossRef]

106. Mehrbod, P.; Abdalla, M.A.; Fotouhi, F.; Heidarzadeh, M.; Aro, A.O.; Eloff, J.N.; McGaw, L.J.; Fasina, F.O. Immunomodulatory properties of quercetin-3-O-alpha-L-rhamnopyranoside from Rapanea melanophloeos against influenza a virus. *BMC Complement. Altern. Med.* **2018**, *18*, 184. [CrossRef]

107. Barthelemy, A.; Ivanov, S.; Fontaine, J.; Soulard, D.; Bouabe, H.; Paget, C.; Faveeuw, C.; Trottein, F. Influenza A virus-induced release of interleukin-10 inhibits the anti-microbial activities of invariant natural killer T cells during invasive pneumococcal superinfection. *Mucosal Immunol.* **2017**, *10*, 460–469. [CrossRef] [PubMed]

108. Zu, M.; Yang, F.; Zhou, W.L.; Liu, A.L.; Du, G.H.; Zheng, L.S. In vitro anti-influenza virus and anti-inflammatory activities of theaflavin derivatives. *Antivir. Res.* **2012**, *94*, 217–224. [CrossRef] [PubMed]

109. Janeway Jr, C.A.; Medzhitov, R. Innate immune recognition. *Annu. Rev. Immunol.* **2002**, *20*, 197–216. [CrossRef] [PubMed]

110. Hur, S. Double-stranded RNA sensors and modulators in innate immunity. *Annu. Rev. Immunol.* **2019**, *37*, 349–375. [CrossRef] [PubMed]

111. Chiou, W.F.; Chen, C.C.; Wei, B.L. 8-Prenylkaempferol Suppresses Influenza A Virus-Induced RANTES Production in A549 Cells via Blocking PI3K-Mediated Transcriptional Activation of NF-kappaB and IRF3. *Evid. Based Complement. Alternat. Med.* **2011**, *2011*, 920828. [CrossRef] [PubMed]

112. Zhu, H.Y.; Han, L.; Shi, X.L.; Wang, B.L.; Huang, H.; Wang, X.; Chen, D.F.; Ju, D.W.; Feng, M.Q. Baicalin inhibits autophagy induced by influenza A virus H3N2. *Antivir. Res.* **2015**, *113*, 62–70. [CrossRef]

113. Abdal Dayem, A.; Choi, H.Y.; Kim, Y.B.; Cho, S.G. Antiviral effect of methylated flavonol isorhamnetin against influenza. *PLoS ONE* **2015**, *10*, e0121610. [CrossRef]

114. Jeong, J.J.; Kim, D.H. 5,7-Dihydroxy-6-Methoxy-Flavonoids Eliminate HIV-1 D3-transfected Cytoprotective Macrophages by Inhibiting the PI3K/Akt Signaling Pathway. *Phytother. Res.* **2015**, *29*, 1355–1365. [CrossRef] [PubMed]

115. Solnier, J.; Fladerer, J.P. Flavonoids: A complementary approach to conventional therapy of COVID-19? *Phytochem. Rev.* **2020**, 1–23. [CrossRef]

116. Russo, M.; Moccia, S.; Spagnuolo, C.; Tedesco, I.; Russo, G.L. Roles of flavonoids against coronavirus infection. *Chem. Biol. Interact.* **2020**, 109211. [CrossRef]

117. Jo, S.; Kim, S.; Shin, D.H.; Kim, M.S. Inhibition of SARS-CoV 3CL protease by flavonoids. *J. Enzyme Inhib. Med. Chem.* **2020**, *35*, 145–151. [CrossRef] [PubMed]

118. Jo, S.; Kim, H.; Kim, S.; Shin, D.H.; Kim, M.S. Characteristics of flavonoids as potent MERS-CoV 3C-like protease inhibitors. *Chem. Biol. Drug. Des.* **2019**, *94*, 2023–2030. [CrossRef] [PubMed]

119. Bhowmik, D.; Nandi, R.; Prakash, A.; Kumar, D. Evaluation of flavonoids as 2019-nCoV cell entry inhibitor through molecular docking and pharmacological analysis. *Heliyon* **2021**, *7*, e06515. [CrossRef]

120. Bora, A.; Pacureanu, L.; Crisan, L. In silico study of some natural flavonoids as potential agents against COVID-19: Preliminary results. *Chem. Proc.* **2020**, *3*, 25. [CrossRef]

121. Gorla, U.S.; Rao, G.K.; Kulandaivelu, U.S.; Alavala, R.R.; Panda, S.P. Lead Finding from Selected Flavonoids with Antiviral (SARS-CoV-2) Potentials against COVID-19: An in-silico Evaluation. *Comb. Chem. High Throughput Screen.* **2020**. [CrossRef]

122. Jain, A.S.; Sushma, P.; Dharmashekar, C.; Beelagi, M.S.; Prasad, S.K.; Shivamallu, C.; Prasad, A.; Syed, A.; Marraiki, N.; Prasad, K.S. In silico evaluation of flavonoids as effective antiviral agents on the spike glycoprotein of SARS-CoV-2. *Saudi J. Biol. Sci.* **2021**, *28*, 1040–1051. [CrossRef]

123. Varughese, J.K.; Joseph Libin, K.; Sindhu, K.; Rosily, A.; Abi, T. Investigation of the inhibitory activity of some dietary bioactive flavonoids against SARS-CoV-2 using molecular dynamics simulations and MM-PBSA calculations. *J. Biomol. Struct. Dyn.* **2021**, 1–16. [CrossRef]

124. Gour, A.; Manhas, D.; Bag, S.; Gorain, B.; Nandi, U. Flavonoids as potential phytotherapeutics to combat cytokine storm in SARS-CoV-2. *Phytother. Res.* **2021**. [CrossRef] [PubMed]

125. Liskova, A.; Samec, M.; Koklesova, L.; Samuel, S.M.; Zhai, K.; Al-Ishaq, R.K.; Abotaleb, M.; Nosal, V.; Kajo, K.; Ashrafizadeh, M. Flavonoids against the SARS-CoV-2 induced inflammatory storm. *Biomed. Pharmacother.* **2021**, 111430. [CrossRef]

126. Muchtaridi, M.; Fauzi, M.; Khairul Ikram, N.K.; Mohd Gazzali, A.; Wahab, H.A. Natural flavonoids as potential angiotensin-converting enzyme 2 inhibitors for anti-SARS-CoV-2. *Molecules* **2020**, *25*, 3980. [CrossRef]

127. Santana, F.P.R.; Thevenard, F.; Gomes, K.S.; Taguchi, L.; Câmara, N.O.S.; Stilhano, R.S.; Ureshino, R.P.; Prado, C.M.; Lago, J.H.G. New perspectives on natural flavonoids on COVID-19-induced lung injuries. *Phytother. Res.* **2021**. [CrossRef]

128. Sanna, C.; Rigano, D.; Corona, A.; Piano, D.; Formisano, C.; Farci, D.; Franzini, G.; Ballero, M.; Chianese, G.; Tramontano, E.; et al. Dual HIV-1 reverse transcriptase and integrase inhibitors from Limonium morisianum Arrigoni, an endemic species of Sardinia (Italy). *Nat. Prod. Res.* **2019**, *33*, 1798–1803. [CrossRef]

129. Liu, A.L.; Shu, S.H.; Qin, H.L.; Lee, S.M.; Wang, Y.T.; Du, G.H. In vitro anti-influenza viral activities of constituents from Caesalpinia sappan. *Planta Med.* **2009**, *75*, 337–339. [CrossRef]

130. Miki, K.; Nagai, T.; Suzuki, K.; Tsujimura, R.; Koyama, K.; Kinoshita, K.; Furuhata, K.; Yamada, H.; Takahashi, K. Anti-influenza virus activity of biflavonoids. *Bioorg. Med. Chem. Lett.* **2007**, *17*, 772–775. [CrossRef]

131. Bang, S.; Li, W.; Ha, T.K.Q.; Lee, C.; Oh, W.K.; Shim, S.H. Anti-influenza effect of the major flavonoids from Salvia plebeia R.Br. via inhibition of influenza H1N1 virus neuraminidase. *Nat. Prod. Res.* **2018**, *32*, 1224–1228. [CrossRef]

132. De Freitas, C.S.; Rocha, M.E.N.; Sacramento, C.Q.; Marttorelli, A.; Ferreira, A.C.; Rocha, N.; de Oliveira, A.C.; de Oliveira Gomes, A.M.; Dos Santos, P.S.; da Silva, E.O.; et al. Agathisflavone, a Biflavonoid from Anacardium occidentale L. Inhibits Influenza Virus Neuraminidase. *Curr. Top. Med. Chem.* **2020**, *20*, 111–120. [CrossRef]

133. Lee, J.S.; Kim, H.J.; Lee, Y.S. A new anti-HIV flavonoid from glucoronide from Chrysanthemum marifolium. *Planta Med.* **2003**, *69*, 859–861. [CrossRef]

134. Li, B.Q.; Fu, T.; Yao, D.Y.; Mikovits, J.A.; Ruscetti, F.W.; Wang, J.M. Flavonoid baicalin inhibits HIV-1 infection at the level of viral entry. *Biochem. Biophys. Res. Commun.* **2000**, *276*, 534–538. [CrossRef]

135. Chang, C.C.; You, H.L.; Huang, S.T. Catechin inhibiting the H1N1 influenza virus associated with the regulation of autophagy. *J. Chin. Med. Assoc.* **2020**, *83*, 386–393. [CrossRef]

136. Song, J.M.; Lee, K.H.; Seong, B.L. Antiviral effect of catechins in green tea on influenza virus. *Antivir. Res.* **2005**, *68*, 66–74. [CrossRef]

137. Muller, P.; Downard, K.M. Catechin inhibition of influenza neuraminidase and its molecular basis with mass spectrometry. *J. Pharm. Biomed. Anal.* **2015**, *111*, 222–230. [CrossRef]

138. Jiang, F.; Chen, W.; Yi, K.; Wu, Z.; Si, Y.; Han, W.; Zhao, Y. The evaluation of catechins that contain a galloyl moiety as potential HIV-1 integrase inhibitors. *Clin. Immunol.* **2010**, *137*, 347–356. [CrossRef]

139. Dao, T.T.; Tung, B.T.; Nguyen, P.H.; Thuong, P.T.; Yoo, S.S.; Kim, E.H.; Kim, S.K.; Oh, W.K. C-Methylated Flavonoids from Cleistocalyx operculatus and Their Inhibitory Effects on Novel Influenza A (H1N1) Neuraminidase. *J. Nat. Prod.* **2010**, *73*, 1636–1642. [CrossRef] [PubMed]

140. Kim, M.; Kim, S.Y.; Lee, H.W.; Shin, J.S.; Kim, P.; Jung, Y.S.; Jeong, H.S.; Hyun, J.K.; Lee, C.K. Inhibition of influenza virus internalization by (-)-epigallocatechin-3-gallate. *Antivir. Res.* **2013**, *100*, 460–472. [CrossRef] [PubMed]

141. Nance, C.L.; Siwak, E.B.; Shearer, W.T. Preclinical development of the green tea catechin, epigallocatechin gallate, as an HIV-1 therapy. *J. Allergy Clin. Immunol.* **2009**, *123*, 459–465. [CrossRef] [PubMed]

142. Williamson, M.P.; McCormick, T.G.; Nance, C.L.; Shearer, W.T. Epigallocatechin gallate, the main polyphenol in green tea, binds to the T-cell receptor, CD4: Potential for HIV-1 therapy. *J. Allergy Clin. Immunol.* **2006**, *118*, 1369–1374. [CrossRef]

143. Kawai, K.; Tsuno, N.H.; Kitayama, J.; Okaji, Y.; Yazawa, K.; Asakage, M.; Hori, N.; Watanabe, T.; Takahashi, K.; Nagawa, H. Epigallocatechin gallate, the main component of tea polyphenol, binds to CD4 and interferes with gp120 binding. *J. Allergy Clin. Immunol.* **2003**, *112*, 951–957. [CrossRef]

144. Huh, J.; Ha, T.K.Q.; Kang, K.B.; Kim, K.H.; Oh, W.K.; Kim, J.; Sung, S.H. C-Methylated Flavonoid Glycosides from Pentarhizidium orientale Rhizomes and Their Inhibitory Effects on the H1N1 Influenza Virus. *J. Nat. Prod.* **2017**, *80*, 2818–2824. [CrossRef]

145. Nguyen, T.T.H.; Kang, H.K.; Kim, Y.M.; Jang, T.S.; Kim, D. Inhibition effect of flavonoid compounds against neuraminidase expressed in Pichia pastoris. *Biotechnol. Bioprocess Eng.* **2014**, *19*, 70–75. [CrossRef]

146. Sauter, D.; Schwarz, S.; Wang, K.; Zhang, R.H.; Sun, B.; Schwarz, W. Genistein as Antiviral Drug against HIV Ion Channel. *Planta Med.* **2014**, *80*, 682–687. [CrossRef] [PubMed]

147. Ortega, J.T.; Serrano, M.L.; Suarez, A.I.; Baptista, J.; Pujol, F.H.; Rangel, H.R. Methoxyflavones from *Marcetia taxifolia* as HIV-1 Reverse Transcriptase Inhibitors. *Nat. Prod. Commun.* **2017**, *12*, 1677–1680. [CrossRef]

148. Bang, S.; Ha, T.K.Q.; Lee, C.; Li, W.; Oh, W.K.; Shim, S.H. Antiviral activities of compounds from aerial parts of *Salvia plebeia* R. Br. *J. Ethnopharmacol.* **2016**, *192*, 398–405. [CrossRef]

149. Connell, B.J.; Chang, S.Y.; Prakash, E.; Yousfi, R.; Mohan, V.; Posch, W.; Wilflingseder, D.; Moog, C.; Kodama, E.N.; Clayette, P.; et al. A Cinnamon-Derived Procyanidin Compound Displays Anti-HIV-1 Activity by Blocking Heparan Sulfate- and Co-Receptor-Binding Sites on gp120 and Reverses T Cell Exhaustion via Impeding Tim-3 and PD-1 Upregulation. *PLoS ONE* **2016**, *11*, e0165386. [CrossRef] [PubMed]

150. Mehla, R.; Bivalkar-Mehla, S.; Chauhan, A. A flavonoid, luteolin, cripples HIV-1 by abrogation of tat function. *PLoS ONE* **2011**, *6*, e27915. [CrossRef]

151. Tewtrakul, S.; Miyashiro, H.; Nakamura, N.; Hattori, M.; Kawahata, T.; Otake, T.; Yoshinaga, T.; Fujiwara, T.; Supavita, T.; Yuenyongsawad, S.; et al. HIV-1 integrase inhibitory substances from *Coleus parvifolius*. *Phytother. Res.* **2003**, *17*, 232–239. [CrossRef] [PubMed]

152. Pasetto, S.; Pardi, V.; Murata, R.M. Anti-HIV-1 activity of flavonoid myricetin on HIV-1 infection in a dual-chamber in vitro model. *PLoS ONE* **2014**, *9*, e115323. [CrossRef]

153. Ortega, J.T.; Suarez, A.I.; Serrano, M.L.; Baptista, J.; Pujol, F.H.; Rangel, H.R. The role of the glycosyl moiety of myricetin derivatives in anti-HIV-1 activity in vitro. *AIDS Res. Ther.* **2017**, *14*, 57. [CrossRef]

154. Ortega, J.T.; Estrada, O.; Serrano, M.L.; Contreras, W.; Orsini, G.; Pujol, F.H.; Rangel, H.R. Glycosylated Flavonoids from Psidium guineense as Major Inhibitors of HIV-1 Replication in vitro. *Nat. Prod. Commun.* **2017**, *12*, 1049–1052. [CrossRef]

155. Li, L.Y.; Li, X.; Shi, C.; Deng, Z.W.; Fu, H.Z.; Proksch, P.; Lin, W.H. Pongamone A-E, five flavonoids from the stems of a mangrove plant, Pongamia pinnata. *Phytochemistry* **2006**, *67*, 1347–1352. [CrossRef]

156. Park, J.Y.; Ko, J.A.; Kim, D.W.; Kim, Y.M.; Kwon, H.J.; Jeong, H.J.; Kim, C.Y.; Park, K.H.; Lee, W.S.; Ryu, Y.B. Chalcones isolated from Angelica keiskei inhibit cysteine proteases of SARS-CoV. *J. Enzyme Inhib. Med. Chem.* **2016**, *31*, 23–30. [CrossRef]

157. Tewtrakul, S.; Nakamura, N.; Hattori, M.; Fujiwara, T.; Supavita, T. Flavanone and flavonol glycosides from the leaves of Thevetia peruviana and their HIV-1 reverse transcriptase and HIV-1 integrase inhibitory activities. *Chem. Pharm. Bull.* **2002**, *50*, 630–635. [CrossRef]

158. Sookkongwaree, K.; Geitmann, M.; Roengsumran, S.; Petsom, A.; Danielson, U.H. Inhibition of viral proteases by Zingiberaceae extracts and flavones isolated from Kaempferia parviflora. *Pharmazie* **2006**, *61*, 717–721.
159. Liu, A.L.; Wang, H.D.; Lee, S.M.Y.; Wang, Y.T.; Du, G.H. Structure-activity relationship of flavonoids as influenza virus neuraminidase inhibitors and their in vitro anti-viral activities. *Bioorg. Med. Chem.* **2008**, *16*, 7141–7147. [CrossRef]
160. Liu, A.L.; Liu, B.; Qin, H.L.; Lee, S.M.Y.; Wang, Y.T.; Du, G.H. Anti-influenza virus activities of flavonoids from the medicinal plant Elsholtzia rugulosa. *Planta Med.* **2008**, *74*, 847–851. [CrossRef]
161. Reutrakul, V.; Ningnuek, N.; Pohmakotr, M.; Yoosook, C.; Napaswad, C.; Kasisit, J.; Santisuk, T.; Tuchinda, P. Anti HIV-1 flavonoid glycosides from Ochna integerrima. *Planta Med.* **2007**, *73*, 683–688. [CrossRef]
162. Ha, S.Y.; Youn, H.; Song, C.S.; Kang, S.C.; Bae, J.J.; Kim, H.T.; Lee, K.M.; Eom, T.L.; Kim, I.S.; Kwak, J.H. Antiviral Effect of Flavonol Glycosides Isolated from the Leaf of Zanthoxylum piperitum on Influenza Virus. *J. Microbiol.* **2014**, *52*, 340–344. [CrossRef]
163. Jeong, H.J.; Ryu, Y.B.; Park, S.J.; Kim, J.H.; Kwon, H.J.; Kim, J.H.; Park, K.H.; Rho, M.C.; Lee, W.S. Neuraminidase inhibitory activities of flavonols isolated from *Rhodiola rosea* roots and their in vitro anti-influenza viral activities. *Bioorg. Med. Chem.* **2009**, *17*, 6816–6823. [CrossRef] [PubMed]
164. Datta, B.K.; Datta, S.K.; Khan, T.H.; Kundu, J.K.; Rashid, M.A.; Nahar, L.; Sarker, S.D. Anti-cholinergic, cytotoxic and anti-HIV-1 activities of sesquiterpenes and a flavonoid glycoside from the aerial parts of *Polygonum viscosum. Pharm. Biol.* **2004**, *42*, 18–23. [CrossRef]
165. Zhang, G.H.; Wang, Q.; Chen, J.J.; Zhang, X.M.; Tam, S.C.; Zheng, Y.T. The anti-HIV-1 effect of scutellarin. *Biochem. Biophys. Res. Commun.* **2005**, *334*, 812–816. [CrossRef] [PubMed]

Article

Elucidation of Antioxidant Compounds in Moroccan *Chamaerops humilis* L. Fruits by GC–MS and HPLC–MS Techniques

Hafssa El Cadi [1], Hajar El Bouzidi [1,2], Ginane Selama [2], Btissam Ramdan [3], Yassine Oulad El Majdoub [4], Filippo Alibrando [5], Katia Arena [4], Miguel Palma Lovillo [6], Jamal Brigui [1], Luigi Mondello [4,5,7,8], Francesco Cacciola [9,*] and Tania M. G. Salerno [8]

[1] Laboratory of Valorization of Resources and Chemical Engineering, Department of Chemistry, Abdelmalek Essaadi University, Tangier 90000, Morocco; hafssa.elcadi@yahoo.fr (H.E.C.); hajarelbouzidi1995@gmail.com (H.E.B.); jamalbrigui@yahoo.fr (J.B.)

[2] Laboratory of Biochemistry and Molecular Genetics, Abdelmalek Essaadi University, Tangier 90000, Morocco; ginane.selama@gmail.com

[3] Department of Biology, Laboratory of Biotechnology and Valorization of Natural Resources, Faculty of Science, University Ibn Zohr, Agadir 80000, Morocco; ramdanbtissam8@gmail.com

[4] Department of Chemical, Biological, Pharmaceutical and Environmental Sciences, University of Messina, 98168 Messina, Italy; youladelmajdoub@unime.it (Y.O.E.M.); arenak@unime.it (K.A.); lmondello@unime.it (L.M.)

[5] Chromaleont s.r.l., c/o Department of Chemical, Biological, Pharmaceutical and Environmental Sciences, University of Messina, 98168 Messina, Italy; filippo.alibrando@chromaleont.it

[6] Department of Analytical Chemistry, Faculty of Sciences, Agrifood Campus of International Excellence (ceiA3), University of Cadiz, IVAGRO, 11510 Cadiz, Spain; miguel.palma@uca.es

[7] Department of Sciences and Technologies for Human and Environment, University Campus Bio-Medico of Rome, 00128 Rome, Italy

[8] BeSep s.r.l., c/o Department of Chemical, Biological, Pharmaceutical and Environmental Sciences, University of Messina, 98168 Messina, Italy; tania.salerno@besep.it

[9] Department of Biomedical, Dental, Morphological and Functional Imaging Sciences, University of Messina, 98125 Messina, Italy

* Correspondence: cacciolaf@unime.it; Tel.: +39-090-676-6570

Citation: Cadi, H.E.; Bouzidi, H.E.; Selama, G.; Ramdan, B.; Majdoub, Y.O.E.; Alibrando, F.; Arena, K.; Lovillo, M.P.; Brigui, J.; Mondello, L.; et al. Elucidation of Antioxidant Compounds in Moroccan *Chamaerops humilis* L. Fruits by GC–MS and HPLC–MS Techniques. *Molecules* **2021**, *26*, 2710. https://doi.org/10.3390/molecules26092710

Academic Editor: Mirella Nardini

Received: 8 April 2021
Accepted: 3 May 2021
Published: 5 May 2021

Publisher's Note: MDPI stays neutral with regard to jurisdictional claims in published maps and institutional affiliations.

Abstract: The aim of this study was to characterize the phytochemical content as well as the antioxidant ability of the Moroccan species *Chamaerops humilis* L. Besides crude ethanolic extract, two extracts obtained by sonication using two solvents with increased polarity, namely ethyl acetate (EtOAc) and methanol-water (MeOH-H$_2$O) 80:20 (*v/v*), were investigated by both spectroscopy and chromatography methods. Between the two extracts, the MeOH-H$_2$O one showed the highest total polyphenolic content equal to 32.7 ± 0.1 mg GAE/g DM with respect to the EtOAc extract (3.6 ± 0.5 mg GAE/g DM). Concerning the antioxidant activity of the two extracts, the EtOAc one yielded the highest value (1.9 ± 0.1 mg/mL) with respect to MeOH-H2O (0.4 ± 0.1 mg/mL). The *C. humilis* *n*-hexane fraction, analyzed by GC–MS, exhibited 69 compounds belonging to different chemical classes, with *n*-Hexadecanoic acid as a major compound (21.75%), whereas the polyphenolic profile, elucidated by HPLC–PDA/MS, led to the identification of a total of sixteen and thirteen different compounds in both EtOAc (major component: ferulic acid: 104.7 ± 2.52 μg/g) and MeOH-H$_2$O extracts (major component: chlorogenic acid: 45.4 ± 1.59 μg/g), respectively. The attained results clearly highlight the potential of *C. humilis* as an important source of bioactive components, making it a valuable candidate to be advantageously added to the daily diet. Furthermore, this study provides the scientific basis for the exploitation of the Doum in the food, pharmaceutical and nutraceutical industries.

Keywords: Arecaceae; polyphenols; volatile content; antioxidant activity; liquid chromatography

1. Introduction

The Moroccan wild palm tree (*Chamaerops humilis* L.), widely called "Doum", is found in six cities of the eastern region of Morocco, namely Oujda, Berkane, Ahfir, Saidia, Nador and Jerrada [1], and represents 7.74% of the total number of Moroccan palm trees [2].

Such a species is cultivated in many Mediterranean countries as an ornament, considering its robustness and decorative features.

Some components of this plant have been used as food as an important source of nutritional energy [3], or in traditional medicine. The husks are eaten in Southern Spain, the fruits in Morocco and the young suckers in Italy. Leaf extracts of *Chamaerops humilis* L. (*C. humilis*) have been commonly used for the treatment of diabetes, digestive disorders, spasms, tone and gastrointestinal disorders [4,5]. Moreover, their fruits have astringent properties thanks to their tannin content, even though, in Morocco, they have been rarely consumed due to their bitter taste [4].

Other studies have shown the beneficial effects of these fruits against hyperlipidemia in an animal model of obesity and hyperglycemia [6]. Thanks to their sedative action, they have been also used to treat insomnia, cough attacks and bronchitis [7]; also, the "Doum" has shown anti-inflammatory, anabolic, antiseptic, urinary, antilithic and diuretic activities [4,7,8]. Leaf extracts have also been reported to possess antioxidant activity and the ability to inhibit lipoxygenase [9,10].

The phytochemical properties of *C. humilis* are so far only little characterized. The analysis of the grain's oil showed higher levels of oleic and linoleic acids than other seed oils, as well as a significant amount of tocopherols and tocotrienols [11].

Several biologically important secondary metabolites such as flavonoids, phenols, saponins, gallic tannins and terpenoids have been detected in the leaves and fruits of *C. humilis* L., which may explain the pharmacological effects mentioned above [4,7,9,12].

With regard to flavonoids, they have been previously reported as constituents of the Arecaceae family of plants, even though the literature lacks detailed information on the phytochemical composition of *C. humilis*. Further, no work has been so far devoted to the analysis of the volatile content of such a species.

The aim of this work was to determine the volatile and polyphenolic content of Moroccan Doum fruits (*C. humilis* L.) by GC–MS and HPLC–PDA/MS. In addition, the evaluation of the physico-chemical properties, and the antioxidant activities of the fruit extracts, was performed as well.

This study represents an effort to provide more reliable information about the antioxidant and beneficial health properties of such a species in order to promote its use in different food, pharmaceutical and supplement industries.

2. Results and Discussion

2.1. Physico-Chemical Parameters

Table 1 reports the physico-chemical parameters for the *C. humilis* fruit under investigation.

Table 1. Physico-chemical parameters of *C. humilis* fruit samples. The results are expressed as mean ± standard deviation.

Fruit	Crude Extract	Solvent Fractions	
		EtOAc	MeOH-H$_2$O
pH	3.0 ± 0.06	–	–
Acidity	1.5 ± 0.28	–	–
RI	1.4 ± 0.10	1.3 ± 0.00	1.3 ± 0.00
TSS	15.2 ± 0.68	0.4 ± 0.01	3.0 ± 0.01
S/A	10.3 ± 0.5	–	–
DM (%)	69.5 ± 0.51	–	–
Ash (%)	3.0 ± 0.31	–	–

Table 1. *Cont.*

Fruit	Crude Extract	Solvent Fractions	
		EtOAc	MeOH-H$_2$O
TS (%)	23.7 ± 0.86	6.4 ± 0.05	4.6 ± 0.10
RS (%)	18.1 ± 0.72	–	–
Lipids(mg/g)	0.70 ± 0.05	–	–
Proteins(mg/g)	5.33 ± 1.5	–	0.6 ± 0.01
Vitamin C (mg/g)	31.4 ± 0.53	13.6 ± 0.45	30 ± 0.28

RI: refractive index; TSS: total soluble solid (°Brix); DM: dry matter; S/A: sugar/acidity; TS: total sugars; RS: reducing sugars.

The percentage of dry matter attained was equal to 69.6 ± 0.5, approximately indicating the presence of 30.4% water in these fruits. The latter value is twice as high than the one reported by Bouhafsoun et al. (17.4 ± 0.12%) [2]. On the other hand, another study showed a higher value (79.6 ± 0.04%) in *Butia odorata*, which belongs to the same family (Arecaceae) [13].

The ash content revealed interesting amounts of minerals (3.0 ± 0.3). Such a value coincides with the mean value of ash content (2.4 to 5.0%) recommended by FAO [14], even though it is lower than that recently reported for the Algerian species (4.2 ± 0.7%) [2].

Concerning the pH measurement, a value of 3 ± 0.06 was attained. This value is lower than the one found by Bouhafsoun et al. (5.0 ± 0.0) [2] and, in general, other species belonging to the Arecaceae family, e.g., date palm (*Phoenix dactylifera* L.) (5.3 ± 0.0) [15] and doum palm (*Hyphaene thebaica*) (4.8 ± 0.0) [16].

The titratable acidity of *C. humilis* L. fruit revealed a percentage of 1.5 ± 0.3%. This value is slightly different from Algerian fruits (0.2 ± 0.0%) [2], but similar to other species, e.g., *Hyphaene thebaica* (0.22%) [16].

The TSS results showed a mean value of 15.2 ± 0.7%. Similar values were found in *Butia odorata* fruits (13.1–14.6%) [17], despite Ferrão et al. (2013) revealing, for the same species, a value of 9.5 ± 0.0% [13]. These results are not in agreement with other studies where values reported were 2.4% in leaves and rachis and 4% in fruits [2]. This can be directly related to the sugar content of the fruit samples, which have higher sugar content than other parts of *C. humilis* L. [2].

The S/A ratio was 10.3 ± 0.5%. Such a ratio is an important biochemical parameter that influences the taste and acceptability of the fruits. The high values of this ratio indicate good technological properties and consumer acceptance of these fruits [18,19]. The result achieved in this study falls within the range found for *Butia odorata* fruits (4.42–14.20%) [13]. On the other hand, the S/A ratio values of the *C. humulis* L. fruits investigated in this work showed higher values compared to those of *B. capitata* reported in the literature, 4.7–5.8% [20].

Results of RS and TS were equal to 18.1 ± 0.7% and 23.7 ± 0.9%, respectively. Vitamin C contents in Doum extracts were determined to be 31.5 ± 0.5 mg/g, which is slightly higher than other research (20.1 ± 0.5 mg/g) [21].

The refractive index values for the *C. humilis* L. in each extract were 1.3 ± 0.0 and 1.34 ± 0.0 for ethyl acetate (EtOAc) and methanol–water (MeOH-H$_2$O), respectively. The ANOVA test ($p > 0.05$) showed that the difference between the fruits in IR was not significant.

With regard to lipid and protein contents, values of 0.7 ± 0.0% and 5.3 ± 1.5% were attained, respectively. A value of 0.6 ± 0.0 mg/g was attained for the MeOH-H$_2$O extract, whereas they were absent in the EtOAc fraction. The low levels of protein content can be caused by the ultrasonic extraction, which leads to protein denaturation, as proven by some researchers [22].

2.2. Phytochemical Screening

The phytochemical screening of *C. humilis* was carried out, for the first time for a Moroccan species. The phytochemical tests revealed the presence of different chemical families, distributed for the studied species according to the solvent concentration used. Anthocyanins were not detected in any of the samples investigated. In the EtOAc extract, unsaturated sterols, terpenes and glycosides, which were absent in the crude extract, were revealed. On the other hand, in the crude extract, catechic tannins, anthracenosides, sterols and steroids were detected in high concentrations.

In the literature, the phytochemical properties of *C. humilis* are not well characterized, although several studies have reported the presence of tannins, flavonoids, saponins, sterols and terpenoids [10]. These results are similar to those found in samples from Algeria [23]. Notably, saponosides, responsible for many pharmacological properties, e.g., anti-inflammatory [24,25], were also detected in the extract of *C. humilis*. From the results achieved, such a species does contain important phytochemical constituents that may contribute to its anti-inflammatory and antioxidant activities (Table 2).

Table 2. Phytochemical screening of *C. humilis* samples.

Compounds Group/Solvent of Extraction		Crude Extract	EtOAc	MeOH-H$_2$O
Alkaloids		+	+	±
Polyphenols	Flavonoids	B	B	A++
	Tannins	+	+	+
	Anthocyanins	−	−	−
	Catechic tannins	++	−	+
	Gallic tannins	−	−	+
	Coumarins	+	+	+
	Anthracenosides	++	−	−
	Anthraquinones	+	−	−
	Anthracenosides and Anthocyanosides	+	−	−
Steroids	Saponosides	++	-	−
	Unsaturated Sterols/Terpenes	−	+	−
	Sterols and Steroids	++	−	−
Sugars	Starch	+	−	−
	Deoxysugars	+	−	−
	Glycosides	−	+	±
	Mucilages	+	−	+

A: Flavones; B: Isoflavones; ++: Abondant; +: Present; −: Absent.

2.3. Phytochemical Content and Antioxidant Ability

The spectrophotometric assays showed an important amount of polyphenols. Comparing the two extracts, the MeOH-H$_2$O one showed the highest total polyphenolic (TPP) content, equal to 32.7 ± 0.1 mg GAE/g DM, with respect to the EtOAc extract, 3.6 ± 0.5 mg GAE/g DM (Table 3). The same considerations can be made for the total flavonoid (TFv) and total tannin (TT) contents.

Table 3. TPP, TFv and TT content in *C. humilis* solvent fractions.

Extract	TPP	TFv	TT	IC$_{50}$
EtOAc	3.6 ± 0.5	6.5 ± 0.1	6.2 ± 0.5	1.9 ± 0.1
MeOH-H$_2$O	32.7 ± 0.1	11.1 ± 0.45	54.3 ± 0.8	0.4 ± 0.1

Statistical analysis (ANOVA) showed that there was a highly significant difference in results ($p < 0.001$) between the different solvent concentrations, thus indicating an effect of the solvent concentration on the extraction of these compounds [26].

Concerning the antioxidant activity of the two fractions (1.9 $\pm$ 0.1 mg/mL and 0.4 $\pm$ 0.1 mg/mL, respectively, for EtOAc and MeOH-H$_2$O), the values attained are higher than those found by other authors, e.g., Belhaoues et al. (2017) [27] and Gonçalves et al. (2018) [10] (0.12 mg/mL and 0.081 mg/mL). Another two studies obtained from a methanolic extract of *C. humilis* reported IC$_{50}$ values of 0.024 mg/mL [28] and 0.455 mg/mL [29]. Our findings are in agreement with previously reported papers on different species [26,30].

2.4. GC–MS Analyses

The *C. humilis* *n*-hexane fraction exhibited 69 compounds belonging to different chemical classes (Figure 1). Similarity ranged from 87% to 96%. The main volatile compound was represented by *n*-Hexadecanoic acid (21.75%), followed by oleic acid (14.66%) (Table 4). Such findings are consistent with other *C. humilis* works [31,32].

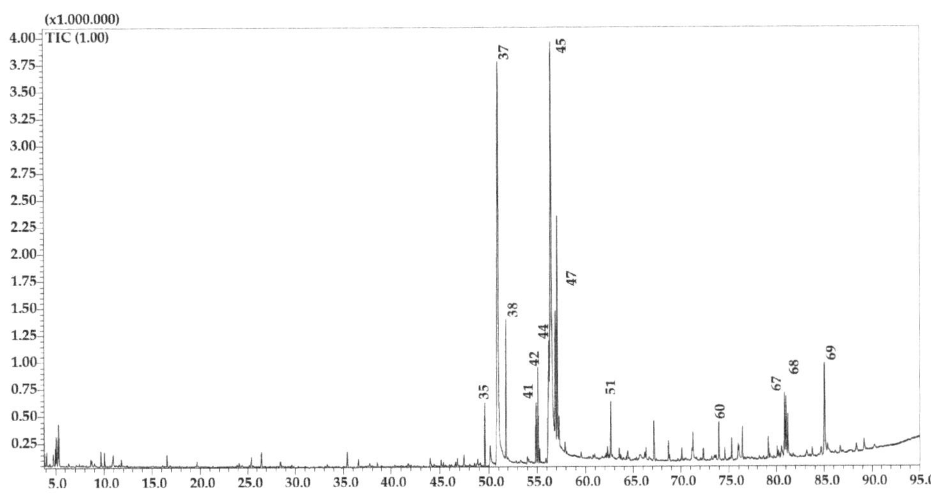

Figure 1. GC–MS profile of the *n*-hexane fraction of *C. humilis*. Main peaks are labeled. Peak assignment as in Table 4.

Table 4. List of compounds identified in the *n*-hexane fraction of *C. humilis* by GC–MS.

No.	Compound	LRI (lib)	LRI (exp)	Similarity	Area(%)	Library
1	*n*-Hexanol	867	867	90	0.04	FFNSC 4.0
2	Acetonylacetone	923	925	90	0.11	FFNSC 4.0
3	*n*-Hexanoic acid	997	977	96	0.31	FFNSC 4.0
4	*n*-Nonanal	1107	1106	96	0.27	FFNSC 4.0
5	*n*-Octanoic acid	1192	1171	94	0.19	FFNSC 4.0
6	*n*-Decanal	1208	1207	91	0.06	FFNSC 4.0
7	(2E)-Decenal	1265	1264	92	0.06	FFNSC 4.0
8	Nonanoic acid	1289	1269	92	0.13	FFNSC 4.0
9	(2E,4E)-Decadienal	1322	1296	93	0.41	FFNSC 4.0
10	*n*-Decanoic acid	1398	1366	93	0.15	FFNSC 4.0
11	ethyl-Decanoate	1399	1395	93	0.08	FFNSC 4.0
12	(E)-, β-Ionone	1482	1482	87	0.07	FFNSC 4.0
13	methyl-Dodecanoate	1527	1524	88	0.03	FFNSC 4.0
14	5,6,7,7a-tetrahydro-4,4,7a-trimethyl-,(R)-2(4H)-Benzofuranone	1532	1533	90	0.46	W11N17

Table 4. *Cont.*

No.	Compound	LRI (lib)	LRI (exp)	Similarity	Area(%)	Library
15	*n*-Dodecanoic acid	1581	1563	94	0.18	FFNSC 4.0
16	ethyl-Dodecanoate	1598	1594	89	0.11	FFNSC 4.0
17	*n*-Hexadecane	1600	1600	87	0.03	FFNSC 4.0
18	*n*-Tetradecanal	1614	1614	91	0.13	FFNSC 4.0
19	1.1'-oxybis-Octane	1657	1663	88	0.10	W11N17
20	*n*-Heptadecane	1700	1700	90	0.17	FFNSC 4.0
21	2-Pentadecanol	1710	1707	92	0.05	W11N17
22	Pentadecanal	1717	1716	90	0.08	W11N17
23	methyl-Tetradecanoate	1727	1725	87	0.03	FFNSC 4.0
24	*n*-Tetradecanoic acid	1773	1762	87	0.23	FFNSC 4.0
25	ethyl-Tetradecanoate	1794	1793	93	0.22	FFNSC 4.0
26	*n*-Octadecane	1800	1800	91	0.09	FFNSC 4.0
27	Hexadecanal	1820	1818	93	0.09	W11N17
28	Pentadecanoic acid, methyl ester	1824	1825	88	0.11	W11N17
29	Neophytadiene	1836	1836	92	0.10	FFNSC 4.0
30	Phytone	1841	1842	94	0.19	FFNSC 4.0
31	Pentadecylic acid	1869	1863	90	0.11	FFNSC 4.0
32	ethyl-Pentadecanoate	1893	1893	91	0.09	FFNSC 4.0
33	*n*-Nonadecane	1900	1900	90	0.13	FFNSC 4.0
34	(Z)-9-Hexadecenoic acid, methyl ester	1895	1903	93	0.14	W11N17
35	methyl-Hexadecanoate	1925	1926	95	1.62	FFNSC 4.0
36	Hexadecanolact-16-one	1938	1943	88	0.77	FFNSC 4.0
37	*n*-Hexadecanoic acid	1977	1969	94	21.75	FFNSC 4.0
38	ethyl-Palmitate	1993	1993	96	3.80	FFNSC 4.0
39	Heptadecanoic acid, methyl ester	2028	2026	90	0.06	W11N17
40	Heptadecanoic acid	2080	2064	94	0.33	W11N17
41	methyl-Linoleate	2093	2093	90	1.57	FFNSC 4.0
42	methyl-Oleate	2098	2098	92	2.61	FFNSC 4.0
43	methyl-Octadecanoate	2127	2127	89	0.11	FFNSC 4.0
44	Linoleic acid	2144	2137	92	6.90	FFNSC 4.0
45	Oleic acid	2142	2145	89	14.66	FFNSC 4.0
46	(Z)-Vaccenic acid	2161	2148	92	4.03	W11N17
47	ethyl-Linoleate	2164	2160	92	5.04	FFNSC 4.0
48	(E)-9-Octadecenoic acid ethyl ester	2174	2173	92	2.10	W11N17
49	ethyl-Stearate	2198	2194	91	0.53	FFNSC 4.0
50	*n*-Tricosane	2300	2300	90	0.22	FFNSC 4.0
51	(Z)-9-Octadecenamide	2375	2362	94	1.69	W11N17
52	*n*-Tetracosane	2400	2400	88	0.12	FFNSC 4.0
53	Behenyl alcohol	2493	2495	89	0.19	FFNSC 4.0
54	*n*-Pentacosane	2500	2500	93	0.31	FFNSC 4.0
55	1-Hexacosene	2596	2595	90	0.54	W11N17
56	*n*-Hexacosane	2600	2599	93	0.19	FFNSC 4.0
57	Heptacos-1-ene	2694	2695	89	0.33	W11N17
58	*n*-Heptacosane	2700	2700	92	0.88	FFNSC 4.0
59	*n*-Octacosane	2800	2799	89	0.16	FFNSC 4.0
60	Squalene	2810	2813	94	1.15	FFNSC 4.0
61	Hexacosanal	2833	2840	93	0.41	W11N17
62	*n*-Nonacosane	2900	2900	90	0.50	FFNSC 4.0
63	Octacosanal	3039	3044	95	0.65	W11N17
64	γ-Tocopherol	3055	3053	88	0.15	W11N17
65	*n*-Hentriacontane	3100	3100	92	0.13	FFNSC 4.0
66	Octacosanol	3120	3109	94	0.68	W11N17
67	2-Nonacosanone	3125	3123	91	2.16	W11N17
68	Vitamin E	3130	3131	93	2.07	W11N17
69	γ-Sitosterol	3351	3321	90	4.13	W11N17
	Tot. identified				**87.29**	
	Tot. not identified				**12.71**	

2.5. HPLC–PDA/MS Analyses

The analysis of the polyphenolic profile, achieved by HPLC–PDA of the EtOAc and MeOH-H$_2$O extracts of *C. humilis*, is reported in Figure 2. A total of sixteen and thirteen different polyphenolic compounds were detected in both extracts, respectively. Tentative identification was based on combined data coming from retention times, PDA, MS and standard co-injection, when available (thirteen in EtOAc vs. twelve in MeOH-H$_2$O extracts (Tables 5 and 6)). Interestingly, the totality of the polyphenolic compounds in both extracts belong to the hydroxycinnamic acids class, whereas only two flavonols were identified in both extracts. Most of the compounds were already reported as constituents of fruits of botanical species, belonging to the same family, e.g., ferulic acid, feruloylquinic acid, ferulic acid hexoside [33], *p*-Coumaric acid, dicaffeoylshikimic acid and isorhamnetin-diglucoside [34]. Notably, 3-Caffeoylquinic acid and 3-Caffeoylquinic acid were reported as constituents of leaf extracts of *C. humilis* [10], whereas quinic acid, *p*-Coumaric, rutin and kaempferol were found in the fruits of the same species [23]. Cinnamoyl glucose and *p*-Coumaric acid ethyl ester are here reported for the first time.

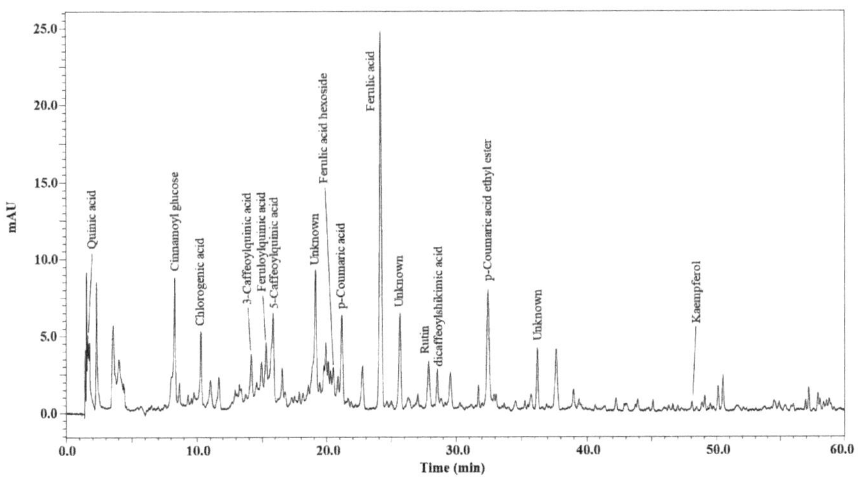

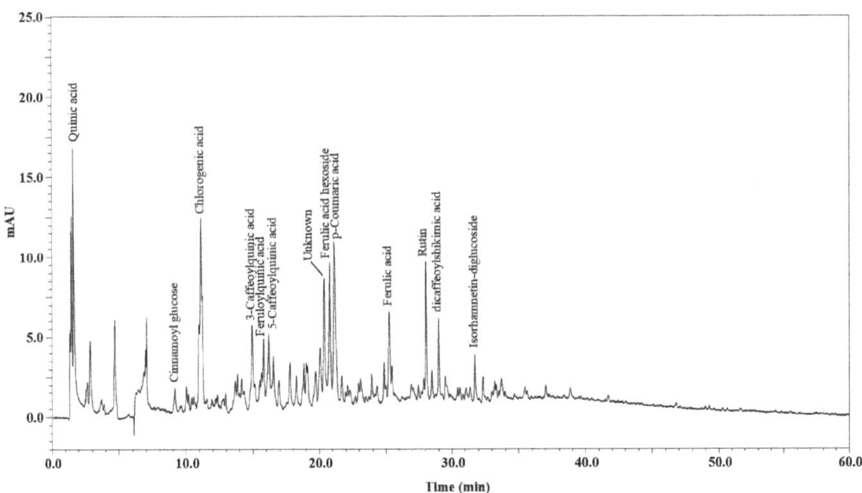

Figure 2. HPLC–PDA polyphenolic profile of the EtOAc (top) and MeOH-H$_2$O extracts of *C. humilis.*

Table 5. Polyphenolic compounds detected in EtOAc extract of *C. humilis* by HPLC–PDA–ESI/MS.

Tentative Identification	t_R (min)	Identification Type	λ_{MAX} (nm)	$[M-H]^-$	Fragments
Phenolic Acid and Derivatives					
Quinic acid	1.64	PDA/MS	–	191	–
Cinnamoyl glucose	8.31	PDA/MS	258–291	309	–
Chlorogenic acid	10.31	PDA/MS	324	353	179
3-Caffeoylquinic acid	14.15	PDA/MS	321	353	–
Feruloylquinic acid	15.29	PDA/MS	324	367	–
5-Caffeoylquinic acid	15.83	PDA/MS	213–324	353	179
Unknown	19.13	PDA/MS	282–325	336	–
Ferulic acid hexoside	20.51	PDA/MS	214–324	355	191
p-coumaric acid	22.74	PDA/MS	288	163	–
Ferulic acid	24.16	PDA/MS	216–321	193	–
Unknown	25.66	PDA/MS	304	193	–
dicaffeoylshikimic acid	28.54	PDA/MS	217–291	497	179
p-Coumaric acid ethyl ester	32.46	PDA/MS	247–291	191	–
Unknown	36.22	PDA/MS	270	345	263
Flavonols					
Rutin	27.86	PDA/MS	352	609	–
Kaempferol	48.18	PDA/MS	219–369	285	–

Table 6. Polyphenolic compounds detected in MeOH-H$_2$O extract of *C. humilis* by HPLC–PDA–ESI/MS.

Tentative Identification	t_R (min)	Identification Type	λ_{MAX} (nm)	$[M-H]^-$	Fragments
Phenolic Acid and Derivatives					
Quinic acid	1.64	PDA/MS	–	191	–
Cinnamoyl glucose	8.31	PDA/MS	258–291	309	–
Chlorogenic acid	10.31	PDA/MS	324	353	179
3-Caffeoylquinic acid	14.15	PDA/MS	321	353	–
Feruloylquinic acid	15.29	PDA/MS	324	367	–
5-Caffeoylquinic acid	15.83	PDA/MS	213–324	353	179
Unknown	19.13	PDA/MS	282–325	336	–
Ferulic acid hexoside	20.51	PDA/MS	214–324	355	191
p-coumaric acid	22.74	PDA/MS	288	163	–
Ferulic acid	24.16	PDA/MS	216–321	193	–
Dicaffeoylshikimic acid	28.54	PDA/MS	217–291	497	179
Flavonols					
Rutin	27.86	PDA/MS	352	609	–
Isorhamnetin-diglucoside	31.75	PDA/MS	353	623	–

The quantification was determined for three repetitions of different extracts of the same sample. As far as quantification is concerned (Table 7), ferulic acid in the EtOAc extract turned out to be the most abundant one (104.7 µg/g), followed by 5-Caffeoylquinic acid (36.5 µg/g). On the other hand, in the MeOH-H$_2$O extract, chlorogenic acid (45.4 µg/g) was predominant, along with quinic acid (37.0 µg/g).

In total, 276.7 µg/g and 262.2 µg/g of polyphenolic compounds for the EtOAc and MeOH-H$_2$O extracts of *C. humilis*, respectively, were attained. Such results are comparable with other Moroccan fruits, e.g., *Ziziphus lotus*, at least for the EtOAc extract (298.5 µg/g) [26].

Table 7. Semi-quantification of polyphenols detected in *C. humilis* fruits in µg/g (*w/w*).

Compounds	EtOAc	MeOH-H$_2$O	Standard Used for Semi-Quantification
Phenolic Acid and Derivatives			
Quinic acid	6.3 ± 0.02	37.0 ± 0.36	Gallic acid
Cinnamoyl glucose	8.1 ± 0.40	0.3 ± 0.03	Cinnamic acid
Chlorogenic acid	18.8 ± 0.90	45.4 ± 1.59	Caffeic acid
3-Caffeoylquinic acid	16.6 ± 0.30	22.4 ± 0.14	Ferulic acid
Feruloylquinic acid	26.3 ± 1.02	12.4 ± 0.07	Ferulic acid
5-Caffeoylquinic acid	36.5 ± 1.05	20.3 ± 0.62	Caffeic acid
Ferulic acid hexoside	12.9 ± 0.82	20.3 ± 0.21	Ferulic acid
p-coumaric acid	11.3 ± 0.50	0.4 ± 0.01	Coumarin
Ferulic acid	104.7 ± 2.52	20.6 ± 0.9	Ferulic acid
Dicaffeoylshikimic acid	7.5 ± 0.10	10.1 ± 0.5	Caffeic acid
p-Coumaric acid ethyl ester	12.7 ± 0.12	–	Coumarin
Flavonols			
Rutin	17.7 ± 0.03	60.2 ± 1.9	Rutin
Isorhamnetin-diglucoside	–	12.8 ± 0.8	Kaempferol
Kaempferol	15.0 ± 0.93	–	Kaempferol

3. Materials and Methods

3.1. Samples and Sample Extraction

Chamaerops humilis L. fruits were harvested in Tangier-Tetouan-Al Hoceima, an area located in the extreme north-west of Morocco. The samples were collected for 4 months (May, June, July and August 2018). All of the harvest areas were between the longitudes 5°94′84106 and the latitudes 35°44′701. The fruit harvesting was carried out at their physiological maturity in the early morning, transported in well-closed boxes and stored at −10 °C in the Materials and Resources Valorization Laboratory, Faculty of Sciences and Technology of Tangier. The extraction method employed was previously described by El Cadi et al. (2020) [26]. Briefly, 5 g of lyophilized powder underwent a defatting step by adding three times 50 mL of *n*-hexane; afterwards, it was dried and homogenized with 50 mL of two solvents with increased polarity, namely EtOAc and MeOH-H$_2$O 80:20 (*v/v*). Each fraction was extracted by using an ultrasound bath (130 kHz) for 45 min. After centrifugation at 5000 g for 5 min, the supernatant was filtered through a paper filter, dried, reconstituted with MeOH-H$_2$O and then filtered through a 0.45 µm Acrodisc nylon membrane (Merck Life Science, Merck KGaA, Darmstadt, Germany) prior to HPLC–PDA-ESI/MS analysis.

3.2. Chemical Reagents and Solvents

Folin-Ciocalteu phenol reagent was obtained from Fluka. Standards (gallic acid, caffeic acid, cinnamic acid, ferulic acid, coumarin, rutin and kaempferol) were obtained from Merck Life Science (Merck KGaA, Darmstadt, Germany). In addition, 2,2-diphenyl-1-picrylhydrazyl (DPPH) and butylated hydroxytoluene (BHT) were purchased from Sigma (St. Louis, MO, USA). LC-MS grade methanol, acetonitrile, acetic acid, EtOAc, acetone and water were purchased from Merck Life Science (Merck KGaA, Darmstadt, Germany). All of the other chemicals were of analytical grade and obtained from Sigma (St. Louis, MO, USA).

3.3. Physico-Chemical Analyses and Phytochemical Screening

Physico-chemical analyses and phytochemical screening were carried out according to a previously published work [26].

3.4. Analysis and Quantification of Phenolic Contents

TPP content was estimated using Folin-Ciocalteu method [35] and was expressed as mg of gallic acid (GAE)/g of dry mass (DM). TFv content was expressed as mg of quercetin (QE)/g of dry mass (DM) and quantified according to the method of Zhishen et al. [36]. TT content was determined by the vanillin method of Julkunen-Tiitto and expressed as mg (+)-catechin/g DW [37].

3.5. Determination of Antioxidant Activity

The DPPH method followed the method described by Braca et al. [38]. Butylated hydroxytoluene (BHT) was used as a positive control and the DPPH radical scavenging activity was calculated according to the equation:

$$\text{DPPH radical scavenging activity: } I\ (\%) = (A\ blank - A\ sample)\ /\ A\ blank \times 100 \qquad (1)$$

The IC_{50} of the DPPH radical was calculated from linear regression (%DPPH remaining radical versus sample concentration).

3.6. GC–MS

GC analyses of the volatile fraction were performed on a GC–MS-QP2020 system (Shimadzu, Kyoto, Japan) with an "AOC-20i" system auto-injector. The analyses were realized on an SLB-5ms column (30 m in length $\times$ 0.25 mm in diameter $\times$ 0.25 µm in thickness of film, Merck KGaA). The initial temperature was set at 50 °C, and afterwards increased up to 350 °C (increase rate: 3 °C/min; holding time: 5 min).

GC–MS parameters were as follows: injection temperature: 280 °C; injection volume: 1.0 µL (split ratio: 10:1); pure helium gas (99.9%); linear velocity: 30.0 cm/s; inlet pressure: 26.7 KPa; EI source temperature: 220 °C; interface temperature: 250 °C. The acquisition of MS spectra was realized in full scan mode, in the mass range of 40–660 m/z, with an event time of 0.2 s.

Relative quantity of the chemical compounds present in each sample was expressed as a percentage based on peak area produced in the GC chromatogram.

Compounds were identified by using the "FFNSC 4.0" (Shimadzu Europa GmbH, Duisburg, Germany) and "W11N17" (Wiley11-Nist17, Wiley, Hoboken, NJ, USA; Mass Finder 3). Each compound was identified applying a MS similarity match and an LRI filter. Linear retention indices (LRI) were calculated by using a C7–C40 saturated alkanes reference mixture (49452-U, MerckKGaA).

Data files were collected and processed by using "GCMS Solution" software, ver. 4.50 (Shimadzu, Kyoto, Japan) [26].

3.7. HPLC–PDA/ESI-MS

LC analyses were performed on a Shimadzu liquid chromatography system (Kyoto, Japan), consisting of a CBM-20A controller, two LC-30AD dual-plunger parallel-flow pumps, a DGU-20A5R degasser, a CTO-20AC column oven, a SIL-30AC autosampler, an SPD-M30A photo diode array detector and an LCMS-8050 triple quadrupole mass spectrometer, through an ESI source (Shimadzu, Kyoto, Japan).

Chromatographic separations were attained on 150×4.6 mm; 2.7 µm Ascentis Express RP C18 columns (Merck Life Science, Merck KGaA, Darmstadt, Germany). The mobile phase was composed of two solvents: water/acetic acid (99.85/0.15 v/v, solvent A) and acetonitrile/acetic acid (99.85/0.15 v/v, solvent B). The flow rate was set at 1 mL/min under gradient elution: 0–5 min, 5% B, 5–15 min, 10% B, 15–30 min, 20% B, 30–60 min, 50% B, 60 min, 100% B. PDA detection was: λ = 200–400 nm (λ = 280 nm) (sampling frequency: 40.0 Hz, time constant: 0.08 s). MS conditions were as follows: scan range and the scan speed were set at m/z 100–800 and 2500 amu sec^{-1}, respectively, event time: 0.3 sec, nebulizing gas (N_2) flow rate: 1.5 L min^{-1}, drying gas (N_2) flow rate: 15 L min^{-1}, interface temperature: 350 °C, heat block temperature: 300 °C, DL (desolvation line) temperature: 300 °C, DL voltage: 1 V, interface voltage: −4.5 kV [26].

3.8. Statistical Analysis

The experiments were carried out in triplicate and the results were expressed as the average of the three measurements ± SD. The comparison of means between groups was performed with one-way analysis of variance (ANOVA) followed by a Tukey test. Differences were considered significant when $p < 0.05$ (Microsoft [®] Office, Santa Rosa, California, CA, USA).

4. Conclusions

The present study aimed to elucidate the bioactive content of *Chamaerops humilis* L. fruits. Considering the two extracts tested, in terms of the antioxidant activity, the EtOAc one turned out to be the most active with respect to the MeOH-H$_2$O. A total of 69 compounds belonging to different chemical classes were positively identified by GC coupled to MS, whereas sixteen and thirteen polyphenolic compounds were detected by HPLC–PDA/MS in both EtOAc and MeOH-H$_2$O extracts, respectively. Such results demonstrate that this fruit can be used for industrial applications in food preparations. In addition, the data attained emphasize an interesting functional composition of the *Chamaerops humilis* L. fruits, which could be considered a valuable new co-product with commercial importance in the food industry.

Author Contributions: Conceptualization, H.E.C. and F.C.; Methodology, H.E.C. and F.C.; Investigation, H.E.C.; H.E.B.; G.S.; B.R.; Y.O.E.M.; F.A.; and K.A.; Writing—Original Draft Preparation, H.E.C.; Writing—Review and Editing, F.C., M.P.L. and T.M.G.S.; Supervision, F.C. and J.B.; Project Administration: L.M. All authors have read and agreed to the published version of the manuscript.

Funding: This research received no external funding.

Institutional Review Board Statement: Not applicable.

Informed Consent Statement: Not applicable.

Data Availability Statement: Data sharing not applicable.

Acknowledgments: The authors thank Merck Life Science and Shimadzu Corporations for their continuous support.

Conflicts of Interest: The authors declare no conflict of interest.

Sample Availability: Samples of the compounds are not available from the authors.

References

1. Tassin, C. *Paysages Végétaux du Domaine Méditerranéen: Bassin Méditerranéen, Californie, Chili Central, Afrique du Sud, Australie Méridionale*; IRD Éditions: Marseille, France, 2012.
2. Bouhafsoun, A.; Boukeloua, A.; Yener, I.; Diare Mohamed, L.; Diarra, T.; Errouane, K.; Mezmaz, R.; Temel, H.; Kaid-Harche, M. Chemical composition and mineral contents of leaflets, rachs and fruits of *Chamaerops humilis* L. *Acad. J. Agric. Res.* **2019**, *20*, 8.
3. Nehdi, I.A.; Mokbli, S.; Sbihi, H.; Tan, C.P.; Al-Resayes, S.I. *Chamaerops humilis* L. var. argentea André Date Palm Seed Oil: A Potential Dietetic Plant Product: Nutritional value of C. humilis seed oil. *J. Food Sci.* **2014**, *79*, C534–C539. [CrossRef] [PubMed]
4. Bnouham, M.; Merhfour, F.Z.; Ziyyat, A.; Aziz, M.; Legssyer, A.; Mekhfi, H. Antidiabetic effect of some medicinal plants of Oriental Morocco in neonatal non-insulin-dependent diabetes mellitus rats. *Hum. Exp. Toxicol.* **2010**, *29*, 865–871. [CrossRef]
5. Hasnaoui, O.; Bouazza, M.; Benali, O.; Thinon, M. Ethno Botanic Study of *Chamaerops humilis* L. Var. argentea Andre (Are-caceae) in Western Algeria. *Agric. J.* **2011**, *6*, 1–6.
6. Gaamoussi, F.; Israili, Z.H.; Lyoussi, B. Hypoglycemic and hypolipidemic effects of an aqueous extract of *Chamaerops humilis* leaves in obese, hyperglycemic and hyperlipidemic Meriones shawi rats. *Pak. J. Pharm. Sci.* **2010**, *23*, 212–219. [PubMed]
7. Benmehdi, H.; Hasnaoui, O.; Benali, O.; Salhi, F. Phytochemical investigation of leaves and fruits extracts of *Chamaerops humilis* L. *J. Mater. Environ. Sci.* **2012**, *3*, 320–337.
8. Blumenthal, M.; Busse, W.; Goldberg, A.; Gruenwald, J.; Hall, T.; Riggins, C.W.; Rister, R.S. *The Complete German Commis-sion E Monographs: Therapeutic Guide to Herbal Medicines*; American Botanical Council: Austin, TX, USA; Integrative Medicine Communications: Boston, MA, USA, 1998.

9. Benahmed-Bouhafsoun, A.; Djied, S.; Mouzaz, F.; Kaid-Harche, M. Phytochemical composition and in vitro antioxidant activity of *Chamaerops humilis* L. extracts. *Int. J. Pharm. Pharm. Sci.* **2013**, *5*, 741–744.
10. Miguel, M.; Bouchmaaa, N.; Aazza, S.; Gaamoussi, F.; Lyoussi, B. Antioxidant, anti-inflammatory and anti-acetylcholinesterase activities of Moroccan plants. *Fresenius Environ. Bull.* **2014**, *23*, 1–14.
11. Fekar, G.; Aiboudi, M.; Bouyazza, L. Composition en acides gras, stérols et tocophérols de l huile végétale non convention-nelle extraite des graines du *Chamaerops humilis* L. du Maroc. *Afrique Sci.* **2015**, *11*, 6.
12. Gonçalves, S.; Medronho, J.; Moreira, E.; Grosso, C.; Andrade, P.B.; Valentão, P.; Romano, A. Bioactive properties of *Chamaerops humilis* L.: Antioxidant and enzyme inhibiting activities of extracts from leaves, seeds, pulp and peel. *3 Biotech* **2018**, *8*, 88. [CrossRef]
13. Ferrão, T.S.; Ferreira, D.F.; Flores, D.W.; Bernardi, G.; Link, D.; Barin, J.S.; Wagner, R. Evaluation of composition and quality parameters of jelly palm (*Butia odorata*) fruits from different regions of Southern Brazil. *Food Res. Int.* **2013**, *54*, 57–62. [CrossRef]
14. FAO. *Street Foods: A Summary of FAO Studies and Other Activities Relating to Street Foods*; Food and Agriculture Organization of the United States: Rome, Italy, 1988.
15. Shaba, E.Y.; Ndamitso, M.M.; Mathew, J.T.; Etsunyakpa, M.B.; Tsado, A.N.; Muhammad, S.S. Nutritional and anti-nutritional composition of date palm (*Phoenix dactylifera* L.) fruits sold in major markets of Minna Niger State, Nigeria. *Afr. J. Pure Appl. Chem.* **2015**, *9*, 167–174.
16. Aamer, R.A. Characteristics of aqueous doum fruit extract and its utilization in some novel products. *Ann. Agric. Sci.* **2016**, *61*, 25–33. [CrossRef]
17. Beskow, G.T.; Hoffmann, J.F.; Teixeira, A.M.; Fachinello, J.C.; Chaves, F.C.; Rombaldi, C.V. Bioactive and yield potential of jelly palms (*Butia odorata* Barb. Rodr.). *Food Chem.* **2015**, *172*, 699–704. [CrossRef]
18. Carvalho Filho, C.D.; Honório, S.L.; Gil, J.M. Qualidade pós-colheita de cerejas cv. Ambrunés utilizando coberturas comestíveis. *Rev. Bras. Frutic.* **2006**, *28*, 180–184. [CrossRef]
19. Chitarra, M.I.F.; Chitarra, A.B. *Pós-Colheita de Frutos e Hortaliças: Fisiologia e, Pós-colheita de Frutos e Hortaliças: Fisiologia e Manuseio*, 1st ed.; Federal University of Lavras: Lavras, Brazil, 1990.
20. Schwartz, E.; Fachinello, J.C.; Barbieri, R.L.; da Silva, J.B. Avaliação de populações de Butia capitata de Santa Vitória do Palmar. *Rev. Bras. Frutic.* **2010**, *32*, 736–745. [CrossRef]
21. Santas, J.; Carbó, R.; Gordon, M.; Almajano, M. Comparison of the antioxidant activity of two Spanish onion varieties. *Food Chem.* **2008**, *107*, 1210–1216. [CrossRef]
22. Lv, S.; Taha, A.; Hu, H.; Lu, Q.; Pan, S. Effects of Ultrasonic-Assisted Extraction on the Physicochemical Properties of Different Walnut Proteins. *Molecules* **2019**, *24*, 4260. [CrossRef]
23. Bouhafsoun, A.; Yilmaz, M.A.; Boukeloua, A.; Temel, H.; Harche, M.K. Simultaneous quantification of phenolic acids and flavonoids in *Chamaerops humilis* L. using LC–ESI-MS/MS. *Food Sci. Technol.* **2018**, *38*, 242–247. [CrossRef]
24. Estrada, A.; Katselis, G.S.; Laarveld, B.; Barl, B. Isolation and evaluation of immunological adjuvant activities of saponins from *Polygala senega* L. *Comp. Immunol. Microbiol. Infect. Dis.* **2000**, *23*, 27–43. [CrossRef]
25. Just, M.J.; Recio, M.C.; Giner, R.M.; Cuéllar, M.J.; Máñez, S.; Bilia, A.R.; Ríos, J.-L. Anti-Inflammatory Activity of Unusual Lupane Saponins from *Bupleurum fruticescens*. *Planta Med.* **1998**, *64*, 404–407. [CrossRef] [PubMed]
26. El Cadi, H.; El Bouzidi, H.; Selama, G.; El Cadi, A.; Ramdan, B.; Oulad El Majdoub, Y.; Alibrando, F.; Dugo, P.; Mondello, L.; Fakih Lanjri, A.; et al. Physico-Chemical and Phytochemical Characterization of Moroccan Wild Jujube "*Zizyphus lotus* (L.)" Fruit Crude Extract and Fractions. *Molecules* **2020**, *25*, 5237. [CrossRef] [PubMed]
27. Belhaoues, S.; Amri, S.; Bensouilah, M.; Seridi, R. Antioxidant, antibacterial activities and phenolic content of organic frac-tions obtained from *Chamaerops humilis* L. leaf and fruit. *Int. J. Biosci.* **2017**, *11*, 284–297.
28. Khoudali, S.; Benmessaoudleft, D.; Essaqui, A.; Zertoubi, M.; Mohammed, A.; Benaissa, B. Study of antioxidant activity and anticorrosion action of the methanol extract of dwarf palm leaves (*Chamaerops humilis* L.) from morocco. *J. Mater. Environ. Sci.* **2014**, *5*, 887–898.
29. Coelho, J.P.; Veiga, J.G.; Elvas-Leitao, R.; Brigas, A.F.; Dias, A.M.; Oliveira, M.C. Composition and in vitro antioxidants ac-tivity of *Chamaerops humilis* L. In Proceedings of the 2017 IEEE 5th Portuguese Meeting on Bioengineering (ENBENG), Coimbra, Portugal, 16–18 February 2017; pp. 1–4.
30. Chung-Weng, P.; Abd Malek, S.N.; Ibrahim, H.; Wahab, N.A. Antioxidant properties of crude and fractionated extracts of Alpinia mutica rhizomes and their total phenolic content. *Afr. J. Pharm. Pharmacol.* **2011**, *5*, 842–852.
31. Mokbli, S.; Sbihi, H.M.; Nehdi, I.A.; Romdhani-Younes, M.; Tan, C.P.; Al-Resayes, S.I. Characteristics of *Chamaerops humilis* L. var. humilis seed oil and study of the oxidative stability by blending with soybean oil. *J. Food Sci. Technol.* **2018**, *55*, 2170–2179. [CrossRef]
32. Giovino, A.; Marino, P.; Domina, G.; Rapisarda, P.; Rizza, G.; Saia, S. Fatty acid composition of the seed lipids of *Chamaerops humilis* L. natural populations and its relation with the environment. *Plant Biosyst. Int. J. Deal. All Asp. Plant Biol.* **2014**, *149*, 767–776. [CrossRef]
33. Khallouki, F.; Ricarte, I.; Breuer, A.; Owen, R.W. Characterization of phenolic compounds in mature Moroccan Medjool date palm fruits (*Phoenix dactylifera*) by HPLC–PDA-ESI-MS. *J. Food Comp. Anal.* **2018**, *70*, 63–71. [CrossRef]
34. Ma, C.; Dunshea, F.R.; Suleria, H.A.R. LC-ESI-QTOF/MS Characterization of Phenolic Compounds in Palm Fruits (Jelly and Fishtail Palm) and Their Potential Antioxidant Activities. *Antioxidants* **2019**, *8*, 483. [CrossRef]

35. Singleton, V.; Rossi, J. Colorimetry of Total Phenolic Compounds with Phosphomolybdic-Phosphotungstic Acid Reagents. *Am. J. Enol. Vitic.* **1965**, *16*, 144–158.
36. Zhishen, J.; Mengcheng, T.; Jianming, W. The determination of flavonoid contents in mulberry and their scavenging effects on superoxide radicals. *Food Chem.* **1999**, *64*, 555–559. [CrossRef]
37. Julkunen-Tiitto, R. Phenolic constituents in the leaves of northern willows: Methods for the analysis of certain phenolics. *J. Agric. Food Chem.* **1985**, *33*, 213–217. [CrossRef]
38. Braca, A.; Sortino, C.; Politi, M.; Morelli, I.; Mendez, J. Antioxidant activity of flavonoids from *Licania licaniaeflora*. *J. Ethnopharmacol.* **2002**, *79*, 379–381. [CrossRef]

Article

Polyphenol Profiling of Chestnut Pericarp, Integument and Curing Water Extracts to Qualify These Food By-Products as a Source of Antioxidants

Gabriella Pinto [1], Sabrina De Pascale [2], Maria Aponte [3], Andrea Scaloni [2], Francesco Addeo [3] and Simonetta Caira [2,*]

1. Department of Chemical Sciences, University of Naples "Federico II", via Cintia, 80126 Naples, Italy; gabriella.pinto@unina.it
2. Proteomics & Mass Spectrometry Laboratory, ISPAAM, National Research Council, via Argine 1085, 80147 Naples, Italy; sabrina.depascale@ispaam.cnr.it (S.D.P.); andrea.scaloni@cnr.it (A.S.)
3. Dipartimento di Agraria, Università degli Studi di Napoli "Federico II", via Università 100, Parco Gussone, 80055 Portici, Italy; maria.aponte@unina.it (M.A.); addeo@unina.it (F.A.)
* Correspondence: simonetta.caira@cnr.it

Abstract: Plant polyphenols have beneficial antioxidant effects on human health; practices aimed at preserving their content in foods and/or reusing food by-products are encouraged. The impact of the traditional practice of the water curing procedure of chestnuts, which prevents insect/mould damage during storage, was studied to assess the release of polyphenols from the fruit. Metabolites extracted from pericarp and integument tissues or released in the medium from the water curing process were analyzed by matrix-assisted laser desorption ionization-time of flight-mass spectrometry (MALDI-TOF-MS) and electrospray-quadrupole-time of flight-mass spectrometry (ESI-qTOF-MS). This identified: (i) condensed and hydrolyzable tannins made of (epi)catechin (procyanidins) and acid ellagic units in pericarp tissues; (ii) polyphenols made of gallocatechin and catechin units condensed with gallate (prodelphinidins) in integument counterparts; (iii) metabolites resembling those reported above in the wastewater from the chestnut curing process. Comparative experiments were also performed on aqueous media recovered from fruits treated with processes involving: (i) tap water; (ii) tap water containing an antifungal *Lb. pentosus* strain; (iii) wastewater from a previous curing treatment. These analyses indicated that the former treatment determines a 6–7-fold higher release of polyphenols in the curing water with respect to the other ones. This event has a negative impact on the luster of treated fruits but qualifies the corresponding wastes as a source of antioxidants. Such a phenomenon does not occur in wastewater from the other curing processes, where the release of polyphenols was reduced, thus preserving the chestnut's appearance. Polyphenol profiling measurements demonstrated that bacterial presence in water hampered the release of pericarp metabolites. This study provides a rationale to traditional processing practices on fruit appearance and qualifies the corresponding wastes as a source of bioactive compounds for other nutraceutical applications.

Keywords: chestnut; water curing; non-targeted MS analysis; antioxidants

Citation: Pinto, G.; De Pascale, S.; Aponte, M.; Scaloni, A.; Addeo, F.; Caira, S. Polyphenol Profiling of Chestnut Pericarp, Integument and Curing Water Extracts to Qualify These Food By-Products as a Source of Antioxidants. *Molecules* **2021**, *26*, 2335. https://doi.org/10.3390/molecules26082335

Academic Editor: Mirella Nardini

Received: 23 March 2021
Accepted: 13 April 2021
Published: 17 April 2021

Publisher's Note: MDPI stays neutral with regard to jurisdictional claims in published maps and institutional affiliations.

1. Introduction

According to FAO statistics (http://www.fao.org/faostat/en/#search/chestnut, accessed on 15 January 2021), Europe is among the three top producers of chestnuts in the world (after Asia and China), with roughly 155 ktons of fruits in 2018, among which more than 30% obtained in Italy. About 65% of the Italian chestnut/marron production is localized in the Campania region (INEA, database on foreign trade), where the typical cultivar, Castagna di Montella, is bred. The latter has been certified as a European Protected Geographical Indication (PGI) product and is exported in the form of fresh whole or peeled fruit, or in its dried forms.

Chestnut is considered a functional food with a rich source of polyphenolic compounds, phenolic acids (such as gallic acid) and tannins (primarily ellagic acid) [1,2]. Recently, a comprehensive review has dealt with the elucidation of the high biological value also of the chestnut pericarp and integument (outer and inner shell), which are discarded as by-products during food manufacturing, as reported by the extensive literature on the corresponding in vitro and in vivo bioactive properties [3]. For example, chestnut shell hydroalcoholic extracts tested on several human cell lines were suggested to display potential anti-angiogenic and anti-inflammatory effects [4]. When assayed on trout blood and intestinal leukocytes, they also elicited a stimulatory response of the immune system against possible infectious agents [5]. Several studies on rat models of diabetes fed with chestnut extracts demonstrated a diet-dependent increase of pancreatic cell viability [6] and a corresponding cytoprotective response against hepatorenal injury [7].

Actually, the bioavailability of polyphenols is an important topic of debate [8], which must consider the possible interaction of these metabolites with other food macromolecules as well as the impact of gastrointestinal digestion of these compounds and their subsequent enterocyte absorption. Indeed, the gastrointestinal digestion of polyphenols seems to be the main limiting physiological factor responsible for the conversion of the glucoside into the aglycone species, especially in the switching from an acid pH value to a basic one, remarkably influencing the corresponding bioavailability process. For example, up to 20% of the initial glucoside polyphenol species were shown to resist the simulated gastrointestinal digestion of coffee grounds [8]. Indeed, only 5–10% of aglycone species was estimated to be absorbed by passive diffusion through the enterocyte membrane of the small intestine when the corresponding glucoside derivatives were digested [9]. On the other hand, the chemical structure of phenolic compounds was demonstrated to be modified from microbiota of the large intestine, thus affecting their bioavailability [10]. These findings suggest that the complexity of physiological events and mechanisms affecting the dietary assumption of polyphenols has not been fully clarified and hasnot yet provided a complete picture of the corresponding molecular bioavailability. Nevertheless, chestnut extracts are nowadays used as natural ingredients for the preparation of various functional foods [1,11,12].

Chestnut pericarp and integument are a rich source of tannins [12–14], which have been subclassified into condensed tannins (CTs) and hydrolysable tannins (HTs). The first compounds are derived from the condensation and polymerization of monomeric flavonoid units [15], while the second ones are polyesters containing a sugar moiety linked to gallic and ellagic acids. HTs, in turn, comprise C-glucosidic ellagitannins, which are further subdivided into vescalagin- and castalagin-types depending on the configuration of the OH group at C-1 position of the glycosidic chain [16], and flavono-ellagitannins that are formed after linkage of a flavonoid molecule to ellagitannins, yielding acutissimin A and B. The latter compounds have been isolated from chestnut wastes [13] and have gained significant interest due to their antitumor activity [17].

A main issue in the chestnut industry is the high perishability of harvested fruits during storage, which is associated with fungal/insect contamination and mould development. This is generally faced with thermo-hydrotherapy or cold-water curing treatment. The latter is a traditional, simple, and inexpensive processing practice commonly used in the Campania region, which is based on soaking chestnuts in cold water at a 1:1 to 1:2 ponderal ratio for 3–9 days [18]. After soaking completion, chestnuts are rinsed with running fresh water, and resulting wastewaters are treated in dedicated waste plants. After chestnut curing in cold water, fruit skin loose its natural luster [19]. The efficiency of the cold water curing treatment depends on associated lactic and alcoholic fermentation processes, which reduce the pH value of the medium [18]. In this context, the increase of CO_2 and CH_3CHO levels in the fermentation medium was reported to play an active role in preserving chestnuts during subsequent storage. Whether the simple treatment of chestnuts with cold water was described to mostly reduce the development of insect larvae of *Curculio elephas* and, depending on concomitant lactic and alcoholic fermentation processes, having a significant effect on fungal contamination during storage [20], the preventive addition of *Lb. pentosus* strain to cold water used for treatment was suggested generating antimicrobial compounds that inhibit fungal growth/germination [20], thus extending shelf-life and preserving the appearance of fruits.

Based on the above, the chestnut by-products that are obtained from the corresponding industry can be different; they may have potential use as a source of polyphenols for food, nutraceutical, leather, and cosmeceutical applications. At first, chestnut pericarp and integument tissues, which are removed during food processing for obtaining fresh and dried products and/or corresponding flour. They represented about 20% of the total weight of the original fruits and were already used for recycling purposes due to their high content of tannins [12,13,21]. However, a characterization of corresponding polyphenols was obtained only for low-mass components. Secondly, the wastewater from the above-mentioned water curing treatments of chestnuts, which should eventually contain polyphenols released from the above-mentioned fruit tissues whose nature, however, has not been characterized yet. The stasis of these dissolved molecules in wastewater should render it susceptible to air oxidation. Thus, knowledge of their molecular nature seems a prerequisite for the possible consideration of reusing these wastewaters as a source of bioactive molecules.

With the aim of considering chestnut by-products for possible recycling purposes, polyphenols from pericarp and integument tissues as well as from wastewater of different fruit water curing treatments were extracted and subjected to qualitative and quantitative experiments based on dedicated assays and detailed MALDI-TOF-MS and ESI-qTOF-MS characterization. This allowed defining different molecular signatures for the analyzed materials, which were peculiar for them and were also prodromal to the rationalization of the mechanisms ongoing during different water curing treatments.

2. Results

2.1. Polyphenol Extraction

The eventual release of polyphenols in wastewater from water curing treatments of chestnuts should hypothetically involve molecules present in the peripheric tissue layers of fruits, namely pericarp and integument, which are imbibed with corresponding aqueous media. In order to evaluate the nature of these phenolic compounds in the above-mentioned fruit districts, sampled chestnuts were peeled to recover pericarp and integument tissues, which were then powdered and extracted in parallel with different solvents. Molecular recovery was evaluated considering overall signal-to-noise ratios measured through dedicated MS procedures (see below). The mixture acetonitrile/methanol/water 2:2:1 *v/v/v* was identified as the optimal one for extraction of polyphenols. Thus, pericarp and integument tissues were extracted under continuous agitation, for 48 h, and recovered material was analyzed with MALDI-TOF-MS and ESI-qTOF–MS procedures.

2.2. MALDI-TOF-MS Profiling of Polyphenols from Chestnut Pericarp and Integument Tissues

In order to provide the maximal representation of molecular species occurring in chestnut epicarp tissues, MALDI-TOF-MS analysis of the corresponding extracts was performed in both linear positive and negative ion mode. This analysis gave mass spectra almost superimposable between different samples, which can be summarized with the polyphenol profiles shown in Figure 1.

Distribution of signals within the MALDI-TOF mass spectrum acquired in positive ion mode (Figure 1A) fitted a Gaussian profile, which enabled to guess the presence of oligomeric CTs differing from each other by units of catechins and esterified catechin with gallic acid residue (Δm = 152 mass units) [22]. Signals recorded by MALDI-TOF-MS analysis were tentatively attributed to CTs, like procyanidins. Increments of about 288, and 441 mass units were tentatively attributed to the presence of (epi)catechin, and (epi)catechin gallate units, respectively, representing the building block of a wide variety of tannins, from simple monomers to multiple oligomers [23]. The precursor compound corresponded to the epicatechin gallate (*m/z* 442.9), with dominant signals at *m/z* 426.9 and 410.9, which were suggestive of the loss of one or two OH groups from the B-ring, respectively. This finding was consistent with the presence of fisetinidin, already identified in chestnut [12], which does not present a -OH group at position C3 of one of the two B-rings.

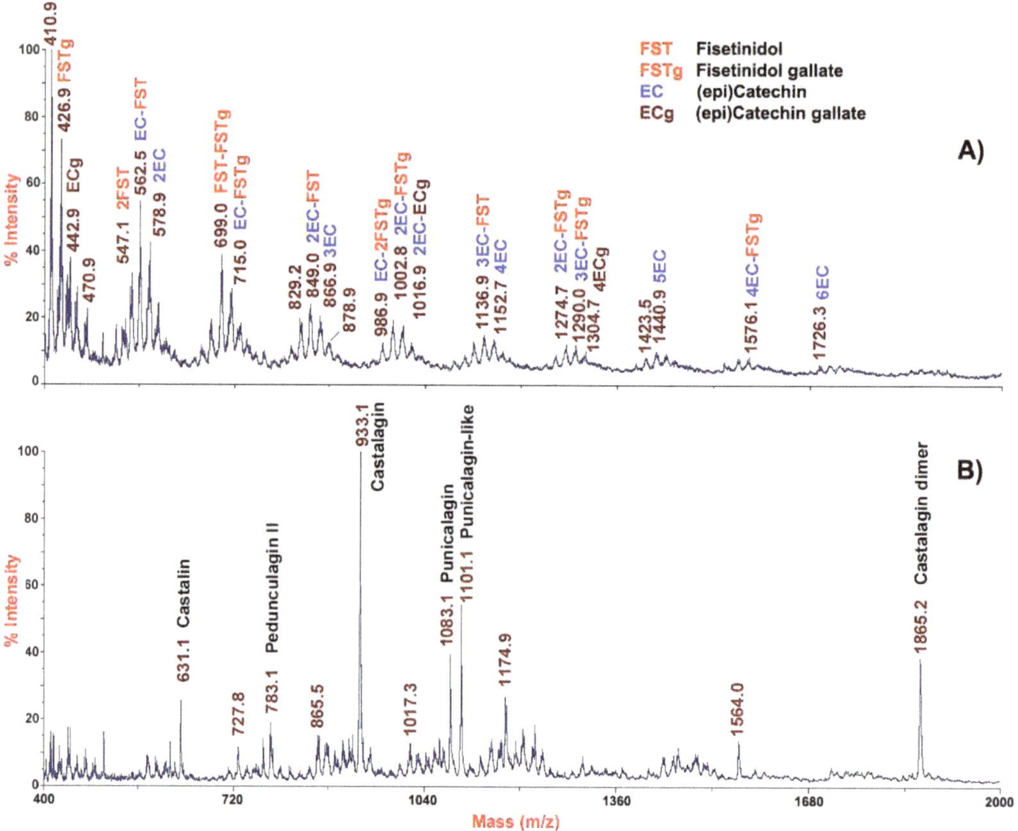

Figure 1. MALDI-TOF mass spectrum of the extract of chestnut pericarp tissues acquired in linear positive (panel (**A**)) and negative (panel (**B**)) ion mode. Condensed polymeric species were tentatively ascribed to gallotannins and ellagitannis derivatives. FST, fisetinidol; FSTg, fisetinidol gallate; EC, (epi)catechin; ECg, (epi)catechin gallate; g, gallic acid.

MALDI-TOF-MS analysis of the same sample carried out in negative ion mode showed a quite different molecular profile (Figure 1B). The dominant mass signal at *m/z* 933.1 indicated the presence of a chestnut-specific HT, namely castalagin, together with its molecular dimer (at *m/z* 1865.2). The occurrence of castalagin in the chestnut pericarp was also confirmed by the co-presence in the mass spectrum of the aglycon castalin (*m/z* 631.1), which is one of the main phenolic compounds present in chestnut shells [16]. The loss of two condensed gallic acid units linked via a galloyl ester bond (Δm = about 152 mass units) between two consecutive product ions likely arose from a different isobaric precursor ion at *m/z* 631.1 (Figure 1B). Other signals present at *m/z* 783.1, 1083.1, and 1101.1 were tentatively assigned to pedunculagin II (bis-HHDP-hex), punicalagin (HHDP-gallagyl-hex), and punicalagin-like species, as already described in other fruits [24,25].

In order to provide the maximal representation of molecular species occurring in chestnut integument tissues, MALDI-TOF-MS analysis of the corresponding extracts was also performed in both linear positive and negative ion mode (Figure 2). Similar to their pericarp counterparts, the mass profiling results for integument extracts showed the occurrence of phenolic polymers.

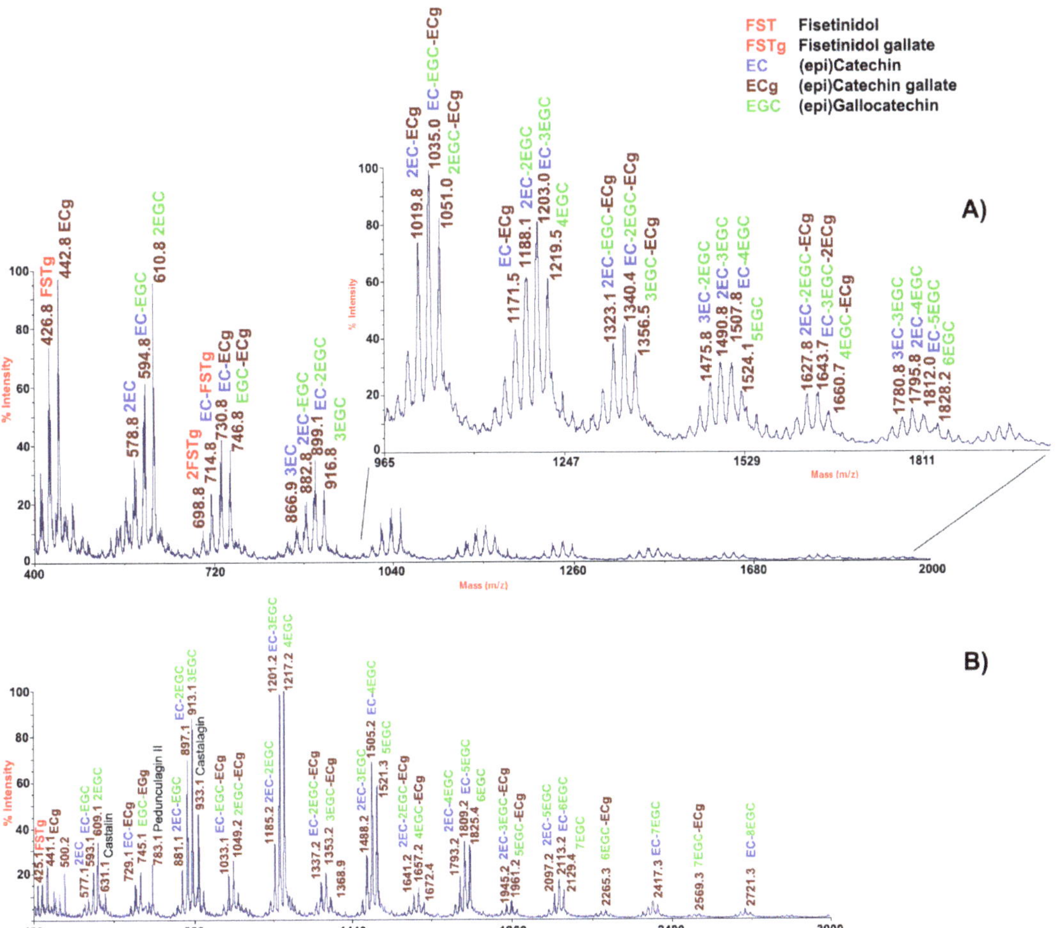

Figure 2. MALDI-TOF mass spectrum of the extract of chestnut integument tissues acquired in linear positive (panel (**A**)) and negative (panel (**B**)) ion mode. FST, fisetinidol; FSTg, fisetinidol gallate; EC, (epi)catechin; ECg, (epi)catechin gallate; EGC, (epi)gallocatechin.

In the case of the MALDI-TOF mass spectrum acquired in positive linear ion mode, corresponding signals differed for the occurrence of repeated units of epigallocatechin and epicatechin gallate residues (Δm = about 304 and 441 mass units) (Figure 2A). Apart from the MH$^+$ signal of epicatechin gallate (m/z 442.8), the most abundant peak was tentatively assigned to prodelphinidin B3 (m/z 610.8), a gallocatechin oligomer. MALDI-TOF-MS analysis of the same sample carried out in negative ion mode highly differed from that recorded under the same experimental conditions for pericarp tissues and confirmed results obtained in positive ion mode (Figure 2B). It reflected the occurrence of prodelphinidin oligomers up to nonamers, together with more complex polymers ascribed to HTs, as already found in chestnut bark [26]. In conclusion, dominant MALDI-TOF-MS signals in integument tissues were associated with oligomeric prodelphinidins, which may have hidden peaks of ellagitannins detected in the chestnut shell as a result of signal ion suppression phenomena.

2.3. ESI-qTOF-MS Analysis of Polyphenols from Chestnut Pericarp and Integument Tissues

High-resolution mass spectrometry and tandem mass spectrometry (MS/MS) has already been used for establishing with high confidence the identity of the polyphenolic species present in various fruit extracts [27]. With the aim of confirming the tentative molecular assignment of polyphenolics obtained by MALDI-TOF-MS, the above-mentioned chestnut pericarp and integument extracts were also analyzed by ESI-qTOF-MS and ESI-qTOF-MS/MS (Figure 3). Corresponding ESI-qTOF-MS profiles acquired in positive ion mode are shown in Figure 3A,B, respectively. They provided another view based on a soft ionization technique of the polyphenolic compounds present in pericarp and integument tissues, which was not dependent on the matrix-assisted laser desorption ionization process. Moreover, the possibility of acquiring fragmentation spectra of selected ions allowed associating the recorded polyphenolic profiles with the structure of single compounds. Tandem mass spectra recorded for the most abundant precursor ions present in Figure 3 are shown in Supplementary Figures S2–S17.

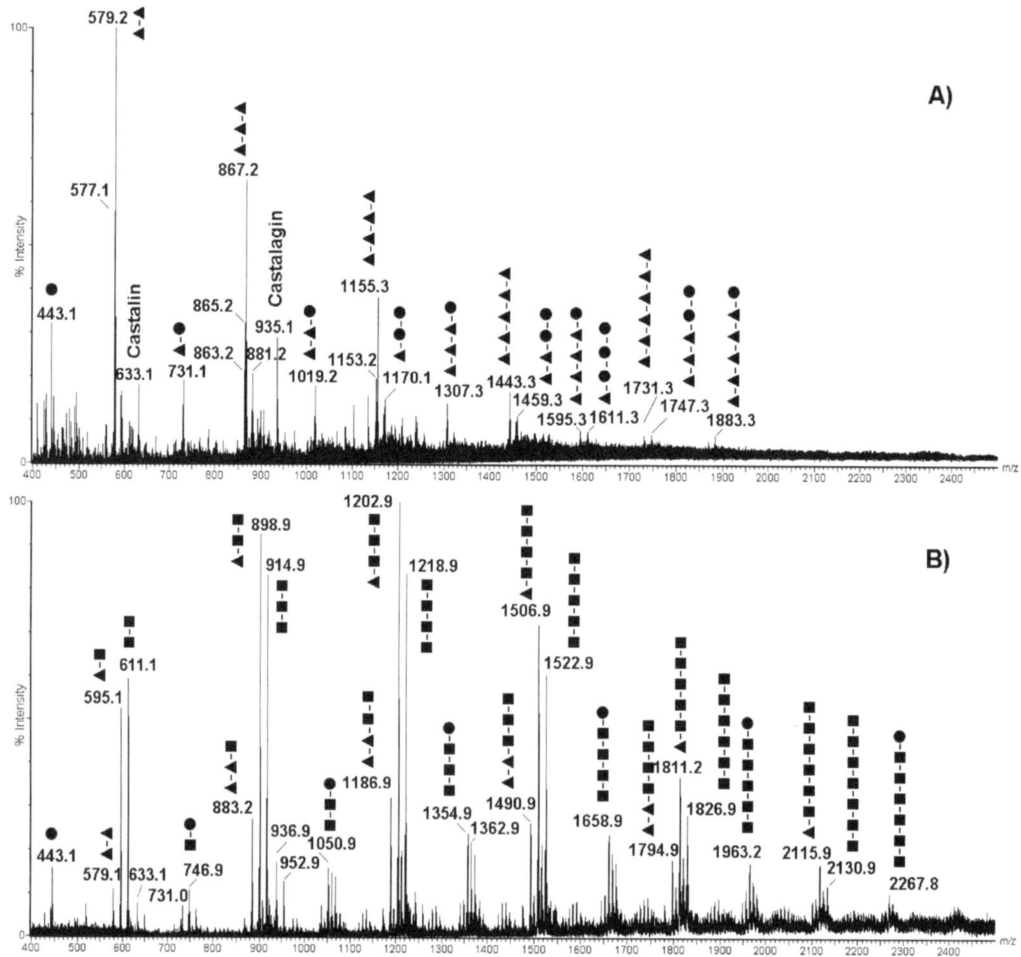

Figure 3. ESI-qTOF mass spectra of extracts from chestnut pericarp (panel (**A**)) and integument (panel (**B**)) tissues acquired in positive ion mode. In panel (**A**), assigned are procyanidins non-galloylated and mono-galloylated oligomers; in panel (**B**), labeled are epigallocatechin and epicatechin gallate oligomers. ▲, epicatechin; ■, epigallocatechin; ●, epicatechin gallate.

When comparing ESI-qTOF-MS profiles of pericarp and integument extracts in the m/z range of 400–2500, evident differences were observed between the analyzed tissues, either regarding the nature of detected polyphenols and the number of linked monomeric units (Figure 3). In particular, the integument extract showed more intense signals for polyphenols containing 3 to 5 monomeric units, while dimeric-trimeric compounds seemed more represented in the pericarp counterpart. This result confirmed the signal distribution observed in the MALDI-TOF mass spectra of integument and pericarp extracts when the acquisition was performed in positive ion mode (Figures 1A and 2A). In addition, a combination of the accurate measurement of detected MH^+ signal values in ESI-qTOF mass spectra, its comparison with polyphenolic structures already detected in chestnut tissues [28] and high-quality fragmentation spectra of most abundant ions present in recorded polyphenolic profiles (Supplementary Figures S2–S17) allowed for the association of a molecular structure to the detected molecular species (Figure 3). The results of this study are summarized in Figure 4, which provides a list of the polyphenols identified in chestnut pericarp and integument extracts by ESI-qTOF-MS, together with their experimental mass value and, in some cases, corresponding diagnostic fragment ions.

Polyphenols Identification	Abbr	Schematic structure	Measured Mass $(M+H)^+$	I	P	MSMS Fragment ions $[M-H]^-$
(epi)Catechin gallate (A-type)	ECg		441.1		X	
(epi)Ccatechin gallate (B-type)	ECg	●	443.1	X	X	
Procyanidin (A-type)	2EC		577.1		X	
Procyanidin (B-type)	2EC	▲-▲	579.2	X	X	425.1; 409.1; 287.1; 271.1; 259.1; 247.1; 165.1; 139.1; 127.0; 123.1
(epi)Catechin-(epi)gallocatechin (A-type)	EC-EGC		593.1	X		467.13; 441.1; 425.11; 355.09;305.09; 289.09; 287.08
(epi)Catechin-(epi)gallocatechin-(B-type)	EC-EGC	▲-■	595.1	X		425.1; 409.1; 317.1; 299.1; 287.0; 275.0; 179.0; 139.0; 127.0; 123.1
Dimeric (epi)gallocatechin (A type)	2EGC		609.1	X		443.13; 425.12; 317.09;307.10;305.08
Dimeric(epi)gallocatechin (B-type) or Prodelphinidin dimer	2EGC	■-■	611.1	X		591.14; 573.13;495.09;441.10;425.11; 423.09; 305.09 287.08;263.07;179.05;153.09;139.05;127.05
Castalin			633.1	X	X	
Procyanidin monogallate	EC-ECg	▲-●	731.2	X	X	579.2; 561.1; 441.1; 409.1; 331.1; 291.1; 289.1; 271.1; 247.1; 153.0; 139.0; 123.1
(epi)Gallocatechin-(epi)catechin gallate	EGC-ECg	■-●	747.1	X		
Procyanidin C1 (A-type)	3EC		863.2		X	
Procyanidin C1 (A-type)	3EC		865.2		X	
Procyanidin C1 (B-type)	3EC	▲ ▲ ▲	867.2	-	X	715.2; 697.2; 679.2; 579.2; 559.2; 535.2; 451.1; 437.1; 425.1; 409.1; 407.1; 397.1; 289.1; 271.1; 247.1; 163.0; 139.1; 127.0; 123.1
Procyanidin-(epi)gallocatechin (A-type)	2EC-EGC		881.1		X	
Procyanidin-(epi)gallocatechin (B-Type)	2EC-EGC	▲-▲-■	883.2	X		
(epi)Catechin-dimer(epi)gallocatechin	EC-2EGC	▲-■-■	898.9	X		731.1; 713.1; 609.1; 595.1; 483.1; 469.1; 443.1; 425.1; 423.1; 317.1; 305.1; 287.0; 263.1; 245.1; 179.0; 139.1; 127.0; 123.1
Trimer(epi)gallocatechin or Prodelphinidin trimer	2EGCg	■-■-■	915.1	X		
Castalagin			935.1		X	633.1; 615.1; 597.1; 553.1; 535.1; 495.1; 469.1; 439.1; 437.1; 345.1; 303.0; 277.1; 153.0; 139.0
Procyanidin-C1 monogallate	2EC-ECg	▲-▲-●	1019.2		X	867.1; 731.1; 579.1; 559.1; 441.1; 289.2; 271.2; 219.0; 203.0
Dimer(epi)gallocatechin-(epi)catechin gallate	2EGC-ECg	■-■-●	1050.9	X		
Cinnamtannin A2 or Procyanidin tetramer	4EC	▲ ▲ ▲ ▲	1155.3		X	1003.1; 985.1; 867.1; 865.1; 713.1; 695.1; 633.0; 577.1; 535.1; 495.0; 425.1; 409.1; 289.1; 247.0; 139.0
(epi)Catechin-dimer(epi)catechin gallate	EC-2ECg	▲-●-●	1170.9	X	X	
Dimer(epi)catechin-dimer(epi)gallocatechin	2EC-2EGC	▲-▲-■-■-	1186.9	X		
(epi)Catechin-trimer(epi)gallocatechin	EC-3EGC	▲-■-■-■	1202.9	X		1035.1; 913.1; 897.1; 899.1; 747.1; 731.1; 729.1; 609.1; 595.1; 591.1; 441.1; 425.0; 423.0; 305.0; 287.0; 263.0; 247.0
Tetramer(epi)gallocatechin or Prodelphinidin tetramer	4EGC	■-■-■-■-	1218.9	X		
Procyanidin tetramer or Cinnamtannin A2 monogallate	4EC-ECg	▲-▲-▲-●	1307.3		X	1153.1; 1019.2; 867.2; 865.1; 847.1; 729.1; 579.2; 577.1; 559.1; 535.1; 495.0; 441.1; 425.1; 409.1; 331.1; 289.1; 271.1; 247.1; 163.0; 139.0; 127.0
Trimer(epi)gallocatechin-(epi)gallocatechin gallate	3EGC-ECg	■-■-■-●	1354.9	X		
Procyanidin pentamer	5EC	▲ ▲ ▲ ▲ ▲	1443.3		X	1155.2; 865.1; 847.1; 713.1; 695.1; 577.1; 575.1; 441.1; 425.1; 409.1; 289.1; 247.0
Procyanidin tetramer digallate i	3EC-ECg	▲-▲-●-●	1459.3	X		
Dimer(epi)catechin-trimer(epi)gallocatechin	2EC-3EGC	▲-▲-■-■-■	1490.9	X		
(epi)Catechin-tetramer(epi)gallocatechin	EC-4EGC	▲-■-■-■-■	1506.9	X		1539.3; 1203.3; 1201.2; 899.2; 897.2; 747.1; 609.1; 595.1; 591.1; 483.1; 441.1; 397.4; 305.1; 287.0; 263.0
Pentamer (epi)gallocatechin Prodelphinidin pentamer	5EGC	■-■-■-■-■	1522.9	X		
Procyanidin pentamer monogallate	4EC-ECg	▲ ▲ ▲ ▲ ●	1595.3		X	1307.2; 1155.2; 1019.2; 867.1; 729.1; 577.1; 559.1; 441.1; 409.1; 271.1; 247.0
(epi)Catechin-trimer(epi)catechin gallate	EC-3ECg	▲-●-●-●	1611.3	X		
Tetramer (epi)gallocatechin-(epi)catechin gallate	4EGC-ECg	■-■-■-■-●	1658.9	X		
Procyanidin hexamer	6EC	▲ ▲ ▲ ▲ ▲ ▲	1731.3		X	
Procyanidin pentamer digallate	5EC-EGC	▲-▲-▲-●-●	1747.3		X	1457.1; 1169.2; 1155.2; 1017.1; 867.1; 865.1; 729.1; 579.1; 577.1; 441.1; 425.1; 289.1; 271.1; 247.0
Dimer(epi)catechin-tetramer(epi)gallocatechin	2EC-4EGC	▲-▲-■-■-■-■	1794.9	X		
(epi)Catechin-pentamer(epi)gallocatechin	EC-5EGC	▲-■-■-■-■-■	1810.9	X		1523.3; 1507.3; 1505.3; 1203.3; 1201.2; 913.2; 899.2; 897.2; 883.2; 761.1; 609.1; 595.1; 593.1; 579.2; 457.1; 441.1; 425.1; 305.1; 287.1; 277.1; 151.0
Hexamer(epi)gallocatechin Prodelphinidin hexamer	6EGC	■-■-■-■-■-■	1826.9	X		
Procyanidin hexamer monogallate	5EC-ECg	▲ ▲ ▲ ▲ ▲ ●	1883.9	X		
Pentamer (epi)gallocatechin-(epi)catechin gallate	5EGC-ECg	■-■-■-■-■-●	1962.9	X		
(epi)Catechin-hexamer(epi)gallocatechin	EC-6EGC	▲-■-■-■-■-■-■	2115.9	X		
Heptamer(epi)gallocatechin Prodelphinidin tetramer	7EGC	■-■-■-■-■-■-■	2130.8	X		
Hexamer(epi)gallocatechin-(epi)catechin gallate	6EGC-ECg	■-■-■-■-■-■-●	2267.8	X		

Figure 4. List of polyphenols identified in chestnut pericarp and integument extracts (Figure 3A,B), along with their measured mass values ascertained by ESI-qTOF-MS analysis in positive ion mode and major diagnostic fragment ions recorded for each precursor ions. I, integument; P, pericarp; EC, (epi)catechin, ▲; EGC, (epi)gallocatechin, ■; ECg, (epi)catechingallate, ●.

In particular, polyphenol polymerization in pericarp tissues was associated with the progressive incorporation of (epi)catechin units, which determined the formation of oligomeric species differing for a $\Delta m = +288$ mass units (Figures 3A and 4). In this case, the most significant oligomeric species were (epi)catechin gallate monomer (m/z 443.1) and (epi)catechin dimer (m/z 579.1) (Supplementary Figure S2), which undergo (epi)catechin addition of up to five molecular units. The assignment of signals at m/z 731.1, 867.2, 1019.2, 1155.3, 1307.3, 1443.3, 1595.3 and 1747.3 reported in Figure 3A was confirmed by the corresponding fragmentation spectra (Figure 4 and Supplementary Figures S5, S6, S9, S10, S12, S13, S15 and S16). Thus, signals at m/z 443.1, 731.1, 1019.2, 1307.3, 1595.3 and 1883.3 (Figure 3A) were attributed to type-B procyanidin monogalloylated species classified in the group of CTs (Figure 4) Interestingly, hop extracts have been reported containing similar procyanidin profiles [29]. The presence of these type-B procyanidin oligomers was paralleled by the A-type counterparts, as revealed by the corresponding signals at m/z 577.1, 865.2 and 1153.2 (Figures 3A and 4). This finding was peculiar of the pericarp extract, which thus showed catechin oligomers having a double type-A inter flavanol linkage at C2→O→C7' (Figure 4). On the other hand, the low-intensity signals at m/z 1170.1, 1459.3 and 1747.3 (Figure 3A) were attributed to oligomers of the type-B digalloylated procyanidin series (Figure 4), as also proved by the fragmentation spectrum of the latter species (Supplementary Figure S16). Finally, the mass signals observed at m/z 935.1 was definitively assigned to castalagin/vescalagin (Figures 3A and 4), as proved by the corresponding MS/MS spectrum that showed fragments at m/z 633 and 303 corresponding to vescalin/castalin and ellagic acid, respectively (Supplementary Figure S8).

Other authors were able to discriminate vescalagin from its isomer, castalagin, by recording fragmented ions in negative ion mode related to galloyl-gallagyl-hexoside and gallagyl species [24,30]. When recorded in positive ion mode, some of the above-cited fragment ions were not detected in the case of the chestnut pericarp extract, while the identification of fragment signals at m/z 277 and 303 definitively proved the assignment of this molecule to castalagin (Supplementary Figure S8), as already demonstrated by other authors [31].

Conversely, the ESI-qTOF mass spectrum of integument extract showed the occurrence of molecular ions at m/z 595.1 and 611.1 (Figure 3B), which were associated with epicatechin-(epi)gallocatechin and gallocatechin-(epi)gallocatechin dimer based on fragmentation spectra, respectively (Supplementary Figures S3 and S4). Their mass difference (Δm = +16 mass units) corresponded to an (epi)catechin to gallocatechin substitution. Starting from these two precursor ions, two series of mass signals occurred in the mass spectrum, which differed from each other in the number of polymerized (epi)gallocatechin units (Δm = +304 mass units), reaching a maximum value of six added molecules for (epi)gallocatechin (m/z 2115.9) and gallocatechin parents (m/z 2130.8) (Figures 3B and 4). Assignment of signals at m/z 898.9, 1202.9, 1506.9 and 1811.2 was confirmed by the corresponding fragmentation spectra (Supplementary Figures S7, S11, S14 and S17). The above-mentioned oligomerization process was also observed starting from (epi)catechin gallate monomer (m/z 443.1) and (epi)catechin dimer (m/z 579.1), and involved the polymerization of up to six (m/z 2267.8) and five (m/z 2098.9) (epi)gallocatechin units, respectively (Figures 3B and 4). In this case, the assignment of parent (epi)catechin dimer (m/z 579.1) was confirmed by the corresponding fragmentation spectrum (Supplementary Figure S2).

In conclusion, ESI-qTOF-MS and ESI-qTOF-MS/MS experiments demonstrated that the pericarp extracts contain CTs composed of (epi)catechin units, also called procyanidins, and HTs such as ellagitannins, while the integument counterparts were rich in condensed tannins characterized by gallocatechin and catechins units condensed with gallate, also known as prodelphinidins.

2.4. Time-Course Analysis of Total Phenols Released in Wastewater from Different Chestnut Water Curing Treatments

With the aim of quantitatively evaluating the extraction of the above-mentioned polyphenols as a result of different water curing processes, comparative experiments were also performed on aqueous media recovered from fruit treatment with: (i) tap water containing the above-mentioned antifungal *Lb. pentosus* strain (process A); (ii) tap water (process B); (iii) wastewater recycled from a previous curing treatment (process C). Kinetics of polyphenols migration from soaked fruits into different water curing media resulting from the water curing processes A, B and C was evaluated by comparative TPC measurements on corresponding wastewater samples at 0, 24, 48, 72 and 96 h. Quantitative results are shown in Figure 5. As expected, the fruit release of polyphenols in wastewater samples was time-dependent and progressively increased along with the duration of the water curing process.

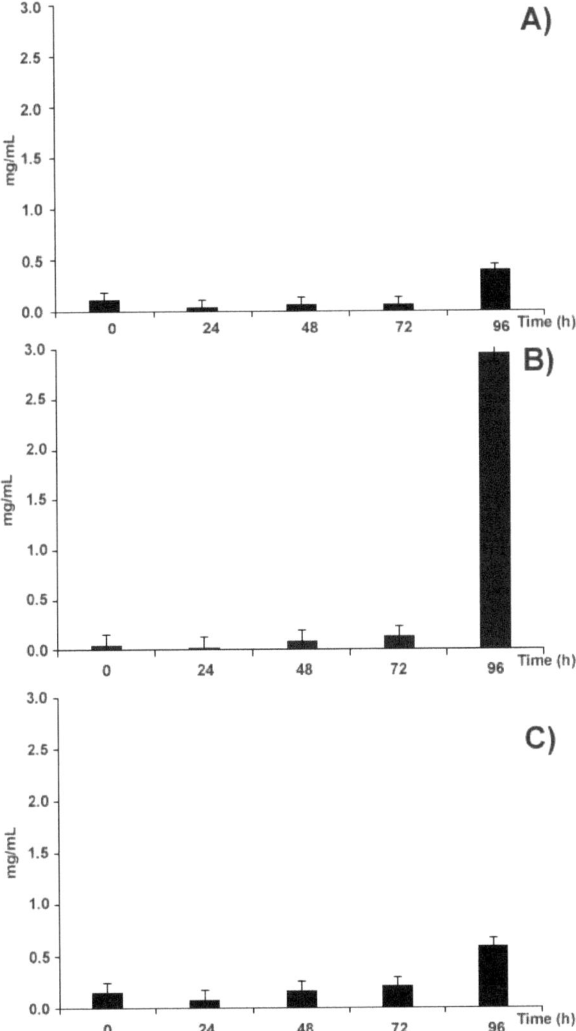

Figure 5. Kinetics of the diffusion of phenolic compounds from the whole fruits into the water-curing media deriving from the chestnut curing processes with: (i) tap water containing cultured *Lb. pentosus* (panel (**A**)); (ii) tap water (panel (**B**)); (iii) wastewater from a previous curing treatment (panel (**C**)). Reported values refer to data from water curing media from different curing processes, which were subtracted from counterparts from blank samples.

The greatest diffusion of polyphenols in wastewater samples occurred for the treatment with tap water (process B) (Figure 5B), for which a polyphenol concentration value of 2.9 mg mL^{-1} was measured at 96 h. In this case, time-course analysis of polyphenols migration into the water curing medium demonstrated that the highest and sudden molecular release occurred at the end of the curing process, namely at 96 h, when more than 90% of the total migration phenomenon was observed.

Conversely, the kinetics of polyphenol migration from soaked fruits into water curing media as result of processes A and C were similar to each other, and highly differed from that of process B (Figure 5); at the end of the treatment, the corresponding molecular recovery values (0.4 and 0.6 mg mL^{-1}, respectively) accounted for about one-sixth of the one observed for process B. The appearance, color and luster of recovered chestnuts from processes A and C provided a rationale to data reported in Figure 5, being more intense than those of counterparts from process B (data not shown), in agreement with a preserved maintenance of polyphenols in corresponding fruits. Their appearance resembled that of untreated chestnuts. This was consistent with the widely accepted consideration that the water curing process performed with tap water makes the treated chestnuts lose their natural luster [32].

A comparison of processes A and B, which were both performed in tap water, suggested that the presence of *Lb. pentosus* (in process A), yielding lactic acid in the curing medium and thus lowering the corresponding pH value (Supplementary Figure S1), was the reason for the corresponding reduced release of polyphenols from the fruit, with respect to the treatment not involving bacterial addition (process B).

Although it was reasonable to speculate that wastewater samples from a previous water curing treatment (process C) at the beginning of treatment (t = 0) should have contained higher levels of polyphenolic compounds than those measurable in counterparts from processes A and B, analysis of corresponding samples showed lower concentration values than those expected (Figure 5). This finding was tentatively associated with a putative molecular precipitation phenomenon occurring during wastewater exposition to air, which involved oxidized polyphenol polymerization and left only a small part of molecules in their native and soluble state. A comparative inspection of wastewater sample centrifugates from processes A, B and C confirmed this hypothesis (data not shown), revealing the formation of evident molecular precipitates only in the latter case. The possible interaction of the above-mentioned polymerizing phenolic compounds with the cell wall of pathogenic microorganisms, eventually determining the formation of metabolite-organism co-precipitates, should represent an interesting process for removing the spoilage microflora from chestnuts [33]. However, this polyphenol-microorganism interaction has to be confirmed by additional experimental evidence, and future studies are needed for this purpose.

2.5. Mass Spectrometric Analysis of Polyphenols Released in Wastewater from Different Chestnut Water Curing Treatments

In order to comparatively evaluate the nature of polyphenols present in water curing media resulting from processes A, B and C, corresponding wastewater samples were subjected to molecular profiling analysis through MALDI-TOF-MS and ESI-qTOF-MS, as already described for pericarp and integument extracts. In particular, MALDI-TOF-MS analysis in linear positive (Supplementary Figure S18) and negative ion mode (Figure 6) of wastewater samples A, B and C showed differences between signal profiles. This finding was particularly evident for spectra acquired in negative ion mode, in which samples from process A showed mass signals that were quite different with respect to that present in counterparts from treatments B and C (Figure 6). A careful inspection of the mass spectra from the latter samples evidenced intense mass signals related to castalagin, castalin, castalagin dimer, punicalagin, punicalagin-like and peduncalagin II species (Figure 6B,C), which were already detected as prominent species in the MALDI-TOF-MS profile acquired in linear negative ion mode of pericarp extracts (Figure 1B). Additional minority signals in the mass spectra of samples B and C (Figure 6B,C) were associated with (epi)catechin-(epi)catechin gallate-(epi)gallocatechin-based polyphenols already detected in the MALDI-TOF-MS profile acquired in linear negative ion mode of integument extracts (Figure 2B). This signal intensity pattern was totally inverted in the case of wastewater samples A, where MH$^+$ peaks associated with typical polyphenol species detected in integument extracts were prominent with respect to counterparts associated with metabolites assigned in the pericarp extracts. Negligible spectral differences observed between samples B and C suggested that the nature of extracted, soluble polyphenols between the two treatments was not affected by the pre-existing occurrence of these metabolites in the water curing medium before chestnut treatment. The results reported above suggest that the contribution of pericarp polyphenols to total metabolites present in wastewater samples from water curing processes B and C was more relevant than that ongoing after treatment A. Thus, the occurrence of bacteria in the water-curing medium did not hamper the extraction of integument polyphenols but had a pronounced effect on the solubilization of pericarp polyphenolics. Although with a less evident effect, the above-mentioned differences observed between samples A and samples B-C were confirmed when the same wastewater specimens were analyzed in linear positive ion mode (Supplementary Material Figure S18). In this case, spectral variations were mostly associated with signals at m/z 605.0 and 893.2 in samples from treatment A, which again highlighted the occurrence of different phenolic polymers between analyzed samples.

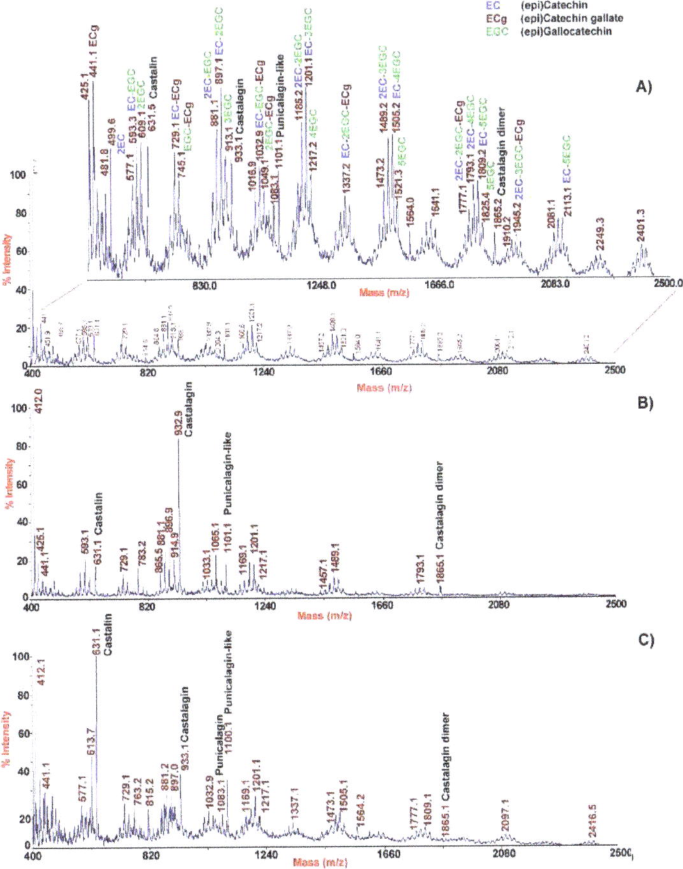

Figure 6. MALDI-TOF mass spectra acquired in linear negative ion mode of wastewater samples resulting from chestnuts subjected to water curing with: (i) tap water containing cultured *Lb. pentosus* (panel (**A**)); (ii) tap water (panel (**B**)); (iii) wastewater from a previous curing treatment with tap water (panel (**C**)). Condensed polymeric species were ascribed to polyphenols based on MALDI-TOF mass spectra acquired in linear negative ion mode of chestnut pericarp and integument extracts.

The above-reported polyphenol mass spectral profile similarities/dissimilarities between wastewater samples were further evidenced when these specimens were analyzed by ESI-qTOF-MS analysis in positive ion mode (Figure 7). Evident signals at about m/z 579.1 (procyanidin B-type), 617.2, 731.1, 867.1 (procyanidin C1) and 1155.2 (procyanidin tetramer) were observed as molecular signatures of wastewater samples from treatments B and C, which were lacking in counterparts from treatment A. Conversely, peculiar signals at m/z 416.9, 514.9, 530.9, 666.9, 682.9, 691.5, 702.2, 818.9, 834.9, 859.2, 995.4 and 1011.4 were typical of the ESI-qTOF-MS profile of wastewater from treatment A, which were absent in the counterparts from treatment B and C.

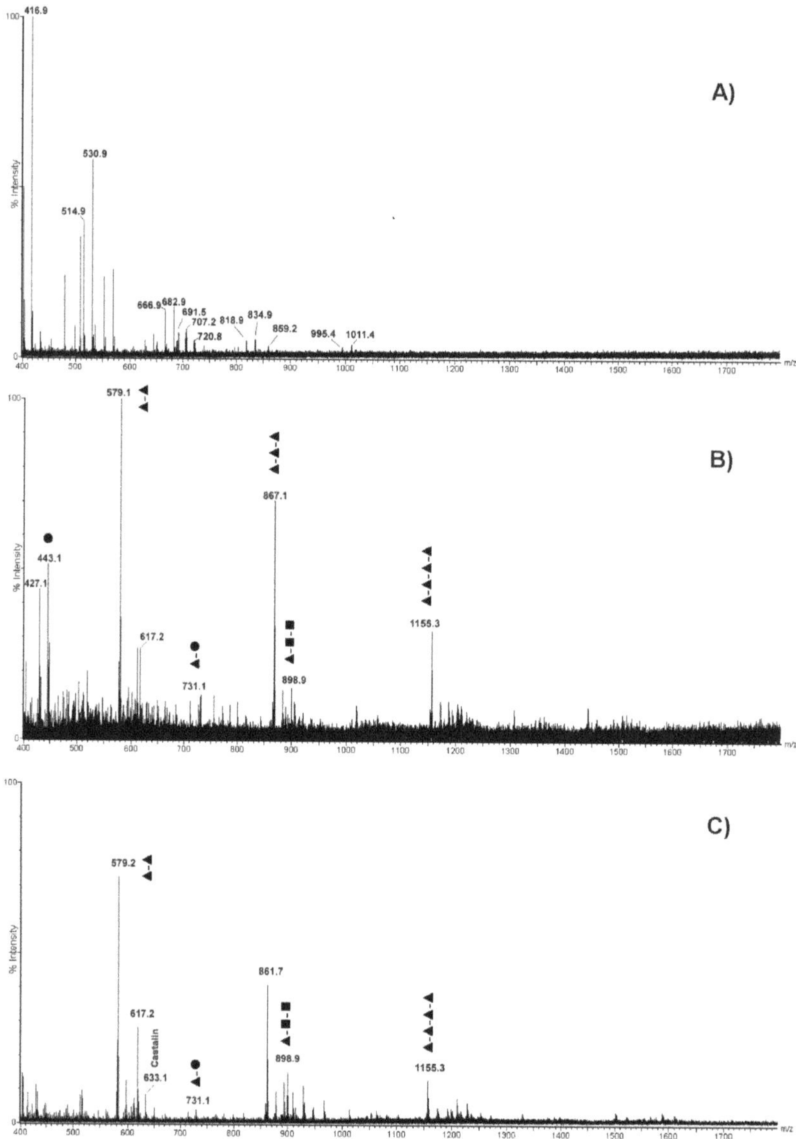

Figure 7. MS spectra of the polyphenol compounds recorded in positive ion mode occurring in the wastewater samples recovered at the end of three curing assays: (**A**) water containing cultured *Lb. pentosus*; (**B**) tap water; (**C**) recycled wastewater.

These signals were not assigned to specific metabolites. These results again demonstrated that polyphenol compounds present in the wastewater samples from treatment B and C represented the most abundant polyphenols (procyanidin oligomers) present in pericarp (Figure 3A), and emphasized the significant molecular migration of these compounds from the outer part of the chestnuts into the wastewater following treatments B and C. This was not the case in treatment A, where this migration was hampered, and no signals attributable to known chestnut polyphenols were detected, based on data reported in Figure 4.

3. Materials and Methods

3.1. Chestnut Sampling and Treatments

Chestnuts from the cultivar Castagna di Montella were subjected to the water curing process based on soaking fruits in water at a 1:2 fruit to water ponderal ratio; treatments were performed at the farm "Malerba Castagne", which is located in Montella (Avellino), Italy. Three identical batches of fresh chestnuts weighing each about 200 kg were soaked in: (A) tap water with adjunct of *Lb. pentosus* strain OM13 added on the second day of the treatment, which was already identified by the agar well-diffusion assay to exhibit a broad-spectrum antifungal activity [20]; (B) tap water; (C) wastewater from a previous water curing treatment, which was recycled for a new consecutive process. All trials were performed in parallel. Samples of wastewater from treatments A, B and C (100 mL) were randomly collected just after the beginning of the water curing process (t = 0) and after 24, 48, 72 and 96 h, and then subjected to the analytical assays reported below. Corresponding pH values are reported in Supplementary Figure S1. Original tap water used for different treatments was also stored for subsequent use as control samples.

3.2. Quantitative Evaluation of Total Phenolic Compounds in Wastewaters

The Folin–Ciocalteu assay was used for the quantification of total polyphenols content (TPC) in wastewater from treatments A, B and C, as well as in original tap water control samples. To this purpose, a 1 mL-aliquot of wastewater was mixed in stopper test tubes with 1 mL of Folin-Ciocalteu reagent (Sigma-Aldrich Co., St. Louis, MO, USA), followed by the addition of 1 mL of Na_2CO_3 (75 g/L) ($\geq$99.9%, Sigma-Aldrich Co., USA). The mixture was kept at room temperature for 90 min, and the corresponding absorbance at 765 nm was measured with a UV-VIS Cary 1-E spectrophotometer (Varian, Palo Alto, USA), with respect to a blank sample. All measurements were compared to a standard curve of 10–100 μg/mL gallic acid (98%, Acros Organics, ThermoFischer Scientific, Waltham, MA, USA), and the results expressed as milligrams of gallic acid equivalents (GAE)/100 g of wastewater extract. All measurements were performed in technical triplicate. Validation of the whole quantification procedure was obtained with two blind samples.

3.3. Extraction of Polyphenols from Chestnut Integument, Pericarp and Curing Wastewaters

One kg of randomly collected fresh chestnuts was manually peeled to recover corresponding pericarp and integument tissues, which were powdered separately using a food processor (model 843, Moulinex, Milan, Italy). Depending on their nature, polyphenols have a variable solubility in various polar organic solvents [3]. Thus, different solvents were assayed to obtain in a single step the best extraction of the different classes of polyphenols present in the chestnut shell and integument. The corresponding molecular recovery was evaluated on a quantitative basis using the Folin-Ciocalteu assay (data not shown), but also according to the overall signal-to-noise ratios measured during dedicated MS analysis. In this context, the best procedure combining both parameters for the extraction of polyphenols from the solid parts of chestnuts was identified as the treatment with the mixture acetonitrile/methanol/water (2:2:1, *v/v/v*), at 20 °C. According to the above-mentioned procedure, 5 g of pericarp and integument tissues were placed in 10 mL-centrifuge tubes with screw cap in triplicate and extracted with 10 mL of the above-mentioned mixture, under continuous agitation, for 48 h. After centrifugation at 5000 rpm, for 20 min, 4 °C, recovered liquid was filtered through a 0.20 μm cellulose acetate filter, transferred to novel tubes and dried under a stream of gaseous N_2. All samples were then solved in 200 μL of 0.1% TFA, which were applied to a desalting step on a C18 solid-phase column (Seppak Oasis, Waters Corporation, Milford, MA, USA); salts were washed away with aqueous 0.1% TFA as eluent, and phenolic acids were finally eluted with 50% *v/v* acetonitrile containing 0.1% TFA. Desalted samples were finally dried under a stream of gaseous N_2. Chestnut pericarp and integument samples were dissolved in 50% *v/v* acetonitrile containing 0.1% TFA (1 μg μL^{-1} final concentration) prior to mass spectrometric analysis.

In parallel, 10 mL-aliquots of the liquid recovered from the chestnut water curing treatments A, B and C were filtered through a 0.20 μm cellulose acetate filter and dried at 30 °C under a stream of gaseous N_2; all samples were then solved, extracted, and processed according to the method described above, prior to mass spectrometry analysis. Wastewater samples were dissolved in 50% *v/v* acetonitrile containing 0.1% TFA (1 μg μL^{-1} final concentration) prior to mass spectrometric analysis.

3.4. MALDI-TOF-MS Analysis

Chestnut pericarp and integument, and wastewater samples were analyzed in parallel by MALDI-TOF-MS analysis, using a Voyager DE-Pro spectrometer (PerSeptive BioSystems, Framingham, MA, USA) equipped with an N_2 laser (k = 337 nm). An identical volume (1 μL) of the samples and the matrix solution, which was prepared by dissolving 10 mg of 2,5-dihydroxybenzoic acid in 1 mL of aqueous 50% *v/v* acetonitrile containing 0.1% *v/v* TFA, were mixed in parallel, placed onto the instrument target and dried at room temperature. The instrument operated with an accelerating voltage of 20 kV, a grid voltage of 95% of the accelerating voltage, a guidewire of 0.05 % and a delayed-ion extraction time of 175 ns. External mass calibration was performed with Sequazyme™ Peptide Mass Standards Kit, Calibration Mixture 1 and Calibration Mixture 2 (Applied Biosystems, Foster City, CA) containing des-Arg1-bradykinin, angiotensin I, Glu1-fibrinopeptide B, neurotensin 0.2, angiotensin I, ACTH (1–17 clip), ACTH (18–39 clip), ACTH (7–38 clip) and bovine insulin, respectively. The mass spectra were acquired in linear positive and negative ion modes. Raw data were elaborated using the software program Data Explorer version 4.0 (Applied Biosystems, Foster, CA, USA).

3.5. ESI-qTOF-MS Analysis

Chestnut pericarp and integument, and wastewater samples were analyzed in parallel by ESI-qTOF-MS analysis, using an ESI q-TOF™ hybrid quadrupole/time-of-flight mass spectrometer (Micromass Ltd., Manchester, UK) equipped with a Z-spray ion source. Dissolved samples were delivered in the instrument source through a syringe pump operating at 0.5 μL min^{-1}. Spectral acquisition was performed in positive ion mode and recording MS and MS/MS spectra. The source and desolvation temperature values were 100 °C and 200 °C, respectively. The TOF operated with an acceleration voltage of 9.1 kV, a cone voltage of 100 V, a cone gas (N_2) of 13 L h^{-1}, and the collision energy in MS mode of 10 eV. Collision-induced dissociation (CID) spectra were acquired in a data-dependent method on the most abundant ions having *m/z* values ranging from 300 to 2500. The collision energy was dependent on the *m/z* value and the charge state of the parent ion; it was generally in the range 25–40 V. The collision cell was pressurized with 10.34 Pa ultrapure Ar (99.999%).3.6. Statistical Analysis

All quantitative measurements reported in the present work were performed in technical triplicate; the calculation of the corresponding mean and standard deviation values were performed with Microsoft Excel 2017 software (Microsoft Corporation, Redmond, WA, USA). All statistical analyses were performed at a significance level of 5% ($p \leq 0.05$) using XLSTAT software Version 2020.1 (Microsoft Corporation, Redmond, WA, USA).

4. Conclusions

This study provides useful indications of the molecular processes associated with the traditional practice of the water curing of chestnuts, which is aimed at preventing insect and mould development during fruit storage. At first, the combined MALDI-TOF-MS and ESI-qTOF-MS analyses characterized different polyphenols present in fruit pericarp and integument tissues. In this context, fragmentation experiments with the latter technique allowed assigning occurring metabolites, demonstrating that CTs and HTs made of repeated catechin/epicatechin and ellagic acid units were highly represented in the fruit pericarp, whereas polymerized gallocatechin or catechin units esterified by gallic acid occurred in the integument tissues. Then, different water curing treatments, namely with tap water, tap water containing *Lb. pentosus* and wastewater from a previous water curing treatment, were compared as it regards to the luster/appearance of recovered chestnuts, as well as the amount and the nature of polyphenols released in the corresponding wastewaters. Results definitively indicated that the wastewater from chestnut water curing treatment can be used as a rich source of polyphenols. It also provided a rationale to the loss of chestnut luster during curing treatment with tap water, demonstrating a strong release of pericarp metabolites in the corresponding wastewaters. This condition was hampered when treatments were performed at conditions involving a drop of the pH value of the medium, namely tap water containing *Lb. pentosus* or wastewater from a previous water curing treatment, which preserved fruit appearance. The nature of the polyphenols detected in wastewaters from treatment with tap water or wastewater from a previous water curing treatment suggested an evident contribution of pericarp polyphenols to metabolites dissolved in wastewaters, which was absent in the case of tap water containing *Lb. pentosus*. In conclusion, this study provides a rationale to different water curing treatments as in terms of the recovered chestnuts but also to optimal conditions for promoting the release of bioactive natural products in wastewaters, thus facilitating their recycling. The latter treatments, favoring the optimal recovery of polyphenols, may find a diffused use when the appearance of treated fruits is not important for commercial purposes, as in the case of raw materials used for production of chestnut flour.

Supplementary Materials: The following are available online. Figure S1—pH profile of medium during the water curing with different treatments. Figure S2—MS/MS spectrum at m/z 595 acquired in positive ion mode. Figure S3—MS/MS spectrum at m/z 579 acquired in positive ion mode. Figure S4—MS/MS spectrum at m/z 609 and 611 acquired in positive ion mode. Figure S5—MS/MS spectrum at m/z 731 acquired in positive ion mode. Figure S6—MS/MS spectrum at m/z 867 acquired in positive ion mode. Figure S7—MS/MS spectrum at m/z 899 acquired in positive ion mode. Figure S8—MS/MS spectrum at m/z 935 acquired in positive ion mode. Figure S9—MS/MS spectrum at m/z 1019 acquired in positive ion mode. Figure S10—MS/MS spectrum at m/z 1155 acquired in positive ion mode. Figure S11—MS/MS spectrum at m/z 1203 acquired in positive ion mode. Figure S12—MS/MS spectrum at m/z 1307 acquired in positive ion mode. Figure S13—MS/MS spectrum at m/z 1443 acquired in positive ion mode. Figure S14—MS/MS spectrum at m/z 1507 acquired in positive ion mode. Figure S15—MS/MS spectrum at m/z 1595 acquired in positive ion mode. Figure S16—MS/MS spectrum at m/z 1747 acquired in positive ion mode. Figure S17—MS/MS spectrum at m/z 1811 acquired in positive ion mode. Figure S18—MALDI—TOF—MS in linear positive ion mode of wastewater samples from treatment with: (A) tap water containing *Lb. pentosus*; (B) tap water; (C) recycled wastewater from a previous curing treatment with tap water.

Author Contributions: S.C. has designed the study. S.C., G.P., S.D.P. and M.A. conducted the experiments. S.C., G.P., A.S. and F.A. analyzed the data and prepared the manuscript. All authors have read and agreed to the published version of the manuscript.

Funding: This research received no external funding.

Institutional Review Board Statement: Not applicable.

Informed Consent Statement: Not applicable.

Data Availability Statement: The data presented in this study are available in supplementary material.

Conflicts of Interest: The authors declare no conflict of interest.

Sample Availability: Samples of the compounds are not available from the authors.

References

1. Barreira, J.C.M.; Ferreira, I.C.F.R.; Oliveira, M.B.P.P.; Pereira, J.A. Antioxidant Activities of the Extracts from Chestnut Flower, Leaf, Skins and Fruit. *Food Chem.* **2008**, *107*, 1106–1113. [CrossRef]
2. Gonçalves, B.; Borges, O.; Costa, H.S.; Bennett, R.; Santos, M. Silva Metabolite Composition of Chestnut (Castanea Sativa Mill.) Upon Cooking: Proximate Analysis, Fibre, Organic Acids and Phenolics. *Food Chem.* **2010**, *122*, 154–160. [CrossRef]
3. Hu, M.; Yang, X.; Chang, X. Bioactive phenolic components and potential health effects of chestnut shell: A review. *J. Food Biochem.* **2021**, e13696. [CrossRef]
4. Sorice, A.; Siano, F.; Capone, F.; Guerriero, E.; Picariello, G.; Budillon, A.; Ciliberto, G.; Paolucci, M.; Costantini, S.; Volpe, M.G. Potential Anticancer Effects of Polyphenols from Chestnut Shell Extracts: Modulation of Cell Growth, and Cytokinomic and Metabolomic Profiles. *Molecules* **2016**, *21*, 1411. [CrossRef] [PubMed]
5. Coccia, E.; Siano, F.; Volpe, M.G.; Varricchio, E.; Eroldogan, O.T.; Paolucci, M. Chestnut Shell Extract Modulates Immune Parameters in the Rainbow Trout Oncorhynchus Mykiss. *Fishes* **2019**, *4*, 18. [CrossRef]
6. Vamanu, E.; Gatea, F.; Pelinescu, D.R. Bioavailability and Bioactivities of Polyphenols Eco Extracts from Coffee Grounds after In Vitro Digestion. *Foods* **2020**, *9*, 1281. [CrossRef]
7. Grgić, J.; Šelo, G.; Planinić, M.; Tišma, M.; Bucić-Kojić, A. Role of the Encapsulation in Bioavailability of Phenolic Compounds. *Antioxidants* **2020**, *9*, 923. [CrossRef]
8. Chiva-Blanch, G.; Visioli, F. Polyphenols and Health: Moving Beyond Antioxidants. *J. Berry Res.* **2011**, *2*, 63–71. [CrossRef]
9. Yin, P.; Zhao, S.; Chen, S.; Liu, J.; Shi, L.; Wang, X.; Liu, Y.; Ma, C. Hypoglycemic and Hypolipidemic Effects of Polyphenols from Burs of Castanea Mollissima Blume. *Molecules* **2011**, *16*, 9764–9774. [CrossRef] [PubMed]
10. Jovanovic, J.A.; Mihailovic, M.; Uskokovic, A.S.; Grdovic, N.; Dinic, S.; Poznanovic, G.; Mujic, I.; Vidakovic, M. Evaluation of the Antioxidant and Antiglycation Effects of Lactarius Deterrimus and Castanea Sativa Extracts on Hepatorenal Injury in Streptozotocin-Induced Diabetic Rats. *Front. Pharm.* **2017**, *8*, 793. [CrossRef]
11. Serafini, M.; Peluso, I. Functional Foods for Health: The Interrelated Antioxidant and Anti-Inflammatory Role of Fruits, Vegetables, Herbs, Spices and Cocoa in Humans. *Curr. Pharm. Des.* **2016**, *22*, 6701–6715. [CrossRef]
12. Vazquez, G.; Pizzi, A.; Freire, M.; Santos, J.; Antorrena, G.; González-Álvarez, J. Maldi-Tof, Hplc-Esi-Tof and 13c-Nmr Characterization of Chestnut (Castanea Sativa) Shell Tannins for Wood Adhesives. *Wood Sci. Technol.* **2013**, *47*, 523–535. [CrossRef]
13. Vasconcelos, M.C.B.M.; Richard, N.B.; Quideau, S.; Jacquet, R.; Rosa, E.A.S.; Ferreira-Cardoso, J.V. Evaluating the Potential of Chestnut (Castanea Sativa Mill.) Fruit Pericarp and Integument as a Source of Tocopherols, Pigments and Polyphenols. *Ind. Crops Prod.* **2010**, *31*, 301–311. [CrossRef]
14. Santos, J.; Antorrena, G.; Freire, M.S.; Pizzi, A.; González-Álvarez, J. Environmentally Friendly Wood Adhesives Based on Chestnut (Castanea Sativa) Shell Tannins. *Eur. J. Wood Wood Prod.* **2017**, *75*, 89–100. [CrossRef]
15. Finch, C.A. Wood adhesives: Chemistry and technology. Edited by A. Pizzi, Marcel Dekkar, New York and Basel, 1983. *Br. Polym. J.* **1984**, *16*, 324. [CrossRef]
16. Comandini, P.; Lerma-García, M.J.; Simó-Alfonso, E.F.; Toschi, T.G. Tannin analysis of chestnut bark samples (Castanea sativa Mill.) by HPLC-DAD-MS. *Food Chem.* **2014**, *157*, 290–295. [CrossRef]
17. Quideau, S.; Jourdes, M.; Lefeuvre, D.; Montaudon, D.; Saucier, C.; Glories, Y.; Pardon, P.; Pourquier, P. The chemistry of wine polyphenolic C-glycosidic ellagitannins targeting human topoisomerase II. *Chem. A Eur. J.* **2005**, *11*, 6503–6513. [CrossRef] [PubMed]
18. Botondi, R.; Vailati, M.; Bellincontro, A.; Massantini, R.; Forniti, R.; Mencarelli, F. Technological Parameters of Water Curing Affect Postharvest Physiology and Storage of Marrons (Castanea Sativa Mill., Marrone Fiorentino). *Postharvest Biol. Technol.* **2009**, *51*, 97–103. [CrossRef]
19. Nazzaro, M.; Barbarisi, C.; La Cara, F.; Volpe, M.G. Chemical and Biochemical Characterisation of an Igp Ecotype Chestnut Subjected to Different Treatments. *Food Chem.* **2011**, *128*, 930–936. [CrossRef]
20. Blaiotta, G.; Di Capua, M.; Romano, A.; Coppola, R.; Aponte, M. Optimization of Water Curing for the Preservation of Chestnuts (Castanea Sativa Mill.) and Evaluation of Microbial Dynamics During Process. *Food Microbiol.* **2014**, *42*, 47–55. [CrossRef] [PubMed]
21. Echegaray, N.; Gómez, B.; Barba, F.J.; Franco, D.; Estévez, M.; Carballo, J.; Marszałek, K.; Lorenzo, J.M. Chestnuts and by-products as source of natural antioxidants in meat and meat products: A review. *Trends Food Sci. Technol.* **2018**, *82*, 110–121. [CrossRef]
22. Pasch, H.; Pizzi, A. Considerations on the macromolecular structure of chestnut ellagitannins by matrix-assisted laser desorption/ionization-time-of-flight mass spectrometry. *J. Appl. Polym. Sci.* **2002**, *85*, 429–437. [CrossRef]
23. Salminen, J.-P. Two-Dimensional Tannin Fingerprints by Liquid Chromatography Tandem Mass Spectrometry offer a new dimension to plant tannin analyses and Help to visualize the tannin diversity in plants. *J. Agric. Food Chem.* **2018**, *66*, 9162–9171. [CrossRef] [PubMed]

24. Abid, M.; Yaich, H.; Cheikhrouhou, S.; Khemakhem, I.; Bouaziz, M.; Attia, H.; Ayadi, M.A. Antioxidant properties and phenolic profile characterization by LC–MS/MS of selected Tunisian pomegranate peels. *J. Food Sci. Technol.* **2017**, *54*, 2890–2901. [CrossRef] [PubMed]

25. Mena, P.; Calani, L.; Dall'Asta, C.; Galaverna, G.; García-Viguera, C.; Bruni, R.; Crozier, A.; Del Rio, D. Rapid and comprehensive evaluation of (poly)phenolic compounds in pomegranate (Punica granatum L.) juice by UHPLC-MSn. *Molecules* **2012**, *17*, 14821–14840. [CrossRef] [PubMed]

26. Barreira, J.C.; Pereira, J.A.; Oliveira, M.B.; Ferreira, I.C. Sugars profiles of different chestnut (Castanea sativa Mill.) and almond (Prunus dulcis) cultivars by HPLC-RI. *Plant. Foods Hum. Nutr.* **2010**, *65*, 38–43. [CrossRef]

27. van der Hooft, J.J.J.; Vervoort, J.; Bino, R.J.; Beekwilder, J.; de Vos, R.C.H. Polyphenol identification based on systematic and robust high-resolution accurate mass spectrometry fragmentation. *Anal. Chem.* **2011**, *83*, 409–416. [CrossRef]

28. Cacciola, N.A.; Cerrato, A.; Capriotti, A.L.; Cavaliere, C.; D'Apolito, M.; Montone, C.M.; Piovesana, S.; Squillaci, G.; Peluso, G.; Laganà, A. Untargeted Characterization of Chestnut (Castanea sativa Mill.) Shell Polyphenol Extract: A Valued Bioresource for Prostate Cancer Cell Growth Inhibition. *Molecules* **2020**, *25*, 2730. [CrossRef]

29. Li, H.J.; Deinzer, M.L. The Mass Spectral Analysis of Isolated Hops a-Type Proanthocyanidins by Electrospray Ionization Tandem Mass Spectrometry. *J. Mass Spectrom.* **2008**, *43*, 1353–1363. [CrossRef] [PubMed]

30. Singh, A.; Bajpai, V.; Kumar, S.; Sharma, K.R.; Kumar, B. Profiling of gallic and ellagic acid derivatives in different plant parts of Terminalia arjuna by HPLC-ESI-QTOF-MS/MS. *Nat. Prod. Commun.* **2016**, *11*, 239–244. [CrossRef]

31. Bowers, J.J.; Gunawardena, H.P.; Cornu, A.; Narvekar, A.S.; Richieu, A.; Deffieux, D.; Quideau, S.; Nishanth, T. Rapid Screening of Ellagitannins in Natural Sources Via Targeted Reporter Ion Triggered Tandem Mass Spectrometry. *Sci. Rep.* **2018**, *8*, 10399. [CrossRef] [PubMed]

32. Jermini, M.; Conedera, M.; Sieber, T.N.; Sassella, A.; Schärer, H.; Jelmini, G.; Höhn, E. Influence of Fruit Treatments on Perishability During Cold Storage of Sweet Chestnuts. *J. Sci. Food Agric.* **2006**, *86*, 877–885. [CrossRef]

33. Migliorini, M.; Funghini, L.; Marinelli, C.; Turchetti, T.; Canuti, S.; Zanoni, B. Study of Water Curing for the Preservation of Marrons (Castanea Sativa Mill., Marrone Fiorentino Cv). *Postharvest Biol. Technol.* **2010**, *56*, 95–100. [CrossRef]

MDPI
St. Alban-Anlage 66
4052 Basel
Switzerland
Tel. +41 61 683 77 34
Fax +41 61 302 89 18
www.mdpi.com

Molecules Editorial Office
E-mail: molecules@mdpi.com
www.mdpi.com/journal/molecules